Sex Hormones and Immunity to Infection

Sabra L. Klein • Craig W. Roberts

Editors

Sex Hormones and Immunity to Infection

 Springer

Editors

Sabra L. Klein
Johns Hopkins Bloomberg
School of Public Health
Dept. Molecular Microbiology &
Immunology
615 N. Wolfe Street
Baltimore MD 21205
USA
saklein@jhsph.edu

C.W. Roberts
Strathclyde Institute of Pharmacy
and Biomedical Sciences
University of Strathclyde
27 Taylor Street
G4 0NR Glasgow
Scotland
UK
c.w.roberts@strath.ac.uk

WK 900 Sex

ISBN 978-3-642-02154-1   e-ISBN 978-3-642-02155-8
DOI: 10.1007/978-3-642-02155-8
Springer Heidelberg Dordrecht London New York

Library of Congress Control Number: 2009930390

Cover design: WMXDesign GmbH, Heidelberg, Germany

Cover picture: Gustav Klimt: Der Kuss. By courtesy of Belvedere, Vienna

Printed on acid-free paper

Springer is part of Springer Science+Business Media (www.springer.com)

Foreword

Why Sex Matters

In the biological sciences, we tend to measure comparative differences rather than absolute values. We customarily compare treated versus untreated, exposed versus unexposed, susceptible versus resistant. Yet biologists too often neglect one of the most fundamental differences, male versus female. This distinction is clearly important; the mere fact that it has been adopted by so many species and preserved over so long a time testifies to its evolutionary advantage. By providing a powerful mechanism of diversity, sexual recombination forms a broadened substitute upon which selection pressure can act.

Surely, the careful study of sexual dimorphism will help to answer many of the questions that biologists ask. How are sex-based differences produced and how are they preserved? How do they affect survival and behavior? Most important, how can they help us to better understand astonishing and ever-changing mosaic of life on this planet?

This book proposes to dig deeply into one aspect of the question of why sex matters. Nothing has proven to be more informative than exploring the determining role that sex plays in host resistance to infection.

Among human and nonhuman animals, the prevalence (i.e., the proportion of individual infected) and intensity (i.e., severity of infection) of infection typically is higher in males than in females. Of course, this reflects differences in exposure as well as inherited differences in susceptibility to pathogens (Klein 2000, 2004; Roberts et al. 2001; Zuk and McKean 1996). Heightened susceptibility to infection is one of the leading explanations of the greater death rates among men than among women reported in several locations around the world. In general, females have more intense immune responses than males (Klein 2000, 2004; Zuk and McKean 1996). The greater immunity among females creates a double-edge sword; it is beneficial as a defense against infectious diseases, but is detrimental in the increased occurrence of autoimmune diseases (Wizemann and Pardue 2001). Sex-based differences typically become apparent after puberty (Klein 2000; Roberts et al. 2001) and several field and laboratory studies link sex differences in immune function with circulating steroid hormones (Klein 2000, 2004; Roberts et al. 2001; Zuk and McKean 1996). Sex hormones change profoundly during pregnancy where they must modulate the immune system to facilitate a successful pregnancy among viviparous animals (Roberts et al. 1996). This task must be accomplished without increasing the general vulnerability of the mother to infection.

This book is especially timely, given the recent scientific report (Simon et al. 2005) by the Society of Women's Health Research showing that less than 3% of the funded research grants at the National Institutes of Health (NIH) in the US are awarded for the study of biological differences between males and females. This report followed the publication of an Institute of Medicine report (Wizemann and Pardue 2001) entitled *Exploring the Biological Contributions to Human Health: Does Sex Matter?* The IOM report concluded that sex differences in susceptibility, prevalence, and severity are apparent for many diseases, including cancers, heart disease, autoimmunity, and infectious diseases. These reports emphasize the need for greater research on women's health issues. Although the inclusion of women and minorities in clinical research has increased, examination of the biological differences between the sexes and how they affect health and disease has lagged (GAO 2000). By focusing on the need for including sex-based studies of infectious diseases, this book emphasizes the value of examining responses in both males and females to improve our understanding about host-pathogen interactions in both sexes. These are issues relevant to the entire scientific community.

The contributors are selected from a variety of disciplines, including microbiology, immunology, genetics, pathology, and evolutionary biology all of them have made important contributions to our knowledge of sex differences and the effects of pregnancy on susceptibility to infection. They then chronic six broad themes to represent the current trends of this diverse body of literature. The book begins with a chapter on the evolution of sex differences in susceptibility to infection. Males and females differ in the selection pressures acting on each sex; therefore, in addition to the genetic and hormonal mechanisms that underlie sex differences in immune function, evolutionary factors must also be considered (Chap. 1). Then follow two chapters dedicated to the direct effects of steroid hormones on the functioning of the immune system. The prevailing hypothesis to explain immunologic differences between the sexes is that sex hormones, in particular, testosterone, 17β-estradiol, and progesterone, influence the immune system (Chap. 2). Circulating concentrations of hormones may not be the only index of steroid hormone effects on immune function. Generally, the effects of hormones depend not only on circulating concentrations but also on the availability and affinity of target-tissue receptors. Accordingly, sex steroid hormone receptors have been identified on several classes of immune cells (Chap. 3).

As noted above, the prevalence and intensity of infectious diseases is generally higher in males than in females (Roberts et al. 2001; Simon et al. 2005; Zuk and McKean 1996). The interactions that exist between the endocrine and immune systems are important in considering why males and females differ in susceptibility to infectious agents. The editors have allotted several chapters to review evidence for sex differences in response to viruses, bacteria, and parasites, with emphasis on the role of sex steroids (Chaps. 4–6). Chapter 7 explores the often expressed belief of "female immunological supremacy". Specifically, although males are more susceptible than females to many infectious agents, males are not more susceptible to all parasites. This phenomenon as well as the underlying mechanisms is fully addressed.

Pregnancy is a reproductive condition during which profound hormonal and immunological changes occur. How pregnancy and the associated rise in sex hormones modulate maternal immune responses and the severity of infections is discussed in Chaps. 8 and 9.

The functional significance of sex differences in immune responses to infectious agents is considered in Chaps. 10 and 11. They provide an epidemiological perspective and raise the possibility that if males and females differ in their immunological responses to pathogens, they may differ in their responses to treatments as well.

In summary, this timely volume critically reviews, in a single publication, the evolutionary origin and the functional mechanisms responsible for sexual differences in response to infection. Surely, it will become a standard reference course for those in this growing field. It brings fresh insight into the management of infectious diseases, delineates areas where knowledge is lacking, and highlights the future avenues of research. It brings us closer to an answer to the question of why sex matters.

Noel R. Rose
Sabra L. Klein
Craig W. Roberts

References

GAO (2000) NIH has increased its efforts to include women in research. United States General Accounting Office, USA

Klein SL (2000) The effects of hormones on sex differences in infection: from genes to behavior. Neurosci Biobehav Rev 24:627–638

Klein SL (2004) Hormonal and immunological mechanisms mediating sex differences in parasite infection. Parasite Immunol 26:247–264

Roberts CW, Satoskar A, Alexander J (1996) Sex steroids, pregnancy-associated hormones and immunity to parasitic infection. Parasitol Today 12:382–388

Roberts CW, Walker W, Alexander J (2001) Sex-associated hormones and immunity to protozoan parasites. Clin Microbiol Rev 14:476–488

Simon VR, Hai T, Williams SK, Adams E, Ricchetti K, and Marts SA (2005) National institutes of health: intramural and extramural support for research on sex differences, 2000–2003. The Society for Women's Health Research, Washington, DC

Wizemann TM, Pardue M (eds) (2001) Exploring the biological contributions to human health: does sex matter? National Academy Press, Washington, DC

Zuk M, McKean KA (1996) Sex differences in parasite infections: patterns and processes. Int J Parasitol 26:1009–1023

Contents

Chapter 1
Sex Differences in Susceptibility to Infection: An Evolutionary Perspective

Marlene Zuk and Andrew M. Stoehr

Abstract Patterns of sex differences in parasite infection and immune responses have been noted for many decades. Although numerous explanations for such differences have been proposed, including hormonal patterns and sex-biased exposure to infective stages of pathogens, these have largely been proximate explanations that address the mechanisms immediately responsible for the findings but do not take a more integrative or ultimate approach. Here, we present an evolutionary framework for understanding the origin and maintenance of sex differences in the incidence and susceptibility to infectious disease, using life history theory and sexual selection to make predictions about when males or females in a particular species are expected to be more or less susceptible to parasites.

1.1 Introduction

Sex differences in incidence and pathogenesis of parasite infections have been of interest to parasitologists for a long time, indeed almost since the systematic study of animal parasites became established near the beginning of the twentieth century. Parasitologists examining animals collected in the field found it natural to note differences in infestations between the host sexes, and their interest was continued in laboratory experiments (Addis 1946, Solomon 1966, Alexander and Stimson 1988). Most of these studies focused on mammals, and during the

M. Zuk (✉)
University of California, Riverside, CaliforniaUSA
e-mail: mzuk@citrus.ucr.edu

S.L. Klein and C.W. Roberts (eds.), *Sex Hormones and Immunity to Infection*,
DOI 10.1007/978-3-642-02155-8_1, © Springer-Verlag Berlin Heidelberg 2010

mid-twentieth century a virtual cottage industry developed in which investigators experimentally infected laboratory rodents with identical doses of parasites and documented any resulting sex differences in the prevalence or intensity of the infection that developed (reviewed in Zuk and McKean 1998). Although exceptions could be found, the majority of research found that males were more likely to harbor parasites or to suffer more intensely from their effects than were females. Furthermore, the persistence of these patterns after experimental infestations of animals in the laboratory suggested that the sex difference was not merely due to differences in exposure to parasites, but also due to males and females behaving differently in the field and hence incurring different risks of infection.

The medical community has also known about sex differences in infectious disease susceptibility for many years. In his 1958 paper, *Biological Sex Differences with Special Reference to Disease, Resistance and Longevity*, the influential physician and medical researcher Landrum Shettles listed ways in which males suffered more from illnesses or were otherwise more fragile than women, concluding, "Females are more resistant to disease, the stress, and strain of life. In general, their biological existence is more efficient, preeminent than of males. In brief, the human male with beard and functioning testes pays the higher price."

More recently, interest and research in sex differences in parasite infections have been expanded in several ways. Firstly, researchers have extended documentation of the parasites themselves to an examination of sex differences in immune response. Here too, at least in most mammals, males tended to be more susceptible to infection, with numerous immune measures suggesting reduced responses in males (Zuk and McKean 1998). Secondly, sex differences in parasite prevalence or intensity were connected to endocrine differences, with a variety of hormones, particularly testosterone and estrogen, implicated in the observed patterns. In particular, testosterone is associated with a suppressed immune system in many mammals, although its action is likely to be mediated by other hormones (see Chaps. 2 and 3 for a much more detailed discussion of this topic). Thirdly, the role of immunity in free-living animals began to attract a great deal of attention, as scientists began to realize that susceptibility to disease was important in an ecological and evolutionary context (Sheldon and Verhulst 1996).

Finally, these observations also were seen to dovetail with another set of findings: males from a variety of mammalian species, including our own, tend to die earlier than females, regardless of the cause. A survey of 227 countries showed that women outlive men in all but a handful of places, whether their lifespan is short, as in Sierra Leone (49 years for women, nearly 44 for men), or long, as in Norway (82 years for women, 76 for men) (Kinsella and Gist 1998). The few countries where men outlive women are almost all in a state of HIV- or conflict-churned crisis, such as Zimbabwe, where women live a scant 35 years to men's 38. The gap between male and female longevity actually increases the longer that both sexes live. Kruger and Nesse compared men's and women's mortality rates for 11 causes of death in men and women from 20 countries, including accidents and homicide as well as infectious and noninfectious diseases (Kruger and Nesse 2006). Men virtually always died earlier than women. They concluded, "Being male is now the single largest demographic risk factor for early mortality in developed countries."

Is there a common thread linking sex differences in parasite prevalence and susceptibility to the higher male mortality that results from all causes? We suggest that an evolutionary approach can unify explanations of sex differences in disease and provide a framework for the research being conducted in this area. Current thinking on the underlying theory behind the evolution of sex differences in many traits, including development of disease, is discussed below. This begins by distinguishing between proximate and ultimate explanations for such differences, as well as for other biological characteristics.

1.2   Levels of Analysis: Proximate and Ultimate Explanations in Biology

Before one can understand why sex differences in parasite susceptibility or immune responses exist, it is important to distinguish between two levels of analysis used for understanding phenomena such as "proximate" and "ultimate". Both are equally valid, but scientists often talk at cross-purposes when they conflate the two.

Proximate explanations are dissections of the mechanism behind a trait, the steps that allow the organism to behave in a particular way or exhibit a characteristic. Proximate causes occur during an individual organism's lifetime, and consist of internal developmental and physiological processes that lead, in the short term, to the phenomenon under consideration.

In contrast, ultimate explanations rely on events that occurred over evolutionary time. Understanding the selection pressures that led to the evolution of certain forms of a trait and not others can help us to understand the adaptive significance of the trait, regardless of the mechanism that makes it happen. Information about the historical sequence of events that took place over the long term, often obtained through a phylogeny of species or other taxa related to the organism in question, can sometimes yield even more information about the evolution of the trait.

Consider, for example, the question of why males of many bird species sing to attract a mate in the springtime rather than at some other time of year. A proximate explanation might invoke hormonal changes triggered by lengthening days that then alter neurochemicals in the vocal center of the bird's brain and prompt it to sing. An ultimate explanation, on the other hand, would seek the benefit that birds confining their singing to such a period would obtain. Presumably, more insects are available in the spring and summer, when the chicks require feeding by their parents, than at other times of year. Individuals that sing, and breed, in the spring are thus more likely to successfully rear their offspring and pass on the genes associated with their responsiveness to the increasing hours of daylight. Both explanations are valid and important to a full understanding of the problem, but they operate at different levels of analysis. Some refer to proximate-level questions as "how" questions and ultimate-level questions as "why" questions, but we think they can both be placed in either format and do not see such a dichotomy as particularly helpful.

With respect to sex differences in susceptibility to parasites, explanations about different hormone levels or the differential exposure of the sexes to the infectious stages of parasites are all proximate explanations. Understanding the interactions among, for instance, testosterone, estrogen, or corticosteroids, and various immune system parameters is important in deciphering the mechanism behind observations or experimental demonstrations of such sex differences, but it does not speak to the selective forces that produced these interactions in the first place. For that, an ultimate explanation is required. Furthermore, focusing at an ultimate level of analysis helps to put "exceptions to the rule" in perspective. If females of a particular species happen to be more susceptible to parasites than are males (as discussed in Chap. 7 of this book), while most other species in the group show the opposite pattern, we can attempt to understand how natural and sexual selection in that species might have produced such a contrary pattern. Discovering that testosterone is not always associated with a suppressed immune system, thus, does not negate the ultimate explanation that males are generally expected to be more susceptible to parasites, though it might call into question the mechanism behind the observation.

1.3   Sexual Selection and Sex Differences in Infection

What, then, is an appropriate framework for addressing the ultimate explanation for sex differences in infection? Here, we briefly review sexual selection theory and current thinking on the evolution of reproductive strategies.

Sexual selection is the counterpart to natural selection, and refers to the differential reproduction of individuals due to competition over mates, as opposed to differential reproduction due to the ability to survive. Like natural selection, sexual selection was originated by Charles Darwin, who distinguished between traits used for survival and those used in acquiring mates. He devoted an entire book, published in 1871, *The Descent of Man and Selection in Relation to Sex*, to the latter. He pointed out that many apparently unusual-appearing traits are actually used in daily life, like the long curved bill on a bird, for example, which may help in feeding. But certain other traits are not so clearly functional, and they are frequently confined to one sex. In some birds of paradise, for instance, the male has a pair of ornamental feathers so long they actually impede his flying ability. Traits such as these are common in the animal kingdom, and include vocal signals like bird and frog song as well as visual signals like elaborate plumage or displays.

Darwin further noted that traits occurring in only one sex could be of two types. First are the primary sexual characters, the basic morphology such as the gonads that enable males to produce sperm and females to produce and nurture eggs. The evolution of these traits is fairly obvious, and requires little special explanation. Other traits, such as the bright colors of many birds or the structures like antlers on male deer, were not so simply understood. Darwin called such traits as secondary sexual characters, and in many cases they are actually detrimental to survival, via an enhanced conspicuousness to predators or other natural enemies or via the energetic cost of producing them.

Darwin proposed that secondary sexual characters could evolve in one of two ways. First, they could be useful to one sex, usually males, in fighting for access to members of the other sex. Hence, the antlers and horns on male ungulates or beetles of some species. These are weapons, and they are advantageous because better fighters get more mates and have more offspring. The second way was more problematic. Darwin noted that females often pay attention to traits like long tails and elaborate plumage during courtship, and he concluded that the traits evolved because the females preferred them. Peahens, thus, were expected to find peacocks with long tails more attractive than those with shorter tails. The sexual selection process, then, consisted of two components: male–male competition, which results in weapons, and female choice, which results in ornaments.

Although the scientific community did not accept sexual selection as readily as natural selection, the theory was finally embraced by the middle of the twentieth century, and research into the evolution of sex differences accelerated. Rather than assuming that females would always be the choosy sex and males the competitive one, however, scientists focused on the ways in which each sex is limited in achieving higher reproductive success.

Evolutionary biologist Robert Trivers (1972) pointed out that females and males usually inherently differ because of how they put resources and effort into the next generation, which he termed parental investment. Female reproductive success is limited by the number of offspring a female can successfully produce and rear. Because they are the sex that supplies the nutrient-rich egg, and often the sex that cares for the young, females will usually leave the most genes in the next generation by having the highest quality young they can; the upper limit to the quantity is usually rather low. Which male they mate with could be very important, because a mistake in the form of poor genes or no help with the young could mean that they have lost their whole breeding effort for an entire year. Ornaments could evolve as indicators of this high quality. Males, on the other hand, can leave the most genes in the next generation by fertilizing as many females as possible. Because each mating requires relatively little investment from him, a male who mates with many females sires many more young than a male mating with only one female.

Variance in male reproductive success is thus expected to be higher, on average, than variance in female reproductive success, which in turn selects for what might be termed a "live hard, die young" overall strategy for males, at least with respect to mating behavior. In elephant seals, for example, a single male may sire more than 90% of the pups in a colony, leaving the vast majority of males with no offspring, while females will virtually always give birth to a single pup. Males battle ferociously among themselves for dominance on the breeding grounds.

With regard to susceptibility to infection, these sex differences in reproductive strategy may provide the ultimate selective force behind increased male vulnerability to infections. If males require, for example, testosterone for aggressive behavior and the development of male secondary sexual characters, selection for winning at the high-stake game that the males play may override the cost in terms of any immunosuppressive effects of the hormone. Sex differences in infection may, thus, simply reflect the larger pattern of differential selection on the sexes.

1.4   The Role of Life History Theory

Testosterone alone, however, is not the sole means by which males and females differ in their physiology. A more general approach to the question of which sex is expected to have evolved greater disease susceptibility comes from life history theory, which examines the evolution of such life "decisions" as how many offspring a species is expected to reproduce and how large those offspring should be at birth or hatching. The underlying assumption is that organisms have a finite pool of energy or resources to draw from, and therefore must allocate that energy to different tasks. Because the resources used for one function are unavailable to another, trade-offs between traits such as growth rate and body size, or between the size and number of offspring, are expected. Life history theory explains many of the apparently maladaptive features of life; animals cannot be good at everything. Along these lines, despite the obvious advantage of being resistant to disease, susceptibility is of course rampant. As with other life history traits, it has seemed logical to conclude that resistance is traded off against the need for investment in other important characters, such as competitive ability or development time (Roff 1992). We assume that animals remain vulnerable to pathogens because being resistant is costly. Evolution has, therefore, not perfected the ability to fend off parasites – i.e., produced organisms that are completely parasite-free – because for most if not all individuals, resources are better expended on other physiological activities or processes.

This view of an animal's reaction to infection as simply another drain on a limited pool of resources provides another kind of ultimate explanation for sex differences in susceptibility to parasites. Combined with sexual selection theory it means that we can begin to ask why we see the patterns that we do, not from the standpoint of an individual species' quirks of immunology, but by examining the way natural and sexual selection are expected to act on life history, including disease resistance.

1.5   Empirical Approaches

One of the earliest discussions of sex differences in disease outcome, from an evolutionary-theoretical perspective was that of Zuk (1990), who emphasized the inherently different means by which males and females maximize reproductive success in many species. In those species where male fitness is heavily dependent upon maximizing mating success (i.e., polygynous species, in which a single male may mate with multiple females), males may benefit from sacrificing immune defense if those resources can, instead, be devoted towards mating efforts. In monogamous species, males typically maximize fitness by assisting in the rearing of offspring, as do the females. Thus, this hypothesis predicts that in monogamous species, males and females will have similarly effective immune defenses, but as the mating system departs further from monogamy towards polygyny (meaning that the strength of sexual selection on males increases), the sex differences in immune defenses, with males showing the less effective defenses, increase (Zuk 1990). Since Zuk (1990), this basic hypothesis and associated predictions have been developed in several other papers (Zuk and McKean 1996; Rolff 2002; Zuk and Stoehr 2002). One of the strengths of this hypothesis, as an "ultimate explanation," is that the predictions apply to taxa other than mammals, including those, such as insects, that lack the hormone testosterone.

A proper test of the hypothesis' primary prediction requires sufficient knowledge of (and variation in) both mating system (or some measure of the strength of sexual selection) and immune defense in a number of species-data that are lacking for many systems, although increasing all the time. Measures of parasitic infections, such as prevalence (proportion of hosts infected) or intensity (number of parasites per host) are typically easier to acquire than more direct measures of immune defense. Nevertheless, the available data on infection levels do highlight interesting patterns, and, not surprisingly, raise more questions. A study examining infection levels across arthropods found no consistent evidence for sex biases in infection prevalence or intensity (Sheridan et al 2000). However, a consistent pattern was lacking not because there were no host taxa for which males were more heavily parasitized, but rather because there were similar numbers of taxa in which females were more heavily parasitized.

Even in vertebrates, where we might expect consistent male-biased infection with parasites because of the immunosuppressive effects of testosterone, things are not so simple. For example, Poulin (1996) found evidence for male-biased parasitic infections in birds when the prevalence of helminth infections was considered, but not when the intensity of infection was considered. McCurdy et al. (1998) found no evidence for an overall sex bias in parasitic infections, but when considered by parasite taxon, the prevalence of *Haemoproteus* infections was female-, not male-biased; this was true even in polygynous species, where the male-biased infections would be most expected. Moore and Wilson (2002) examined the relationship between sexual selection and parasitic infection across mammals. Using methods that controlled for correlations between traits due to shared ancestry, Moore and Wilson (2002) used two measures of the strength of sexual selection – mating system and sexual size dimorphism – to determine if sexual selection was associated with sex differences in infection with parasites. As predicted, increases in polygyny or greater male size were associated with greater sex differences in parasitic infection. One of the most interesting findings of the study was that in those species where females are the larger sex, parasitic infection was female-biased (i.e., females had more parasites). However, in these species, larger female size is not thought to be due to sexual selection on females – thus, the cause and effect relationships among sexual selection, sex differences in parasitic infection, and body size appear complex indeed.

To the best of our knowledge, no large comparative (i.e., multiple species, phylogenetic controls, and sexual selection measures) study utilizing more direct measures of immune defense to address sex differences in immune defenses, rather than parasites themselves, has been conducted. However, an alternative and increasingly popular approach to empirically testing the hypothesis that sexual selection influences sex differences in immune defenses is to experimentally manipulate, in a single species, factors such as the strength of sexual selection, mating history and resource abundance. These studies, too, are revealing that the relationship between sexual selection and immune defense is complex. Indeed, in both invertebrates and vertebrates, the direction or presence of sex differences in immune function may depend upon not only the factors manipulated in the experiment, but also which component(s) of immunity were assessed (Klein 2000; Adamo et al. 2001; Hosken 2001; Fedorka et al. 2005; McGraw and Ardia 2005; McKean and Nunney 2005; Rolff et al. 2005; McKean and Nunney 2008). For example, in crickets, sex differences with phenoloxidase activity, one measure of potential immune defense, were apparent in later stages, but not in earlier stages of development. However, no sex differences were found at any stage for hemocyte number (a count of one of the cell types involved in arthropod immune defense) (Adamo et al. 2001).

1.6 Theoretical Approaches

Given these complex patterns, what are we to make of the underlying evolutionary, i.e., ultimate, reasons for sex differences in immune defense? Were the original formulations of the hypothesis, such as those by Zuk (1990) or Rolff (2002) incorrect? Here, we briefly discuss some of the more recent theoretical investigations into the problem of how sex differences in immune defense might have evolved.

All models, verbal or quantitative, make assumptions. Often, these assumptions are less than obvious; this is particularly true in the case of verbal models. The model as articulated by Zuk 1990, Rolff 2002, and others makes two assumptions that may be important for understanding variation in the magnitude and direction of sex differences in susceptibility to parasitic infection. The first assumption is that female fitness is more dependent upon longevity than is male fitness. The second assumption, which is probably the more important of the two, is that the most important benefit of immunocompetence is to increase survival, or, if one likes, that the primary cost of parasitic infection is death. From the perspective of a resource allocation problem, the model with these assumptions in place can be represented graphically, as in Fig. 1.1a. It is clear that with these assumptions in place, the sex

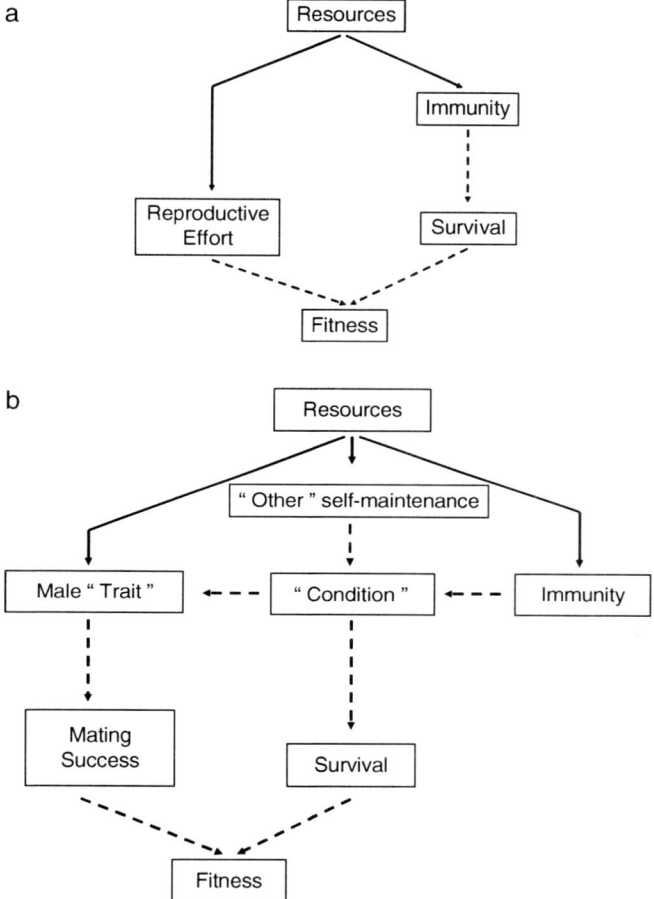

Fig. 1.1 (**a**) Resource allocation to immunity and reproductive effort, assuming that the benefits of immunity only affect survival. In this case, it is clear that the sex that invests the most in survival must necessarily invest more in immunity (solid arrows represent resource allocations; dashed arrows are causal relationships) (**b**) Resource allocation when immunity can affect both survival and mating effort, due to the benefits of immunity for "condition". Shown here is the male case; in females, reproductive effort is simply fecundity

that values survival (typically argued to be females) will be the sex that invests in immune defense.

However, it is not clear how broadly these assumptions apply. For polygynous mammals, it appears that, indeed, longevity is more important for male fitness than for female fitness. But long-term studies in several bird species show that longevity accounts for approximately 60% of the variation in fitness for both females and males, and ranges from about 30 to 80% for both sexes (summarized in Newton, 1989). Longevity may account for a considerable proportion of variation in fitness for both sexes in many insects, as well (Clutton-Brock 1988).

Even in species where longevity is of less importance to males than to females, should we always expect males to invest less in immune defense? Parasites may kill their male hosts, but many infections may reduce the general health or condition of their hosts, which, in turn, may affect traits important for mating success such as bright coloration or energetically expensive courtship behavior, without being lethal. It could be argued such a cost of parasitic infection could be even more detrimental to males than to females, because while a sublethal infection may reduce female fecundity, it may not necessarily prevent her from being mated and rearing some offspring. In some mating systems, however, a parasitized, unhealthy (and therefore less attractive) male may have zero fitness. Thus, the (second) assumption that the primary cost of parasite infection is death, and its implicit accompanying assumption – that the sublethal effects of parasitic infection (e.g., development of disease) are the same for each sex – may not always be true. (This is addressed later – see the reference Blanco et al. 2001 and Tseng 2004)

Stoehr and Kokko (2006) examined the importance of these assumptions by constructing a model of resource allocation to various fitness components, including disease resistance, that would not only allow survival to play an important role in the fitness of both sexes, but more importantly, acknowledge that parasites have sublethal effects, and that these may not be the same for the sexes. In addition, the model incorporates these ideas by also allowing the effects of parasitic infection (and therefore the benefits of immunity) to be realized through the effects of "condition," on the traits that are important to fitness. For the purposes of the model, condition can be defined as that attribute of an organism that is not only affected by resource allocation to it, but also in turn affects other traits such as survival and fecundity; that is, in this model "condition" is what we might generally refer to as the "health" of the organism.

The graphical representation of this model is shown for males in Fig. 1.1b. (The female case is basically the same, except that instead of the male trait and mating success, these are collapsed into female reproductive effort, or fecundity). In the model, resources are allocated to immune defense, reproductive effort (e.g., a male's extravagant plumage or courtship song), and other forms of basic self-maintenance. Immunity, along with other forms of self-maintenance, has positive effects on "condition," and condition in turn has positive effects on survival and on male reproductive effort (i.e., the male "trait"). In this scenario, immunity does have costs, in that immunity and male reproduction compete for limited resources. However, we do not necessarily expect males to simply maximize fitness by investing all resources into reproductive effort, because if immunity is sacrificed entirely, condition, and therefore both survival and reproductive effort, are compromised (the mathematical details of the model, which are explained in Stoehr and Kokko (2006), insure that if no resources are invested in immune defense, then condition, and therefore survival, is zero). Thus, this formulation more realistically represents what we know to be the more general effects of resistance to infection on survival and reproductive effort – i.e., it does not assume that immune defense only evolved in the context of increasing survival.

Stoehr and Kokko (2006) then explored the implications of this model by first constructing a series of mathematical equations that expressed the relationships between these different components of the model and allowed these relationships to take varying shapes. Of primary interest to us for understanding sex differences in immune function are three particular relationships. One is the relationship between the male "trait" and his mating success; this is a measure of the strength of sexual selection. Also of interest is the relationship between immunity and condition. While this could reflect details of the immune system, in the model of Stoehr and Kokko (2006) this is constructed more generally and can be thought of as the impact of parasites and disease outcome on condition. In this manner, it incorporates not only details of immune defense but also variation in parasite combinations, parasite virulence, and behavior that leads to differences in host exposure to parasites, etc. Such a broad approach is important, because the impact of parasites may differ between the sexes; for example, males may be exposed to more (or fewer) parasites because of their courtship behaviors (Tinsley 1989, Zuk and Kolluru 1998). Finally, there is the relationship between condition and reproductive effort. This is, for males, the condition-dependence of traits such as bright coloration, elaborate courtship dances, or loud or complex calls and dances: males in better condition produce more vigorous displays. For females, this is the condition-dependence of fecundity: females in better condition produce more or healthier offspring. Given how different the forms of reproductive effort take for males and females, it would seem highly unlikely that condition would have identical effects on reproductive effort for both sexes. Thus, by varying the shapes of the relationships between immune defense and condition, and condition and reproductive effort, the potential importance of the assumption that the nonlethal effects of parasites are similar (and negligible) for the sexes can be assessed.

Stoehr and Kokko (2006) examined these assumptions numerically, through an evolutionarily stable strategy (ESS) approach. An evolutionarily stable strategy is one that would persist in a population even if a mutant form pursuing an alternative strategy were to enter the population. Stoehr and Kokko (2006) began with an arbitrary resource allocation strategy for a population, given certain parameter values for the strength of sexual selection, the impact of parasites on condition, and the condition-dependence of reproductive effort. Then new resource allocation strategies were explored, and any that resulted in higher fitness could "invade" and replace the old strategy; when the best strategy to adopt is the existing strategy, the evolutionarily stable (i.e., "best") strategy has been achieved.

Recall that the primary prediction of the hypothesis for sexual dimorphism of immune defense is that as the strength of sexual selection increases, the magnitude of the difference between sexes, with males showing an inferior immune response, is expected to increase. Stoehr and Kokko (2006) found that, indeed, this prediction is supported provided that (a) the impact of parasites on condition is the same for the sexes; (b) the condition-dependence of reproductive effort is the same for the sexes; and (c) neither of these effects is particularly strong. If instead parasites are

highly detrimental to condition and/or reproductive effort is highly dependent on
condition, then males cannot afford to sacrifice immune defense to improve
mating success, even in the face of very strong sexual selection. As a result,
both sexes invest in immune defense equally. More importantly, the model shows
that if the impact of parasites on condition is greater for males than for females,
males should invest more of their resources into immune defense than
should females, even in the face of strong sexual selection (Fig. 1.2). A similar,
though not quite as dramatic, effect is found if male reproductive effort is more
condition-dependent than is female reproductive effort. In other words, even if the
effects of sexual selection are to diminish male investment in immunity below
that which would occur in the absence of sexual selection altogether, this dimin-
ishment may still not be sufficient to cause males to invest less in immunity than
do females (Fig. 1.2; *upper thin solid line*).

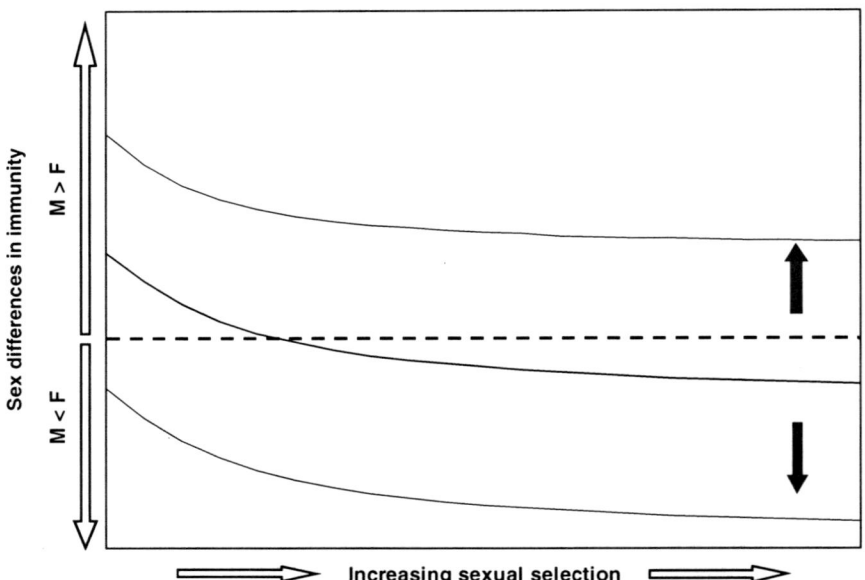

Fig. 1.2 Sex differences in immunity as a function of sexual selection. The thick solid line
represents the case when the condition-dependence of reproduction and the effect of immunity
on condition are equal for the sexes; when sexual selection is absent or weak males should invest
more in immune defense than should females (i.e., thick, solid black line is above the dashed line,
in the region of M>F investment in immunity). As the strength of sexual selection increases, the
female bias in investment in immunity increases. However, if parasites have particularly strong
negative effects on condition in males, and/or if male reproductive success is highly dependent on
condition, relative to those same effects in females, males should invest more in immunity than
should females, even when sexual selection is strong (thin solid line raised above the thick solid
line, and never crossing dashed line). Of course, the converse situation may mean that males never
invest more in immunity than do females (lower thin solid line)

The results of the simulation by Stoehr and Kokko (2006) suggest that the validity of the assumptions implicit in the verbal models arguing for inferior male immune defenses when sexual selection is strong may be very important. We not only know that in many cases male secondary sexual traits are condition-dependent, but, in fact, theory suggests that we should expect these traits to be condition-dependent (Andersson 1994). Of course, we also expect female fecundity to be condition-dependent, so the question, for our purposes, becomes "When do we expect male fitness to be more condition-dependent than female fitness?" because these are the cases where we might (given certain other assumptions) expect males to invest more in their immune defense than do females. Unfortunately, as any biologist who has ever tried to quantify (or even define!) condition will realize immediately, comparing condition and condition-dependence between the sexes is hardly trivial. It would not simply be enough to examine the correlation between some measure of condition and secondary sexual trait (for males) and fecundity (for females) because ultimately, we would also need to know something about how that male secondary sex trait expression translates into fitness. However, there may be some well-studied systems where such a comparison might be possible.

Perhaps, a slightly more tractable question is whether similar parasitic infections affect the condition of the sexes equally. This question is not free from the inherent difficulties of measuring condition, but there is at least some evidence to suggest that, when such a comparison can be made, the answer is that parasites do not always have the same effects on male and female condition (Blanco et al. 2001; Tseng 2004). For example, in magpies, there is a negative correlation between lice infestation and nutritional condition (in this case, body mass adjusted for skeletal size) in both sexes, but the relationship is stronger for males (Blanco et al. 2001). And in mosquitoes, infection with parasites reduces male body size more than it does female body size when the mosquito larvae are reared at high density, but at low larval densities, parasites have a greater impact on female body size (Tseng 2004). Furthermore, because the model of Stoehr and Kokko (2006) includes potential exposure differences as part of "parasitic impact," behaviors that bias exposure in one sex may also be important, and such behaviors have been found (Tinsley 1989; Zuk and Kolluru 1998; Riemchen and Nosil 2001). Finally, it must be remembered that these two important effects – i.e., the impact of parasites on condition and the condition-dependence of reproductive effort – may interact in concert, to increase the magnitude of sex differences in immunity, or in opposition, to diminish or erase sex differences in immunity.

Like all models, Stoehr and Kokko's (2006) make its own assumptions and has its own limitations. The primary purpose of this model was to examine the logic of the basic arguments (or, put another way, the importance of the implicit assumptions) put forth in earlier less quantitative treatments of the sexual selection versus male immune defense hypothesis. As such, the model is successful as it reveals that these assumptions may be crucial in understanding how sexual selection and immune defense interact to produce or eliminate sexual dimorphism in immune defense. However, it is not a detailed model of immune defense. For example, Stoehr and Kokko (2006) ignore potentially important factors such as the complex and multifaceted nature of immune defenses, host–parasite coevolution, and the genetics of resistance. In addition, the model ignores the possibility that individuals (or the sexes) may differ in the amount of resources they acquire.

Although it seems unlikely that incorporating any of these factors will reveal that things are more simple than they appear, these are certainly factors that should be incorporated, in as much as is possible, in future theoretical and empirical approaches to understanding sexual dimorphism in immune function. Indeed, several recent models addressing optimal allocation of resources to immune defense raise several interesting points. None of these models addressed sex differences in immunity, but their findings should be incorporated into future theoretical treatments of this problem. For example, one of the underlying assumptions of earlier treatments of sex differences in susceptibility to infection and the manifestation of disease was that females would invest more in immune defense because they are often the longer-lived sex; that is, it was assumed that inherently long-lived organisms would favor immune defense greater than short-lived organisms. This assumption is challenged in models by van Boven and Weissing (2004) and Miller et al. (2007). Both of these studies found that, under some conditions, optimal investment in immune defense is maximal at intermediate lifespans, not at the longest lifespans. One of the reasons this appears to be so is because of demographic processes: long-lived species do not have high demographic turnover, and therefore do not supply the "fuel," i.e., susceptible individuals, necessary to support some species of parasites (van Boven and Weissing 2004; Miller et al. 2007). As a result, there is less benefit to investing in costly immune defenses in these species. Not surprisingly, however, these conclusions depended on certain assumptions as well; for example, if immunity was innate, instead of acquired, then optimal investment increased with lifespan (Miller et al. 2007).

As mentioned above, Stoehr and Kokko's (2006) model did not consider that males and females might start with differently sized resource pools. Sex differences in resource acquisition might occur, however, if one sex is forced, to a greater degree than the other, to sacrifice, say, foraging effort in order to invest in reproduction. In a model of optimal resource allocation to immune defense, Medley (2002) found that optimal allocation of resources to immune defense calls for little to no allocation in starved individuals, peaks in those individuals with intermediate levels of resources, but then falls again in "well-fed" individuals. Hosts with more resources, i.e., "in better condition," may be better able to tolerate some level of infection, such that the relationship between parasite loads and condition or "quality" may be complex (Medley 2002). A similar problem was addressed by Houston et al. (2007), who modeled optimal allocation of efforts to foraging versus immune defense. In addition, Houston et al. (2007) show that whether individuals of a given state invest primarily in foraging or immune defense is not simply a matter of current nutritional state, but of environmental predictability. In more stable environments, food availability and allocation to immune defense tend to be positively related, but as the environment becomes more unpredictable, this relationship no longer holds.

1.7 Future Directions

Comparative studies of parasite infections in many different kinds of animals, as well as experimental studies of immune defense in single species and theoretical explorations of the role of resource allocation in the evolution of immunity, all suggest that it is simplistic to expect one sex to routinely have an inferior immune ability, even in species in which sexual selection has been intense. The original hypothesis that males were likely to have evolved a greater susceptibility to parasites was on the right track, in that it identified a useful way of thinking about the evolution of such sex differences. A more general perspective on the problem of resource allocation to defense against parasites as well as other outlets should prove even more valuable. The collective findings, both empirical and theoretical, clearly support the idea that life history differences between the sexes matter in understanding sex differences in disease, and that these differences can be most profitably understood in an evolutionary framework. The challenge now is to understand exactly how the differences matter; when we understand the details and mechanisms, we will be able to see why sex differences in immunity are sometimes male-biased and at other times female-biased.

To achieve this understanding, we suggest that a number of issues should be addressed. More large-scale comparative studies, conducted in a phylogenetic context, which examine immunity across species in a variety of taxa to uncover important correlates of sex differences in immunity, will be invaluable. These types of studies can reveal broad, consistent patterns and identify potentially important causal factors that can then be addressed experimentally. However, note that the evidence to date suggests that sex differences in immunity are dynamic, and may change over the course of the life history of an organism, due to changes in external factors such as resource abundance, and may vary with different components of immune defense or different parasites. For example, in *Drosophila melanogaster*, female larvae are more resistant than male larvae to a larval parasite, there are no sex differences in resistance to a pupal parasite, whereas in adult flies, there are sex differences in resistance to a microsporidian, but not to a fungal, infection (Kraaijeveld et al. 2008). Furthermore, sex differences in resistance to bacterial infection in adult *Drosophila* are highly labile: sexual activity reduces male but not female resistance, whereas resource deprivation reduces female but not male resistance, resulting in variation in the direction of sex differences in immunity depending upon how these factors are manipulated (McKean and Nunney 2005).

A relatively unexplored but potentially fruitful area of research is the intersection between population dynamics and sex differences in parasite resistance. For example, in free-living yellow-necked mice, antihelminthic treatment of a dominant parasitic helminth in males reduces infections in females in the population as well, but removal of the same parasite from females has no effect on infections in males (Ferrari et al. 2004). There is also ample evidence from a variety of species that immunity varies seasonally (Nelson and Demas 1996; Altizer et al. 2006; Martin et al. 2008). In the future, we hope to see these kinds of ecological factors considered alongside the life history perspective we have outlined here, and these, in turn, combined with approaches that consider the multifaceted nature of the immune system (Lee 2006). The result should be a much greater, integrative understanding of sex differences in immunity than could be achieved by any single approach alone.

References

Adamo SA, Jensen M, Younger M (2001) Changes in lifetime immunocompetence in male and female *Gryllus texensis* (formerly *G. integer*): trade-offs between immunity and reproduction. Anim Behav 62:417–425

Addis CJJ (1946) Experiments on the relations between sex hormones and the growth of tapeworms *(Hymenolepis diminuta)* in rats. J Parasitol 32:574–580

Alexander J, Stimson WH (1988) Sex hormones and the course of parasitic infection. Parasitol Today 4:189–193

Altizer S, Dobson A, Hosseini P, Hudson P, Pascual M, Rohani P (2006) Seasonality and the dynamics of infectious diseases. Ecol Lett 9:467–484

Andersson M (1994) Sexual selection. Princeton University Press, Princeton NJ

Blanco G, De la Puente J, Corroto M, Baz T, Colas J (2001) Condition-dependent immune defence in the Magpie: how important is ectoparasitism? Biol J Linn Soc 72:279–286

Clutton-Brock TH (ed) (1988) Reproductive success. University of Chicago Press, Chicago IL

Darwin C (1871) The descent of man and selection in relation to sex. Modern Library, New York

Fedorka KM, Zuk M, Mousseau TA (2005) Natural selection drives the link between male immune function and reproductive potential. Can J Zool 83:1012–1014

Ferrari N, Cattadori IM, Nespereira J, Rizzoli A, Hudson PJ (2004) The role of host sex in parasite dynamics: field experiments on the yellow-necked mouse *Apodemus flavicollis*. Ecol Lett 7:88–94

Hosken DJ (2001) Sex and death: microevolutionary trade-offs between reproductive and immune investment in dung flies. Curr Biol 11:R379–R380

Houston AI, McNamara JM, Barta Z, Klasing KC (2007) The effect of energy reserves and food availability on optimal immune defence. Proc R Soc B 274:2835–2842

Kinsella K, Gist YJ (1998) Mortality and health. International brief: gender and aging. US Department of Commerce, Economics and Statistics Administration, Bureau of the Census, Washington DC, USA

Klein SL (2000) Hormones and mating system affect sex and species differences in immune function among vertebrates. Behav Proc 51:149–166

Kraaijeveld AR, Barker CL, Godfray HCJ (2008) Stage-specific sex differences in *Drosophila* immunity to parasites and pathogens. Evol Ecol 22:217–228

Kruger DJ, Nesse RM (2006) An evolutionary life-history framework for understanding sex differences in human mortality rates. Hum Nat 17:74–97

Lee KA (2006) Linking immune defenses and life history at the levels of the individual and the species. Integr Comp Biol 46:1000–1015

Martin LB, Weil ZM, Nelson RJ (2008) Seasonal changes in vertebrate immune activity: mediation by physiological trade-offs. Philos Trans R Soc B 363:321–339

McCurdy DG, Shutler D, Mullie A, Forbes MR (1998) Sex-biased parasitism of avian hosts: relations to blood parasite taxon and mating system. Oikos 82:303–312

McGraw KJ, Ardia DR (2005) Sex differences in carotenoid status and immune performance in zebra finches. Evol Ecol Res 7:251–262

McKean KA, Nunney L (2005) Bateman's principle and immunity: phenotypically plastic reproductive strategies predict changes in immunological sex differences. Evolution 59:1510–1517

McKean KA, Nunney L (2008) Sexual selection and immune function in *Drosophila melanogaster*. Evolution 62:386–400

Medley GF (2002) The epidemiological consequences of optimisation of the individual host immune response. Parasitol 125:S61–S70

Miller MR, White A, Boots M (2007) Host life span and the evolution of resistance characteristics. Evolution 61:2–14

Moore SL, Wilson K (2002) Parasites as a viability cost of sexual selection in natural populations of mammals. Science 297:2015–2018

Nelson RJ, Demas GE (1996) Seasonal changes in immune function. Q Rev Biol 71:511–548

Newton I (1989) Synthesis. In: Newton I (ed) Lifetime reproduction in birds. Academic Press, San Diego, pp 441–469

Poulin R (1996) Sexual inequalities in helminth infections: a cost of being male? Am Natur 147:287–295

Riemchen TE, Nosil P (2001) Ecological causes of sex-biased parasitism in three spine stickleback. Biol J Linn Soc 73:51–63

Roff DA (1992) The evolution of life histories: theory and analysis. Chapman & Hall, New York

Rolff J (2002) Bateman's principle and immunity. Proc R Soc B 269:867–872

Rolff J, Armitage SAO, Coltman DW (2005) Genetic constraints and sexual dimorphism in immune defense. Evolution 59:1844–1850

Shettles LB (1958) Biological sex differences with special reference to disease, resistance and longevity. J Obstet Gyn Brit Empire 65(2):288–295

Sheldon BC, Verhulst S (1996) Ecological immunology: costly parasite defences and trade-offs in evolutionary ecology. Trends Ecol Evol 11:317–321

Sheridan LAD, Poulin R, Ward DF, Zuk M (2000) Sex differences in parasitic infections among arthropod hosts: is there a male-bias? Oikos 88:327–334

Solomon GB (1966) Development of *Nippostrongylus brasiliensis* in gonadectomized and hormone-treated hamsters. Exp Parasitol 18:374–396

Stoehr AM, Kokko H (2006) Sexual dimorphism in immunocompetence: what does life-history theory predict? Behav Ecol 17:751–756

Tinsley RC (1989) The effects of host sex on transmission success. Parasitol Today 5:190–195

Trivers RL (1972) Parental investment and sexual selection. In: Campbell B (ed) Sexual selection and the descent of man, 1871–1971. Heinemann, London, pp 136–179

Tseng M (2004) Sex-specific response of a mosquito to parasites and crowding. Proc R Soc B 271:S186–S188

van Boven M, Weissing FJ (2004) The evolutionary economics of immunity. Am Nat 163:277–294

Zuk M, McKean KA (1996) Sex differences in parasite infections: patterns and processes. Int J Parasitol 26:1009–1024

Zuk M (1990) Reproductive strategies and sex differences in disease susceptibility: an evolutionary viewpoint. Parasitol Today 6:231–233

Zuk M, Stoehr AM (2002) Immune defense and life history. Am Nat 160:s9–s22

Zuk M, Kolluru GR (1998) Exploitation of sexual signals by predators and parasitoids. Q Rev Biol 73:415–438

Chapter 2
Effects of Sex Steroids on Innate and Adaptive Immunity

S. Ansar Ahmed, Ebru Karpuzoglu, and Deena Khan

Abstract Estrogens and androgens are classically recognized as reproductive sex steroid hormones because of their well-documented effects on reproductive tissues. However, extensive research in diverse biological disciplines have clearly established that reproductive hormones have broad physiological effects on nonreproductive tissues, including the immune, central nervous, cardiovascular, and skeletal systems. Thus, the term "sex/reproductive hormones" describes only a narrow (albeit important) aspect of biological effects of sex steroids. In this chapter, the effects of sex hormones on the innate and adaptive immune system are highlighted. Generally, estrogens upregulate proinflammatory cytokines (e.g., IFNγ) and IFNγ-inducible molecules (nitric oxide, NOS2, and COX2), whereas androgens suppress proinflammatory responses. Immunomodulation by sex steroids may have both physiological and pathological implications (e.g., sex differences in immune capabilities and in inflammatory diseases, respectively).

2.1 Sources of Sex Steroids: Physiological and Exogenous

Estrogens are produced in gonadal and extra-gonadal tissues. 17β-estradiol, is principally produced by theca and granulosa cells in the ovaries of premenopausal women (Simpson 2003). In theca cells, androstenedione is converted into testosterone by aromatase. Testosterone and androstenedione are then taken up by granulosa

S. Ansar Ahmed (✉)
Center for Molecular Medicine and Infectious Diseases, Department of Biomedical Sciences and Pathobiology, Virginia-Maryland Regional College of Veterinary Medicine, Virginia Tech, Blacksburg, VA 24060, USA
e-mail: ansrahmd@vt.edu
Current Address: Institute of Genes and Transplantation, Baskent University, Ankara, Turkey

S.L. Klein and C.W. Roberts (eds.), *Sex Hormones and Immunity to Infection*, DOI 10.1007/978-3-642-02155-8_2, © Springer-Verlag Berlin Heidelberg 2010

cells and converted by the aromatase enzyme to 17β-estradiol in premenopausal women (Williams et al. 1998). The levels of estrogen in women vary physiologically during menstrual cycle stages, pregnancy, and with age. In premenopausal women, the physiological range of estrogen during the menstrual cycle is between 40 and 400 pg ml^{-1} (Ruggiero and Likis 2002). Estrogen levels are markedly increased during pregnancy. In the postmenopausal stage, estrogen levels drop significantly (Akhmedkhanov et al. 2001). In mice, the physiological levels of 17β-estradiol found in the serum are as follows: diestrus 20–30 pg ml^{-1}, estrus 100–200 pg ml^{-1}, and during pregnancy 5,000–10,000 pg ml^{-1} (Bebo et al. 2001). Extragonadal estrogen synthesis occurs in mesenchymal cells of adipose tissue, breast, osteoblasts and chondrocytes of bone, vascular endothelium, aortic smooth muscle cells, and several sites in the brain (Simpson 2003). These sites are important sources of estrogen in men and postmenopausal women. Estrogens produced at these sites, unlike that secreted by ovaries, tend to act locally at high concentrations (Labrie et al. 1998).

In addition to natural estrogens, synthetic estrogens (e.g., 17α-ethinyl estradiol) are extensively prescribed as oral contraceptives to premenopausal women and as estrogen replacement therapy for postmenopausal women (Yin et al. 2002). Additionally, a third category of estrogenic compounds, referred to as environmental estrogens, is now recognized. Environmental estrogens account for a large component of endocrine disrupting chemicals (EDC). A number of EDC, including estrogens also have been shown to affect the immune system (Ansar Ahmed 2000). These estrogens can mimic or block natural hormones. Due to their ability to accumulate in adipose tissue and the fact that they are biologically active at very low concentrations, EDCs can accumulate and act cumulatively to alter the immune and reproductive systems (Soto et al. 1995). In this chapter, we detail the effects of estrogens (in particular, 17β-estradiol or E2) and androgens (in particular, testosterone) on the immune system; progesterone, another sex steroid that has profound effects on the functioning of immune cells, is discussed in detail in Chaps. 9 and 10 pertaining to pregnancy.

Testosterone is the principal androgen secreted from Leydig cells in testes of males and in small quantities from theca cells in ovaries of females. Importantly, testosterone is essential not only in sexual development and other reproductive processes but also for modulating immune responses. Males are, in general, more prone to infectious diseases, both in terms of prevalence and intensity, over females partly because of the suppressive effects of testosterone and its metabolite dihydro-testosterone (DHT) on the immune system (Choudhry et al. 2006; Easterbrook et al. 2007; Schuurs and Verheul 1990). Castration has beneficial effects on the immune system following trauma and hemorrhage (Yokoyama et al. 2002). Consistent correlations have been observed between endogenous testosterone levels and the burden of parasites, such as *Babesia microti* and *Plasmodium vivax* (Barnard et al. 1996; Muehlenbein et al. 2005), which is detailed in Chap. 6 of this book. In addition to affecting responses to parasites, androgens affect the development of the immune system as castration results in increased thymus size in mice (Olsen et al. 1991).

2.2 Sex Steroid Regulation of Innate Immune Cells

Antigen invasion into the body is largely prevented by physical barriers that act as a first line of defense. These physical barriers (such as skin, mucosal tissue of gastrointestinal, respiratory, reproductive, and urogenital tracts) are usually fortified by chemical barriers (e.g., mucus, saliva, and tears which contain protease enzymes). Cellular innate immune cells such as neutrophils, macrophages, natural killer (NK) cells, and dendritic cells (DCs) act as the second line of defense. The nature of this response enables effector cells to recognize a number of molecules widely expressed by groups of microbes and to clear or curtail their multiplication by various mechanisms such as phagocytosis and lysis of infected cells. Innate immune cells, predominantly macrophages and DCs, produce cytokines that aid in the activation and influence the nature of the adaptive immune system.

2.2.1 Neutrophils

Neutrophils or polymorphonuclear cells (PMNs) are the "first cellular responders" to counter antigenic invasion. These cells kill pathogens by two complementary effective mechanisms (1) phagocytosis, and (2) release of potent toxic oxygen-free radicals generated by a respiratory burst. Neutrophils contain a population of primary and secondary granules. Primary granules are composed of enzymes, including myeloperoxidase, acid hydrolases (i.e., cathepsins), lysozymes, and neutral proteases. Secondary granules consist of lactoferrin, lysozymes, and collagenases. Neutrophils migrate to sites of injury or inflammation in response to chemoattractants released by damaged tissues where they trap antigen in a vacuole called a phagosome. Phagosomes fuse with primary granules to form phagolysosomes (Faurschou and Borregaard 2003). The phagolysosome is a hostile environment capable of destroying many, but not all, pathogens. Concurrently, neutrophils aid in destroying microbial pathogens by another mechanism called respiratory burst by sequential conversion of oxygen to toxic superoxide anion, hydrogen peroxide, and hypochloride ion in the presence of NADPH oxidase, superoxide dismutase, and myeloperoxidase, respectively (Hampton et al. 1998).

Estrogen regulates both the number and function of neutrophils. For example, estrogen has been shown to suppress bone marrow production of leukocytes including PMNs (Josefsson et al. 1992; Wessendorf et al. 1998). This possibly is in part due to estrogen effects on the bone (osteopetrosis), which tends to occlude the bone marrow cavity. 17β-estradiol, ethinyl estradiol, and idoxifene, a selective estrogen receptor modulator (SERM), but not 17α-estradiol, significantly reduce neutrophil chemotaxis (Delyani et al. 1996; Ito et al. 1995) as well as adherence to the vascular endothelium (Delyani et al. 1996). Although 17β-estradiol prevents neutrophil infiltration and organ damage following trauma-hemorrhage, the mechanism by which it inhibits neutrophil transmigration remains unknown. Estrogens

can alter neutrophil chemotaxis and function by modulating the release of chemoattractants such as CXCL8 from monocytes (Pioli et al. 2007), and CXCL8,
CXCL10, CCL5 from keratinocytes (Kanda and Watanabe 2005). Furthermore,
estrogens decrease chemotaxis of neutrophils by altering the expression of adhesive
proteins, such as intracellular adhesion molecule-1 (ICAM-1) and therefore, protect
against myocardial ischemia-reperfusion injury and myeloperoxidase activity
(Squadrito et al. 1997).

17β-estradiol and weak estrogenic analogs (e.g., estrone and estriol) significantly
reduce neutrophil function as indicated by decreased superoxide anion production (Abrahams et al. 2003; Bekesi et al. 2007). Interestingly, neutrophils in females
have increased resistance to activation by burn or trauma hemorrhage compared
with those in males (Deitch et al. 2006). 17β-estradiol is capable of limiting
neutrophil activation, as reflected by decreased CD11b expression and respiratory
burst activity in response to trauma-hemorrhagic shock or burn injury (Deitch et al.
2006). The salutary effects of estrogen on attenuation of inflammatory responses
are mediated by decreased neutrophil infiltration at sites of injury/inflammation,
improved injury markers (e.g., myleoperoxidase activity), decreased cytokine production (e.g., TNF-α, IL-6, and IL-1β), reduced chemokine levels (e.g., cytokine
induced neutrophil chemo-attractants (CINC1, CINC2, and CINC3), reduced
monocyte chemoattractant protein-1 (MCP1 or CCL2)) and decreased expression
of inflammatory mediators (e.g., P-selectin and intercellular adhesion molecule
(ICAM1)) (Cuzzocrea et al. 2008; Hsu et al. 2007; Yu et al. 2007).

Despite the fact that estrogens can affect neutrophil-mediated immune
responses, some reports suggest that estrogen does not reduce neutrophil infiltration
into cardiac muscle (Cavasin et al. 2006; Tiidus et al. 2002), myeloperoxidase
activity (Tiidus et al. 2002), or neutrophil degranulation and oxidation (Cave et al.
2007; Chiang et al. 2004). However, during endometriosis, an estrogen-dependent
autoimmune disorder affecting women of reproductive age, estrogen enhances the
responsiveness of cells to IL-1β, which acts directly to upregulate CXCL8 (i.e.,
IL-8), a chemokine involved in active angiogenesis and recruitment of neutrophils
(Akoum et al. 2001). This suggests that estrogen may not attenuate inflammation in
all cases, and a number of variables such as the dose of estrogen, tissue, and type of
injury/inflammation may influence the immunomodulatory effect of estrogen.

2.2.2   Macrophages

The term macrophage is derived from the Greek words: macros: large, great, and
phagein: eat; "large eating cell." Macrophages tend to follow neutrophils to sites of
injury/inflammation. These cells, unlike neutrophils, are capable of repeated phagocytosis and have the ability to secrete copious amounts of inflammatory proteins,
including cytokines. Selected macrophages also have the ability to process and
present antigens. Macrophages are key innate immune cells and are one of the

important targets of estrogen within the immune system (Ansar Ahmed et al. 1999). Estrogen increases murine and human macrophage phagocytic activity (Baranao et al. 1992). Further, in mice, the percentage of macrophages in the endometrial stromal and myometrial connective tissues of the cycling uterus changes relative to the stages of the estrus cycle (De and Wood 1990).

Generally, androgens inhibit the function of macrophages in vivo and in vitro (Miller and Hunt 1996). Androgen receptors (AR) have been identified in primary cultured macrophages (Cutolo et al. 1996). Stimulation of murine macrophages with testosterone in vitro reduces the synthesis of proinflammatory products, including TNF-α and nitric oxide synthase (D'Agostino et al. 1999). Testosterone also reduces toll-like receptor (TLR) 4 expression on macrophages (Rettew et al. 2008). Testosterone attenuates *Leishmania donovani*-mediated p38MAPK activation of macrophages, which is considered to be the cause of testosterone-enhanced *L. donovani* survival in macrophages (Liu et al. 2006). Furthermore, androgens such as testosterone, DHT, mesterolone, and danazol modulate the clearance of IgG-sensitized erythocytes by decreasing macrophage FcγR expression (Gomez et al. 2000).

2.2.3 Dendritic Cells

DCs are highly potent APCs, which activate naive T lymphocytes and assist in regulation of Th1 and Th2 development. As also discussed in Chap. 3 of this book, 17β-estradiol promotes differentiation of functional DCs from precursor cells. In vitro estrogen exposure of splenic DCs from rats with experimental allergic encephalomyelitis (EAE) (Zhang et al. 2004), neural DCs from mice with EAE (Liu et al. 2002) or murine bone marrow-derived DCs (Siracusa et al. 2008) increases the expression of markers of DC activation, including major histocompatibility complex II (MHCII), CD80 (B7.1), CD86 (B7.2), and CD40. Estrogen receptor (ER) antagonists, ICI 182 780 and tamoxifen, inhibit DC differentiation (Paharkova-Vatchkova et al. 2004), which is restored by the addition of physiological concentrations of estrogen (Paharkova-Vatchkova et al. 2004). The activity of IFN-producing killer DCs (IKDCs) is increased in spleens from estrogen-treated as compared with ovariectomized C57BL/6 female mice (Siracusa et al. 2008). Estrogen-treated DCs induce IKDCs and increase nitric oxide (Zhang et al. 2004). Treatment of murine splenic DCs with estrogen increases intracellular IL-6 and IL-10 expression, but not IL-12 or TNF-α (Yang et al. 2006). Human monocyte-derived immature DCs have increased IL-6, CXCL8, and CCL2 secretion after short-term in vitro estrogen treatment (Bengtsson et al. 2004). In contrast to estrogen, testosterone decreases the production of inflammatory cytokines (including IL-1β, IL-6, and TNFα) from DCs (Corrales et al. 2006).

During pregnancy, when high levels of estrogen are evident, maturation of monocytes to mature DCs as determined by expression of CD80, CD86,

HLA-DR, and IL-12p70 is diminished. However, the levels of IL-10 are increased from pregnant women when compared with nonpregnant women (i.e., in women with lower physiological levels of estrogen) (Bachy et al. 2008). Splenic DCs stimulated with GM-CSF and IL-4 and exposed to high pregnancy levels of estrogen (500–2,000 pg ml^{-1}) have decreased secretion of IFN-γ, IL-12, and TNF-α (Liu et al. 2002).

2.2.4   Natural Killer Cells

NK cells (CD3$^-$CD16$^+$CD56$^+$ or CD3$^-$CD16$^-$CD56$^+$ in humans and CD3$^+$CD49b$^+$ in mice) are large granular lymphocytes with the ability to lyze pathogen-infected cells and certain tumor cells. Estrogen decreases NK cell activity both in vivo and in vitro (Baral et al. 1995; Ferguson and McDonald 1985; Screpanti et al. 1987). In vivo 17β-estradiol treatment upregulates the number of NK cells and expression of CD69, NKp46, NKG2DL, and 2B4 (CD244), NK cell-activating receptors, granzyme B, and soluble FasL (Hao et al. 2007). Estrogen induces the granzyme B inhibitor, SerpinB9/proteinase inhibitor 9 (PI9) in NK cell lines (Jiang et al. 2007). Although 17β-estradiol increases the number of NK cells (Hao et al. 2007), exposure to 17β-estradiol reduces their cytotoxicity in a dose-dependent manner. This effect can be observed in many mouse strains, but to varying degrees. While estrogen suppresses (>50%) NK cell cytotoxicity in C3H/N, DBA/1, and NZB/W strains of mice, this effect is less pronounced (<30%) in C57BL/6 and MRL lpr/lpr mice (Nilsson and Carlsten 1994). The duration of estrogen treatment also differentially affects NK cell activity. Short-term in vivo treatment with estrogen enhances NK cell activity, whereas prolonged exposure to estrogen decreases NK activity (Screpanti et al. 1987). Decreased NK cell activity occurs in the presence of 17β-estradiol in both WT and ERα knockout mice, suggesting that the effects on NK cells are through the ERβ (Curran et al. 2001). In postmenopausal women, NK cell cytotoxic activity is increased compared with premenopausal women (Albrecht et al. 1996). Oral and transcutaneous 17β-estradiol treatment of postmenopausal women decreases NK cell cytotoxicity (Albrecht et al. 1996). Although 17β-estradiol diminished NK cell activity, 17β-estradiol exposure of activated NK cells, and invariant NKT (iNKT) cells from in vivo α-galactosylceramide (α-GalCer)-stimulated female mice increases IFN-γ production (Gourdy et al. 2005; Nakaya et al. 2006). On the other hand, decreased IFN-γ secretion is observed, from phorbol 12-myristate 13-acetate and ionomycin stimulated NK cells, in pregnant women compared with nonpregnant women. This could be mediated by the high levels of 17β-estradiol present during pregnancy (Bouman et al. 2001).

   Testosterone administration to TA3 (Tianjin Albino 3) mice for 2 weeks causes substantial reduction in NK cell activity (Hou and Zheng 1988). In humans, however, dihydroepiandrosterone supplementation in postmenopausal women decreases CD4$^+$ T cells, but increased CD8$^+$/CD56$^+$ NK cells (Casson et al. 1993).

2.3 Sex Steroid Regulation of Proinflammatory Cytokines and Chemokines

2.3.1 TLR Signaling

Innate immune cells, predominantly macrophages, express toll-like receptors (TLRs) which recognize specific sequences on microbes called pathogen-associated molecular patterns (PAMPs) (Miyake 2007). Therefore, these molecules serve as pattern recognition receptors that sense and initiate an innate immune response to pathogens or other antigenic stimuli. So far, 13 TLRs have been identified in mammals (Roach et al. 2005), of which from TLR1 to TLR9 are conserved among both humans and mice. Some of the TLRs are located on the cell membrane (e.g., TLR1, TLR2, TLR4, TLR5, and TLR6), while others are found intracellularly in endosomes (e.g., TLR3, TLR7, TLR8, TLR9) (Krishnan et al. 2007). TLR1, found on the plasma membrane, can form a heterodimeric complex with TLR2, a receptor that recognizes a wide variety of fungal and protozoal products, including peptidoglycans (Wang et al. 2001a), mycobacterial lipoarabinomannan (Tapping and Tobias 2003) as well as a structurally different LPS from *Bacteroides fragilis, Chlamydia trachomatis,* and *Pseudomonas aeruginosa* (Erridge et al. 2004). TLR2, in addition to endogenous ligands such as heat shock protein 70 (Hsp70) (Vabulas et al. 2002), can recognize diacylated lipopeptides in combination with TLR6 (Nakao et al. 2005). TLR5 is basolaterally expressed on the surface of intestinal epithelial cells and activated by bacterial flagellin (Tallant et al. 2004). The intracellular TLRs are important for recognition of intracellular pathogens such as viruses and include TLR3 (recognizes double-stranded (ds) RNA), TLR7/TLR8 (recognizes single-stranded (ss) RNA) (Takeshita et al. 2004), and TLR9 (recognizes unmethylated bacterial and viral CpG DNA motifs) (Hemmi et al. 2000).

TLR4 is the best characterized TLR (Rock et al. 1998), which is activated by PAMPs such as lipopolysaccharide (LPS) on Gram negative bacteria (Beutler et al. 2001), taxol (which mimics LPS) (Kawasaki et al. 2003), and endogenous ligands (Beg 2002). LPS activation of TLR4 requires the association of three different proteins, CD14, LPS binding protein (LBP), and MD-2 (Shin et al. 2007). LBP, a soluble protein, binds directly to LPS and initiates the binding between LPS and CD14 (Finberg et al. 2004). CD14, a glycosylphosphatidylinositol-anchored protein, is critical in modulating specific recognition of LPS (Muta and Takeshige 2001). MD-2, a secreted soluble glycoprotein, acts as an extracellular adaptor by binding to TLR4 and helps TLR4 in ligand recognition (Miyake 2004). Activation of TLR4 can result in initiation of two distinct signaling mechanisms: a myeloid differentiation factor 88 (MyD88)-dependent and MyD88-independent pathway. MyD88-dependent pathway-induced genes are mostly proinflammatory molecules such as IL-1β, IL-1α, COX2, CXCL1, TNF-α, CCL3 (MIP-1β), IL-6, and IL-12 (Bjorkbacka et al. 2004) (Fig. 2.1). Among the genes induced through the MyD88-independent

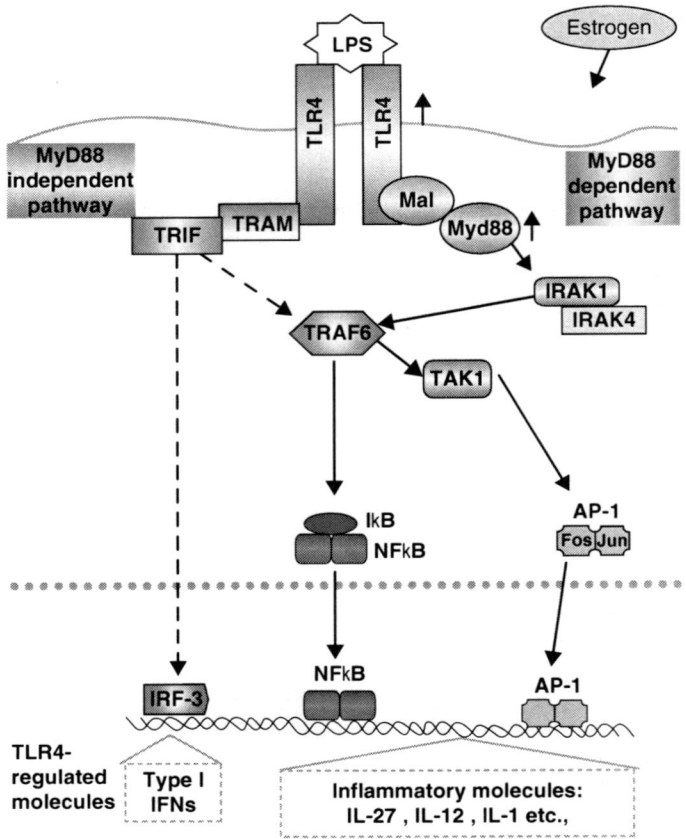

Fig. 2.1 In the MyD88-dependent pathway, upon LPS recognition, TLR4 homodimers recruit MyD88 and MyD88-adaptor-like protein (Mal, also known as TIRAP). MyD88 cooperates with IL-1 receptor-associated kinase-4 (IRAK4), which phosphorylates IRAK1. Phosphorylated IRAK1 then binds with TRAF6 (tumor necrosis factor (TNF) receptor-associated factor-6), followed by the phosphorylation and activation of the kinase, TAK1. TAK1 can activate the inhibitory κB (IκB) kinase (IKK) complex, which then results in the activation of NFκB (Kawai and Akira 2007). Phospho-TAK1 can also induce MAP kinases, p38 through MKK3/6 (Mitogen-activated protein Kinase Kinase3/6) (Wang et al. 2001b) or c-Jun N-terminal kinase (JNK) through MKK7 inducing AP-1 and AP-1-induced genes (Tournier et al. 2001). The TLR4-stimulated MyD88-independent pathway is regulated by TRAM (TRIF-related adapter molecule, also known as TICAM-2) and TRIF (TIR-containing adapter molecule, also known as TICAM-1) (O'Neill and Bowie, 2007). TRAM recruits TRIF to the plasma membrane, which associates with LPS-induced TLR4 (Tanimura et al. 2008). The MyD88-independent TLR4 signaling pathway through TRAM was shown to activate IRF-3 and NFκB (Fitzgerald et al. 2003). In addition, the MyD88-independent pathway can also stimulate NFκB through the association of TRIF with TRAF6 (Palsson-McDermott and O'Neill 2004). LPS induction of the MyD88-independent pathway results in the increased expression of IFN-β, IRF-1, RANTES, TRAF1, and IP-10 (Bjorkbacka et al. 2004). Even though the MyD88-dependent and MyD88-independent pathways regulate the expression of different molecules through different adaptor molecules, both pathways are involved in the modulation of innate immunity.

pathway are *Irf*1, *Cxcl10*, TNF receptor-associated factor 1 (*Traf1*), and *Ccl5* (RANTES) (Bjorkbacka et al. 2004)

There is some evidence that estrogen can alter TLR4-related proteins. In vitro 17β-estradiol-pretreated (10^{-9} M) and then LPS-stimulated macrophage cell line, RAW 264.7 cells have decreased *Tlr4* and *Cd14* mRNA expression (Vegeto et al. 2004). LPS-stimulated rat microglial cells treated with 10 nM 17β-estradiol do not demonstrate a marked change in *Tlr4* and *Cd14* mRNA expression compared with controls (Vegeto et al. 2004). On the other hand, in vivo estrogen and LPS exposure of mice increases CD14 and LBP expression in Kupffer cells (Ikejima et al. 1998). Treatment of DCs with estrogen antagonists (e.g., toremifene or tamoxifen) decreases expression of CD14 (Komi and Lassila 2000). In postmenopausal women (65–80 year), the expression of TLR4 and CD14 from PBMC is lower than in younger women (18–30 year), which might explain the reduction in LPS-stimulated IL-6, IL-1β, and TNF-α expression in postmenopausal women (Flynn et al. 2003; Teranishi et al. 2001). Estrogens and androgens can alter adaptor molecules downstream of TLR4 activation. The precise molecular mechanism of how estrogen modulates TLR4-mediated signaling is reviewed in detail in Chap. 5 of this book.

2.3.2 Proinflammatory Cytokines

The innate immune system has profound effects on adaptive immunity, primarily through cytokine production. These cytokines include tumor necrosis factor-α (TNF-α), Interleukin-1 (IL-1), Interleukin-12 (IL-12), Interleukin-18 (IL-18), and Interleukin-27 (IL-27). In addition, other inflammatory biomolecules, such as nitric oxide (NO) produced as a consequence of nitric oxide synthase 2 (NOS2) expression, provide further connections between innate and adaptive immunity.

2.3.2.1 TNF-α and IL-1α/β

TNF-α is produced by macrophages, monocytes, B- and T-lymphocytes, NK cells, Langerhans cells, and Kupffer cells in response to bacterial and inflammatory stimulants (Tracey and Cerami 1993). Increased levels of TNF-α have been implicated in septic shock (Lin and Yeh 2005) and systemic lupus erythematosus (SLE) (Aringer and Smolen 2008). Although estrogen regulates TNF-α, the effects (i.e., either enhanced or decreased) depend upon whether the studies are in vivo or in vitro, the dose of estrogen used, the species or tissue type examined, and the stimuli used (Mori et al. 1992; Ralston et al. 1990; Shanker et al. 1994; Straub 2007). For example, in vitro estrogen pretreatment of LPS-stimulated murine bone marrow-derived macrophages (BMM) decreases TNF-α protein (Zhang et al. 2001), while in vitro incubation of phorbol-myristate-acetate (PMA)-stimulated human monoblastic U937 cells with 17β-estradiol increases TNF-α production

(Carruba et al. 2003). Furthermore, exposure of mice to estrogen downregulates secretion of TNF-α from splenic macrophages (Hildebrand et al. 2006).

IL-1α/β, produced by monocytes and macrophages, regulates a broad range of immune responses. Similar to TNF-α, estrogen-related differences in IL-1 expression have been observed. In vitro 17β-estradiol-cultured and LPS-stimulated HL-60 promyelocytic leukemia cells demonstrate elevated expression of IL-1β compared with cells stimulated only with LPS (Mori et al. 1992). 17β-estradiol induces mRNA in rheumatoid fibroblast-like cell line and in primary synovial cells from patients with rheumatoid arthritis (RA) (Itoh et al. 2007). LPS-stimulated splenic lymphoid cells from 17β-estradiol-treated mice have increased IL-1α and IL-1β levels compared with control animals. These levels are markedly decreased when cultures are exposed to an NFκB inhibitor, suggesting that NFκB regulates 17β-estradiol-induced IL-1 expression (Dai et al. 2007). Conversely, estrogen-treated ovariectomized rats with acute endoluminal arterial injury have lowered expression of IL-1β as compared with control-ovariectomised rats (Miller et al. 2004). Similarly, LPS-induced IL-1β production by PBMC from postmenopausal women given estrogen implants for 6 months is reduced by estrogen treatment (Rogers et al. 2007).

The androgen, DHT, inhibits LPS and TNFα-mediated mRNA expression of *Ccl2*, *Cd40*, *Tlr4*, *Pai1*, and *Cox2* (Norata et al. 2006). Furthermore, DHT also suppresses the LPS and TNFα-mediated release of cytokines and chemokines such as GM-CSF, VCAM1, and ICAM1 protein expression from human endothelial cells (Norata et al. 2006).

2.3.2.2   IL-12

IL-12 is an immunomodulatory cytokine that is critical in linking innate immunity and protective cell-mediated immune responses. IL-12p70 (formed of p40 and p35 subunits) is produced by activated monocytes, macrophages, neutrophils, and DCs in response to bacteria, fungi, or by LPS stimulation (Bekeredjian-Ding et al. 2006; Fernandez-Lago et al. 1999; Hermann et al. 1998). IL-12 is a major inducer of IFN-γ and stimulates the differentiation/activation of helper T cell type 1 (Th1) cells (Trinchieri 2003).

Modulation of IL-12 by estrogen has been shown to fluctuate with reproductive phases, dose of estrogen, and stimuli utilized. During the third trimester of human pregnancy, IL-12 production from LPS-stimulated monocytes (Elenkov et al. 2001) and peripheral blood mononuclear cells (PBMCs) (Sakai et al. 2002) is lower compared with cells from nonpregnant women. IL-12 also is decreased in supernatants of PHA-activated lymphocytes from the third trimester of pregnancy in women with RA compared with healthy pregnant women (Tchorzewski et al. 2000). Increased IL-12 is observed from LPS and IFNγ-stimulated PMBCs from pregnant women in their first and second trimester compared with similarly treated PBMC obtained from nonpregnant women (Germain et al. 2007; Sacks et al. 2003).

PBMC from women with preeclampsia spontaneously produce more IL-12 when cultured in vitro compared with healthy pregnant women (Sakai et al. 2002).

Sex differences in IL-12 levels are evident in peritoneal exudate cells from Swiss Jim Lambert (SJL) mice (Wilcoxen et al. 2000). Macrophages from female SJL mice produce more IL-12 protein compared with male SJL mice in response to LPS (Wilcoxen et al. 2000). Expression of IL-12 is decreased in splenic lymphocytes from males compared with females (Bao et al. 2002) and in lymphocytes from DHT-treated female mice compared with placebo-treated female mice (Liva and Voskuhl 2001). Physiological sex hormone concentrations may regulate IL-12 production. However, there are only limited studies that directly examine the effects of estrogens and androgens on IL-12.

2.3.2.3   IL-18

IL-18 is produced by activated murine splenic macrophages, peritoneal exudate cells, and alveolar macrophages (Puren et al. 1999). This cytokine plays an important role in induction of innate immunity (Nakanishi et al. 2001). IL-18, in concert with IL-12, stimulates macrophages and DCs to augment levels of IFN-γ and differentation of Th1 cells (Dinarello 1999).

In humans, PBMCs from women in the first trimester of pregnancy produce more IL-18 following IFNγ-pretreatment or costimulation with IFN-γ and LPS compared with PBMC from nonpregnant women (Germain et al. 2007). Mice given 17β-estradiol implants for 7 days (providing 1,500–2,000 pg ml^{-1} in serum, 20–50% of pregnancy levels) before the induction of experimental autoimmune encephalomyelitis (EAE) have increased expression of *Il18* mRNA in their splenocytes (Matejuk et al. 2002). The expression of *Il18* in the uteri of mice is regulated by estrogen. *Il18* mRNA levels are upregulated during sexual maturation and upregulation of *Il18* mRNA expression is noted in stromal cells from uteri of estrogen-treated mice (Kusumoto et al. 2005). During estrus, female mice have more *Il18* mRNA expressed in endometrial epithelial and stromal cells compared with females in diestrus (Murakami et al. 2005). A higher level of 17β-estradiol treatment of ovariectomized mice markedly reduces uterine *Il18* expression (Murakami et al. 2005). To date, data on IL-18 and 17β-estradiol are insufficient to determine the immunoregulatory effects of estrogen. A recent study demonstrated decreased IL-18 levels in sexually mature male mice when compared with sexually immature male mice (Abu Elhija et al. 2008). Existing studies imply that sex hormones may modify IL-18 expression and production and further work in this area is warranted.

2.3.2.4   IL-27

In addition to IL-12, which is thought to be the principle inducer of Th1 polarization, the newly discovered cytokine, IL-27, is important in the initiation of Th1 responses through the induction of the Th1 transcriptional factor, T-bet (Hibbert et al. 2003). IL-27 can initiate the mRNA expression of *T-bet* and (currently considered a Th1-specific marker) in the presence of IL-12 (Hibbert et al. 2003). IL-27 is a heterodimer composed of a p40 protein that is called Epstein–Barr virus (EBV)-induced gene 3 (EBI3) and p28, an IL-12p35-related protein. LPS-activated APCs, and monocyte-derived DCs secrete IL-27 through a TLR/MyD88-dependent pathway (Liu et al. 2007; Pirhonen et al. 2007). Additional immunomodulatory roles of IL-27 have become apparent. IL-27 can suppress Th17-type cells (Neufert et al. 2007), induce Th2 cytokines (e.g., IL-10) (Batten et al. 2008), downregulate development of regulatory T cells (Huber et al. 2008), and exert protective effects against autoimmunity (i.e., EAE (Batten et al. 2006)).

There are very limited data on the regulatory role of estrogen on IL-27. Incubation of mature DCs (stimulated with GM-CSF and IL-4) with 17β-estradiol increases IL-27 production compared with nonestrogen-treated control cells (Kyurkchiev et al. 2007). The primary role of IL-27 is to prepare T lymphocytes for early Th1 commitment, to induce the expression of Th1 transcription factors such as T-bet, and to increase responsiveness to IL-12 through upregulation of IL-12Rβ2. The addition of recombinant IL-27 augments T-bet protein levels in splenocytes from estrogen-treated mice when compared with placebo-treated control mice (Karpuzoglu et al. 2007). These findings suggest that in vivo estrogen exposure has a stimulatory effect on IL-27, which might modulate the estrogen-regulated induction of IFN-γ. Future work is needed to decipher estrogen regulation of IL-27 and the molecular pathways involved in IL-27-induced responses. Currently, there is no data pertaining to potential effects of androgens on IL-27.

2.3.2.5 IFN-γ

Although IFN-γ production is a hallmark of Th1 cells, it is also released from other immune cells such as NK cells, NKT cells, and CD8 T cells. IFN-γ was considered a proinflammatory cytokine, but now it is evident that IFN-γ has antiinflammatory action as well (Cabantous et al. 2005). The immunomodulatory effect of estrogen on IFN-γ production is implied when comparing males and females. IFN-γ protein is increased in cultures of spleen or lymph nodes from female BALB/c, C57BL/6J, (NZBxNZW)F1, and CBA/Ca mice in response to various stimulants such as mycobacteria and viruses (Huygen and Palfliet 1984; McMurray et al. 1997; Satoskar and Alexander 1995). Sex differences in IFN-γ production also are observed in a murine lupus model in which female B/W mice have increased concentrations of IFN-γ compared with male mice (Haas et al. 1998; McMurray et al. 1997). In vitro estrogen treatment upregulates *Ifnγ* expression from antigen-specific T cell clones of patients with multiple sclerosis (Gilmore et al. 1997). Treatment with physiological levels of 17β-estradiol can alter the mRNA expression as well as protein production of IFN-γ in humans and mice (Fox et al. 1991; Grasso and Muscettola, 1990; Karpuzoglu-Sahin et al. 2001a,b). Estrogen-treated female mice have enhanced antigen-specific CD4$^+$ T cell responses as well as increased numbers of IFNγ-producing cells from lymph nodes (Maret et al. 2003). In vivo estrogen treatment of mice significantly upregulates IFN-γ protein and mRNA expression from ConA or antiCD3-stimulated splenocytes as well as purified T cells (Karpuzoglu et al. 2006; Karpuzoglu-Sahin et al. 2005; Karpuzoglu-Sahin et al. 2001). In mice, estrogen can modulate the transcription of IFN-γ through the direct interaction of the ER with an estrogen response element (ERE) in the 5′-prime flanking region of the *Ifnγ* gene (Fox et al. 1991), suggesting that estrogen can directly induce synthesis of IFN-γ. In vitro exposure of PHA and LPS-stimulated PMBCs to preovulatory (physiological) levels of estrogen increases IFN-γ, whereas exposure to pregnancy levels of estrogen downregulates IFN-γ secretion (Matalka 2003). *Leishmania*-infected pregnant C57BL/6 mice secrete less IFN-γ from splenocytes, when stimulated with *Leishmania* antigens compared with infected and nonpregnant C57BL/6 mice (Krishnan et al. 1996). In humans, decreased IFN-γ expression is observed during the third trimester of pregnant women with RA compared with healthy pregnant women (Tchorzewski et al. 2000).

2.3.3 Proinflammatory Chemokines

Chemokines (chemotactic cytokines) are proinflammatory cytokines with the ability to induce inflammatory-site-directed chemotaxis of specific immune cells (Rollins 1997). Chemokines such as CCL2 (MCP1), CCL12 (MCP5), and CCL5 (RANTES) activate several lymphocyte populations such as monocytes, eosinophils, basophils, and NK cells (Kopydlowski et al. 1999). In vitro estrogen exposure of splenic endothelial cells increases CCL2 expression, which is downregulated when cultured with the ER antagonists, tamoxifen or ICI182780, indicating that this effect is mediated through the ER (Murphy et al. 2004). CCL2 and CCL5, which are expressed in the uterus during early pregnancy in mice, are involved in macrophage recruitment (Wood et al. 1997). Estrogen and progesterone induce mRNA expression of *Ccl2*, *Ccl3*, *Ccl5*, and *Cxcl1* from uterine tissue (Mackay et al. 1999; Wood et al. 1997). ConA-activated splenic lymphocytes from 17β-estradiol-treated mice produce higher levels of CCL2 and CCL12 than do lymphocytes from placebo-treated mice (Lengi et al. 2006). Estrogen upregulation of CCL2 and CCL12 is mediated through IFNγ because estrogen treatment of IFNγ-knockout mice does not increase production of these chemokines (Lengi et al. 2006). These data suggest that IFNγ is a notable contributor in the upregulation of estrogen-modulated CCL2 and CCL12. While estrogen demonstrates a stimulatory role in induction of CCL2 and CCL12, the CCL5 protein from splenocytes is unaltered by in vivo estrogen treatment. CCL5 protein levels are diminished in estrogen-treated TNFα and IL-1-induced keratinocytes as compared with nonestrogen treated cells (Kanda and Watanabe 2003) and in mononuclear cells isolated from spinal cords of mice treated to induce EAE (Matejuk et al. 2001). Finally, there is no effect of estrogen on mRNA expression in preovulatory rat ovary tissue (Wong et al. 2002).

2.3.4   NOS2-Derived Nitric Oxide

Nitric oxide is produced by the action of NOS on L-arginine in the presence of NADPH-derived electrons and O_2. Direct in vitro exposure of various cell types, such as peritoneal macrophages (Hong and Zhu, 2004), the macrophage cell line RAW 264.7 (Azenabor et al. 2004), and rat peritoneal macrophages (You et al. 2003), to estrogen increases *Nos2* expression and/or NOS2-derived nitric oxide levels following LPS stimulation. Previous studies have demonstrated that in vivo estrogen treatment of C57BL/6 mice upregulates NOS2 protein and mRNA expression as well as NOS2-derived nitric oxide levels from ConA and antiCD3-stimulated whole splenocytes (Karpuzoglu et al. 2006). Estrogen-upregulation of NOS2 and nitric oxide is dependent on IFN-γ because ConA activated cells from IFNγ-knockout mice (unlike wild-type (WT) mice) do not upregulate production of NOS2 or nitric oxide in response to estrogen treatment (Karpuzoglu et al. 2006). However, direct addition of rIFNγ markedly upregulates NOS2 and nitric oxide in cells from estrogen-treated mice only (Karpuzoglu et al. 2006). Short-term and relatively low doses of subcutaneous estrogen treatment can induce NOS2-derived nitric oxide from splenocytes of CD1 outbred mice (Karpuzoglu-Sahin et al. 2005).

2.3.5   NFκB

Nuclear factor of κB (NFκB) family members plays a key role in the immunomodulation of proinflammatory responses. The NFκB family consists of NFκB1 (p50, its precursor p105), NFκB2 (p52, its precursor p100), p65 (RelA), cRel, and RelB. All of the members of the NF-κB family have a Rel homology domain (RHD), which is important for dimer formation, nuclear translocation, and DNA binding (Ghosh et al. 1998; Hayden and Ghosh, 2004). NFκB can be found as homo or heterodimers. In nonstimulated cells, heterodimers (or homodimers) are retained in the cytoplasm by an inhibitory protein family known as the IκB, which diminishes the DNA binding activity of the RHD (Ghosh et al. 1998; Yamamoto and Gaynor, 2004). When lymphocytes are stimulated, IκB becomes phosphorylated, ubiquitinated, and subsequently degraded by the 26S proteosome, which results in the release of NFκB. Free NFκB dimers translocate to the nucleus and activate transcription of target genes (Hayden and Ghosh 2004; Yamamoto and Gaynor 2004). The most commonly studied form of NFκB is a heterodimer that consists of two proteins, p50 (NFκB1) and p65 (RelA), which activate NFκB-dependent genes. The p50 and p52 lack transactivation domains and hence NFκB homodimers such as p50–p50 or p52–p52 are known to inhibit rather than activate NFκB-induced genes. However, when p50 binds to another IκB family member, B cell lymphoma 3 (Bcl3) activation is conferred (Watanabe et al. 1997).

Estrogen downregulates p65 NFκB expression from murine CD11c^{+} splenic DCs (Yang et al. 2006), macrophages (Ghisletti et al. 2005), and neurons

(Bethea et al. 2006), as well as NFκB DNA-binding activity in human retinal pigment epithelial cells (Paimela et al. 2007). Estrogen markedly inhibits the nuclear translocation of the classical NFκB protein, p65 and RelB, as well as cRel. In addition, estrogen preferentially allows p50 and p52 translocation into nuclei of splenocytes (Dai et al. 2007). Furthermore, estrogen treatment increases Bcl3 expression as well as Bcl3 binding to the *Nos2* promoter with p50 NFκB protein, when compared with control samples (Dai et al. 2007). The activation of NFκB p50–p50 homodimer/Bcl3 complexes may explain the estrogen upregulation of proinflammatory NFκB target genes even in the absence of the classical NFκB pathway (p65/p50).

In mammalian testis, spermatogenesis requires testosterone-mediated activation of AR expressed on cells. The expression of AR depends on p50-p65 NFκB proteins (Zhang et al. 2005). TNF-α further promotes NFκB-mediated AR expression (Delfino et al. 2003). Increasing evidence suggests that androgen signaling modulates canonical p65–p50 and noncanonical p52 activity in androgen-sensitive cells such as in prostrate cancer cell (Lessard et al. 2007). Androgens have been reported to enhance prostrate carcinogenesis by increasing oxidative stress and by prolonging AP1 and NFκB DNA binding activity (Ripple et al. 1999). Other studies have shown that androgens have opposite effects on NFκB. For example, testosterone blocks IL-1β-induced promatrilysin expression by inhibiting NFκB transactivation activity in prostrate carcinoma cells (Stratton et al. 2002). Increased apoptosis in AR-expressing prostrate tumors, in vivo, has been reported to be due to suppression of nuclear localization and transcriptional activity of RelA, accompanied with diminished expression of antiapoptotic targets, Bcl2 and IL-6 (Nelius et al. 2007). In addition, exposure of a human monocytic/macrophage cell line to testosterone increases apoptosis and decreases cell growth accompanied with an increase of IκBα and a decrease of the IκBα phosphorylated form (ser32) (Cutolo et al. 2005). DHT decreases TNF-α and LPS-induced VCAM1 and ICAM1 protein expression, and *Il6, Mcp1, Cd40, Tlr4, Pai1, Cox2, Gro*, and *Gmcsf* mRNA expression mediated by NFκB in human endothelial cells (Norata et al. 2006).

2.4 Sex Steroid Regulation of T Cell Responses

$CD4^+$ T cells are currently categorized into four distinct subtypes based on transcriptional factors, cytokine production, and function. These include (1) Th1, which primarily secretes IFN-γ; (2) Th2, which primarily secretes IL-4; (3) Th17, which secretes a potent inflammatory cytokine IL-17, and (4) regulatory T cells, which secrete antiinflammatory cytokines (e.g., IL-10 and TGF-β) that downregulate immune responses. IFNγ-secreting Th1 cells are known to promote cellular immune responses, especially, in defense against viral and bacterial intracellular pathogens. Overexpression of Th1-type cytokines can be observed in many proinflammatory conditions including several autoimmune diseases. Th17 cells secrete IL-17, which plays a key role in inflammatory and autoimmune conditions. Hence, it is critical to understand how estrogen can modulate CD4$^+$ T cells. Conversely, testosterone reduces functions of CD4$^+$ T cells (Maurer et al. 2001; Olsen and Kovacs 1996; Wunderlich et al. 2002). Androgens generally reduce Th1 cytokine release (Angele et al. 2001) and induce Th2 profile (Dalal et al. 1997).

2.4.1   Th1 Cells and T-Bet

A noteworthy discovery in the field of molecular immunology is the identification of IFNγ-inducing specific transcription factor, T-bet (Tbx-21) (Szabo et al. 2000). The expression of T-bet is limited to immune cells, where it is expressed at very low levels in unstimulated naive CD4$^+$ T cells but upregulated upon activation of T and splenic NK cells (Szabo et al. 2000, 2002). The induction of T-bet is regulated by several cytokines including, IL-27, IL-15, and IFN-γ through activating signal transducers and activators of transcription (STATs), in particular STAT1 and STAT4 (Lighvani et al. 2001; Strengell et al. 2003; Szabo et al. 2000). IL-27 induces *Tbet* expression, which in turn induces chromatin remodeling of the *Ifnγ* gene (Hibbert et al. 2003; Takeda et al. 2003), which takes place prior to stimulation of naive T cells by IL-12 and STAT4 activation to differentiate into Th1 cells (Mullen et al. 2001). Elevated T-bet upregulates IL-12Rβ2 on lymphocytes making them more responsive to IL-12 and STAT4 activation (Mullen et al. 2001). The IL-12/STAT4 signaling induces IFN-γ production and Th1 differentiation (Afkarian et al. 2002; Matsuoka et al. 2004; Mullen et al. 2001).

Splenic lymphocytes and purified T cells from 17β-estradiol-treated mice have upregulated T-bet expression compared with control mice (Karpuzoglu et al. 2007). Estrogen-induced T-bet protein expression is augmented in the presence of IL-27, but not in the presence of IL-12. Thus, estrogen may be able to prime splenic lymphocytes for Th1 differentiation via increasing T-bet expression and inducing responsiveness of the splenocytes to immediate Th1-inducing cytokines, i.e., IL-27 (Karpuzoglu et al. 2007). In summary, estrogen-induced T-bet can influence the generation of Th1 and proinflammatory cytokines through regulating the commitment of Th1 cells to produce IFN-γ and maintaining immune homeostasis via activation of several molecular pathways. To date, there are no data on the effects of testosterone on T-bet.

2.4.2   Th2 Cells and IL-4

Although the influence of sex and estrogen on of Th1-type cytokines is evident, the role of sex and/or sex hormones on Th-2 cytokine production, particularly IL-4, is unclear. 17β-estradiol has no marked effect on either IL-4 mRNA or protein from ConA-activated splenic lymphocytes (Karpuzoglu-Sahin et al. 2001a,b) and does not induce *Il4* mRNA expression from spinal cords of mice with EAE

(Matejuk et al. 2001). Estrogen treatment also has no effect on the mRNA expression of *Il4* in auricles of uteri after allergy induction in mice (Sakazaki et al. 2008). mRNA and protein levels are not changed in chondroblasts from the mandibular condyles cultured with different concentrations of 17β-estradiol compared with untreated control cultures (Yun et al. 2008). There are, however, studies using different experimental models and/or cell types, that illustrate that estrogen modulates IL-4 levels. For example, estrogen has a stimulatory effect on IL-4 secretion from bone marrow cells from in vivo ovalbumin-challenged ovarectomized allergic rats compared with control rats (de Oliveira et al. 2007). The increase in IL-4 production from CD4$^+$ T cells positively correlates with cyclical variations in estrogen levels in humans (Verthelyi and Klinman 2000). Similar results also are evident from murine CD4$^+$ T cells exposed to 17β-estradiol and antiCD3/CD28 in vitro (Lambert et al. 2005). With regards to IL-4 levels in studies employing delayed-type hypersensitivity (DTH) response to *C. albicans*, although there is no sex difference in cytokine concentrations, estrogen treatment increased the IL-4 levels (Ma et al. 2007).

Other studies have shown that increased levels of estrogen, such as during pregnancy as well as administration of estrogen at levels higher than physiological dose may skew the cytokine profile towards Th2 (Matalka 2003; Sabahi et al. 1995; Shimaoka et al. 2000). Increased IL-4 secretion as well as impaired resistance to *Leishmania major* infection is observed in pregnant C57BL/6 mice (Arinola et al. 2005). It is not, however, clear whether this is due to increased levels of estrogen or other hormones that also are elevated during pregnancy. PHA-stimulated PBMCs from pregnant women are biased towards a Th2 phenotype, with increased IL-4 protein or mRNA expression and decreased IFN-γ production when compared with PBMC in nonpregnant controls (Marzi et al. 1996; Matthiesen et al. 2003). Interestingly, antiCD3 antibody-stimulated splenic lymphocytes from male mice produce more IL-4 than that of female mice, but treatment of female mice with DHT does not alter IL-4 production when compared with IL-4 production in placebo-treated female mice (Liva et al. 2001). In vitro testosterone exposure of ConA-activated human PBMC decreases IL-4 secretion in the presence of cortisol (Janele et al. 2006). The discrepancies in these reports on the effects of estrogen and testosterone on IL-4 may partly be due to in vivo versus in vitro treatment, duration of exposure, species as well as differential stimulation and incubation procedures.

2.4.3 Th17 Cells and IL-17

IL-17 acts on target cells to activate key signaling molecules associated with inflammation that include: PI3K, NFκB, p38MAPK, and JNK1/JNK2 pathways (Huang et al. 2007). It promotes inflammation by several mechanisms including; (1) recruiting inflammatory cells neutrophils, monocytes, and macrophages to the site of inflammation; (2) acting on target cells (such as fibroblasts, epithelial cells, and lymphocytes) to stimulate a broad host of strong inflammatory molecules such as CXCL1, CXCL2, CXCL3, CXCL5, CXCL6 (Huang et al. 2007), IL-6, CXCL8, MCP1 (Hata et al. 2002), CCL1, CCL2, CCL17, CCL20, epithelial cell-derived neutrophils-activating protein 78 (ENA78) (Kawaguchi et al. 2003), and (3) cosynergizing with TLR ligands, IFN-γ, IL-1β, and TNF-α (Yu and Gaffen 2008). IL-17A induces NOS2 and nitric oxide production, which may augment development of autoimmune and inflammatory disorders. Expression of human βdefensin2 and CCL20 (MIP3) are stimulated by IL-17A, conferring its antimicrobial activity and chemotactic effects on immature DCs and memory T cells. Infection of mice with *Klebsiella pneumoniae* (Ye et al. 2001), *Porphyromonas gingivalis* (Oda et al. 2003), and *Toxoplasma gondii* (Kelly et al. 2005) upregulates production of IL-17. In addition to bacterial and viral infections, IL-17 plays an important role in autoimmune diseases such as multiple sclerosis (Lock et al. 2002), EAE (Aranami and Yamamura 2008), and RA (Yamada et al. 2007). Though IL-17 is protective in infection, overproduction of IL-17 may aggravate disease conditions and contribute to tissue injury. Our recent preliminary data show that exposure of IL-6 and TGF-β to splenocytes from estrogen-treated mice markedly induce IL-17 compared with placebo-treated mice (Ahmed, et al. unpublished data).

2.4.4 Regulatory T Cells

T regulatory (Treg) cells are a population of T cells, which suppress T cell-mediated immune responses and self-reactive T cells in autoimmune diseases (Itoh et al. 1999; Sakaguchi et al. 2001; Shevach 2002). There are two forms of Treg cells. Naturally occurring CD4$^+$CD25$^+$ T cells, termed nTreg cells, which acquire their regulatory capacity during thymopoesis constitute around 1–10% of total CD4$^+$ T cells in thymus, peripheral blood, and lymphoid tissues (Itoh et al. 1999; Sakaguchi et al. 2001; Shevach 2002). Induced Tregs (iTregs) are derived from the activation and differentiation of naive CD4$^+$ T cells in the periphery under unique stimulatory condition such as antigenic stimulation, exposure to cytokines like IL-10, or by blockage of Th1- (i.e., IFN-γ, IL-12p70) and Th2- (i.e., IL-4) type responses (Levings et al. 2001; O'Garra and Barrat 2003). Treg cells express CD25, cytotoxic T-lymphocyte antigen 4 (CTLA4), programmed death-1 (PD-1), and glucocorticoid-induced TNF receptor family-related protein (GITR) on their surface. A forkhead winged helix family transcriptional regulator (Foxp3) is essential for the genetic programming of nTreg cell development and function (Itoh et al. 1999; Sakaguchi et al. 2001; Shevach 2002). The induction of Foxp3 can convert naive CD4$^+$CD25$^-$ T cells to Tregs cells (Fontenot et al. 2003; Hori et al. 2003; Marie et al. 2005). Tregs utlilize a broad range of suppressive mechanism, including the release of the immunosuppressive cytokines, TGF-β and IL-10 (Chen and Wahl 2003; Gorelik and Flavell 2002; Joosten and Ottenhoff 2008; Kingsley et al. 2002; Nakamura et al. 2001). Tregs also suppress in an antigen-specific manner by utilizing cell–cell contact mechanism (Nishioka et al. 2008; O'Garra et al. 2003; Sakaguchi et al. 2006; Salomon et al. 2000; Shimizu et al. 2002). Interestingly, stimulation of naive cells with TGF-β and IL-6 promotes Th17 cell differentiation (Bettelli et al. 2006). However, inhibition of IL-6-driven Th17 induction (e.g., by IL-2 and retinoic acid), but exposure to TGF-β promotes the differentiation of CD4$^+$Foxp3$^+$ Treg cells from naive CD4 T cell precursors (Benson et al. 2007; Laurence et al. 2007). IL-10 and TGF-β secreted by Tregs are critical in dampening immune responses in allergy by skewing production of IgE towards the noninflammatory isotypes, IgG4 and IgA, respectively (Akdis et al. 2006). Furthermore, Tregs are essential in suppressing effector cells in burns (Ni Choileain et al. 2006), cancer (Liyanage et al. 2006), viral diseases (Jiao et al. 2008), and autoimmune diseases (i.e., type 1 diabetes, RA, and multiple sclerosis) (Chatenoud et al. 2001; Kukreja et al. 2002; Leipe et al. 2005). Excessive activity of Tregs can inhibit normal immune response against tumor and pathogen.

Increasing evidence suggests that estrogen is a potential regulatory factor in expansion of CD4$^+$CD25$^+$ T cells at both physiological and pharmacological range. Exogenous estrogen exposure of mice as well as pregnancy results in increased frequency of CD4$^+$CD25$^+$ T cells in the spleen, lymph nodes, and blood compared with nonestrogen or nonpregnant controls (Aluvihare et al. 2004; Polanczyk et al. 2005; Tai et al. 2008). Estrogen upregulates expression of *Foxp3 Il10*, and *Pd1* (Polanczyk et al. 2007; Tai et al. 2008). Interestingly, estrogen-mediated upregulation of *Pd1*, and *Ctla4* expression involves ERα. It has been shown that in ERα$^{-/-}$ mice or in mice treated with the ER antagonist, ICI 182 780, the frequency and suppressive activity of CD4$^+$CD25$^+$ Treg cells as well as *Foxp3* mRNA expression is reduced (Offner 2004; Polanczyk et al. 2004, 2005). The combined effect of estrogen-mediated Treg expansion and activation plays an important role in suppression of effector cells, concomitantly preventing autoreactive responses in EAE and RA (Ito et al. 2001; Offner and Polanczyk 2006; Polanczyk et al. 2003, 2004). The medical castration of men downregulates the frequency of CD4$^+$CD25$^+$ T cells, whereas androgen therapy replenishes the number of Treg cells to controls levels (Page et al. 2006). The molecular mechanisms in the maintenance of Treg cells by both estrogens and androgens merit further attention and indepth analysis.

2.5 Conclusions

The foregoing studies demonstrate that sex steroid hormones markedly regulate various facets of immune cells. It is not uncommon to notice that sex steroid hormones may have contradictory effects. Several factors are known to influence the outcome of sex hormone-mediated immune responses. These include: age, sex, immune status, states of estrus cycle, genetic background, dose of the hormone, pregnancy, duration and route of exposure, type of cells, and activation status of the cells. Even within a single species, sex hormones may have differential effects based on tissue sites. Thus, it is possible that estrogens and androgens may have dissimilar consequences in various autoimmune diseases that may markedly differ in initiating and pathogenic mechanisms. The cell signaling pathways in response to sex steroids may vary with the type of disease or immune status. Another very important variable is that the effects of sex steroid hormones on purified cells may be quite different compared with mixed whole-cell populations. In the body, the outcome of immune responses is dependent on the close interactions and cross regulation of various subsets of cells of innate and adaptive immune system (e.g., cross regulation through costimulatory molecules and cytokines on APC with lymphocytes). Studying sex steroid effects on isolated and purified cells may yield different results than that of natural mixed populations of cells. The complexity of sex hormone effects on biological tissues needs to be recognized. One aspect is unequivocal, that sex hormones can regulate the immune system. It is important, however, not to generalize the effects of sex hormones, but rather each situation or disease needs to be explored independently.

Acknowledgments We thank Dr. R. Dai and Ms. Rebecca Phillips from our lab for their comments. Supported by ROI-A1051880 NIH/NIAID, and IRC grants.

References

Abrahams VM, Collins JE, Wira CR, Fanger MW, Yeaman GR (2003) Inhibition of human polymorphonuclear cell oxidative burst by 17-beta-estradiol and 2, 3, 7, 8-tetrachlorodibenzo-p-dioxin. Am J Reprod Immunol 50(6):463–472

Abu Elhija M, Lunenfeld E, Eldar-Geva T, Huleihel M (2008) Over-expression of IL-18, ICE and IL-18 R in testicular tissue from sexually immature as compared to mature mice. Eur Cytokine Netw 19(1):15–24

Afkarian M, Sedy JR, Yang J, Jacobson NG, Cereb N, Yang SY, Murphy TL, Murphy KM (2002) T-bet is a STAT1-induced regulator of IL-12R expression in naive CD4+ T cells. Nat Immunol 3(6):549–557

Akdis M, Blaser K, Akdis CA (2006) T regulatory cells in allergy. Chem Immunol Allergy 91:159–173

Akhmedkhanov A, Zeleniuch-Jacquotte A, Toniolo P (2001) Role of exogenous and endogenous hormones in endometrial cancer: review of the evidence and research perspectives. Ann NY Acad Sci 943:296–315

Akoum A, Lawson C, McColl S, Villeneuve M (2001) Ectopic endometrial cells express high concentrations of interleukin (IL)-8 in vivo regardless of the menstrual cycle phase and respond to oestradiol by up-regulating IL-1-induced IL-8 expression in vitro. Mol Hum Reprod 7(9):859–866

Albrecht AE, Hartmann BW, Scholten C, Huber JC, Kalinowska W, Zielinski CC (1996) Effect of estrogen replacement therapy on natural killer cell activity in postmenopausal women. Maturitas 25(3):217–222

Aluvihare VR, Kallikourdis M, Betz AG (2004) Regulatory T cells mediate maternal tolerance to the fetus. Nat Immunol 5(3):266–271

Angele MK, Knoferl MW, Ayala A, Bland KI, Chaudry IH (2001) Testosterone and estrogen differently effect Th1 and Th2 cytokine release following trauma-haemorrhage. Cytokine 16(1):22–30

Ansar Ahmed S (2000) The immune system as a potential target for environmental estrogens (endocrine disrupters): a new emerging field. Toxicology 150(1–3):191–206

Ansar Ahmed S, Hissong BD, Verthelyi D, Donner K, Becker K, Karpuzoglu-Sahin E (1999) Gender and risk of autoimmune diseases: possible role of estrogenic compounds. Environ Health Perspect 107(Suppl 5):681–686

Aranami T, Yamamura T (2008) Th17 Cells and Autoimmune Encephalomyelitis (EAE/MS). Allergol Int 57(2):115–120

Aringer M, Smolen JS (2008) The role of tumor necrosis factor-alpha in systemic lupus erythematosus. Arthritis Res Ther 10(1):202

Arinola OG, Louis JS, Tacchini-Cottier F, Aseffa A, Salimonu LS (2005) Pregnancy impairs resistance of C57BL/6 mice to Leishmania major infection. Afr J Med Med Sci 34(1):65–70

Azenabor AA, Yang S, Job G, Adedokun OO (2004) Expression of iNOS gene in macrophages stimulated with 17beta-estradiol is regulated by free intracellular Ca2+. Biochem Cell Biol 82 (3):381–390

Bachy V, Williams DJ, Ibrahim MA (2008) Altered dendritic cell function in normal pregnancy. J Reprod Immunol 78(1):11–21

Bao M, Yang Y, Jun HS, Yoon JW (2002) Molecular mechanisms for gender differences in susceptibility to T cell-mediated autoimmune diabetes in nonobese diabetic mice. J Immunol 168(10):5369–5375

Baral E, Nagy E, Berczi I (1995) Modulation of natural killer cell-mediated cytotoxicity by tamoxifen and estradiol. Cancer 75(2):591–599

Baranao RI, Tenenbaum A, Sales ME, Rumi LS (1992) Functional alterations of murine peritoneal macrophages during pregnancy. Am J Reprod Immunol 27(1–2):82–86

Barnard CJ, Behnke JM, Sewell J (1996) Environmental enrichment, immunocompetence, and resistance to Babesia microti in male mice. Physiol Behav 60(5):1223–1231

Batten M, Kljavin NM, Li J, Walter MJ, de Sauvage FJ, Ghilardi N (2008) Cutting Edge: IL-27 Is a Potent Inducer of IL-10 but Not FoxP3 in Murine T Cells. J Immunol 180(5):2752–2756

Batten M, Li J, Yi S, Kljavin NM, Danilenko DM, Lucas S, Lee J, de Sauvage FJ, Ghilardi N (2006) Interleukin 27 limits autoimmune encephalomyelitis by suppressing the development of interleukin 17-producing T cells. Nat Immunol 7(9):929–936

Bebo BF Jr, Fyfe-Johnson A, Adlard K, Beam AG, Vandenbark AA, Offner H (2001) Low-dose estrogen therapy ameliorates experimental autoimmune encephalomyelitis in two different inbred mouse strains. J Immunol 166(3):2080–2089

Beg AA (2002) Endogenous ligands of Toll-like receptors: implications for regulating inflammatory and immune responses. Trends Immunol 23(11):509–512

Bekeredjian-Ding I, Roth SI, Gilles S, Giese T, Ablasser A, Hornung V, Endres S, Hartmann G (2006) T cell-independent, TLR-induced IL-12p70 production in primary human monocytes. J Immunol 176(12):7438–7446

Bekesi G, Tulassay Z, Racz K, Feher J, Szekacs B, Kakucs R, Dinya E, Kiss E, Magyar Z, Rigo J Jr (2007) The effect of estrogens on superoxide anion generation by human neutrophil granulocytes: possible consequences of the antioxidant defense. Gynecol Endocrinol 2:1–4

Bengtsson AK, Ryan EJ, Giordano D, Magaletti DM, Clark EA (2004) 17-{beta}-estradiol (E2) modulates cytokine and chemokine expression in human monocyte-derived dendritic cells. Blood 104(5):1404–1410

Benson MJ, Pino-Lagos K, Rosemblatt M, Noelle RJ (2007) All-trans retinoic acid mediates enhanced T reg cell growth, differentiation, and gut homing in the face of high levels of co-stimulation. J Exp Med 204(8):1765–1774

Bethea CL, Reddy AP, Smith LJ (2006) Nuclear factor kappa B in the dorsal raphe of macaques: an anatomical link for steroids, cytokines and serotonin. J Psychiatry Neurosci 31(2):105–114

Bettelli E, Carrier Y, Gao W, Korn T, Strom TB, Oukka M, Weiner HL, Kuchroo VK (2006) Reciprocal developmental pathways for the generation of pathogenic effector TH17 and regulatory T cells. Nature 441(7090):235–238

Beutler B, Du X, Poltorak A (2001) Identification of Toll-like receptor 4 (Tlr4) as the sole conduit for LPS signal transduction: genetic and evolutionary studies. J Endotoxin Res 7 (4):277–280

Bjorkbacka H, Fitzgerald KA, Huet F, Li X, Gregory JA, Lee MA, Ordija CM, Dowley NE, Golenbock DT, Freeman MW (2004) The induction of macrophage gene expression by LPS predominantly utilizes Myd88-independent signaling cascades. Physiol Genomics 19 (3):319–330

Bouman A, Moes H, Heineman MJ, de Leij LF, Faas MM (2001) Cytokine production by natural killer lymphocytes in follicular and luteal phase of the ovarian cycle in humans. Am J Reprod Immunol 45(3):130–134

Cabantous S, Poudiougou B, Traore A, Keita M, Cisse MB, Doumbo O, Dessein AJ, Marquet S (2005) Evidence that interferon-gamma plays a protective role during cerebral malaria. J Infect Dis 192(5):854–860

Carruba G, D'Agostino P, Miele M, Calabro M, Barbera C, Bella GD, Milano S, Ferlazzo V, Caruso R, Rosa ML, Cocciadiferro L, Campisi I, Castagnetta L, Cillari E (2003) Estrogen regulates cytokine production and apoptosis in PMA-differentiated, macrophage-like U937 cells. J Cell Biochem 90(1):187–196

Casson PR, Andersen RN, Herrod HG, Stentz FB, Straughn AB, Abraham GE, Buster JE (1993) Oral dehydroepiandrosterone in physiologic doses modulates immune function in postmenopausal women. Am J Obstet Gynecol 169(6):1536–1539

Cavasin MA, Tao ZY, Yu AL, Yang XP (2006) Testosterone enhances early cardiac remodeling after myocardial infarction, causing rupture and degrading cardiac function. Am J Physiol Heart Circ Physiol 290(5):H2043–H2050

Cave NJ, Backus RC, Marks SL, Klasing KC (2007) Modulation of innate and acquired immunity by an estrogenic dose of genistein in gonadectomized cats. Vet Immunol Immunopathol 117(1–2):42–54

Chatenoud L, Salomon B, Bluestone JA (2001) Suppressor T cells–they're back and critical for regulation of autoimmunity!. Immunol Rev 182:149–163

Chen W, Wahl SM (2003) TGF-beta: the missing link in CD4+CD25+ regulatory T cell-mediated immunosuppression. Cytokine Growth Factor Rev 14(2):85–89

Chiang K, Parthasarathy S, Santanam N (2004) Estrogen, neutrophils and oxidation. Life Sci 75(20):2425–2438

Choudhry MA, Bland KI, Chaudry IH (2006) Gender and susceptibility to sepsis following trauma. Endocr Metab Immune Disord Drug Targets 6(2):127–135

Corrales JJ, Almeida M, Burgo R, Mories MT, Miralles JM, Orfao A (2006) Androgen-replacement therapy depresses the ex vivo production of inflammatory cytokines by circulating antigen-presenting cells in aging type-2 diabetic men with partial androgen deficiency. J Endocrinol 189(3):595–604

Curran EM, Berghaus LJ, Vernetti NJ, Saporita AJ, Lubahn DB, Estes DM (2001) Natural killer cells express estrogen receptor-alpha and estrogen receptor-beta and can respond to estrogen via a non-estrogen receptor-alpha-mediated pathway. Cell Immunol 214(1):12–20

Cutolo M, Accardo S, Villaggio B, Barone A, Sulli A, Coviello DA, Carabbio C, Felli L, Miceli D, Farruggio R, Carruba G, Castagnetta L (1996) Androgen and estrogen receptors are present in primary cultures of human synovial macrophages. J Clin Endocrinol Metab 81(2):820–827

Cutolo M, Capellino S, Montagna P, Ghiorzo P, Sulli A, Villaggio B (2005) Sex hormone modulation of cell growth and apoptosis of the human monocytic/macrophage cell line. Arthritis Res Ther 7(5):R1124–R1132

Cuzzocrea S, Genovese T, Mazzon E, Esposito E, Di Paola R, Muia C, Crisafulli C, Peli A, Bramanti P, Chaudry IH (2008) Effect of 17beta-estradiol on signal transduction pathways and secondary damage in experimental spinal cord trauma. Shock 29(3):362–371

D'Agostino P, Milano S, Barbera C, Di Bella G, La Rosa M, Ferlazzo V, Farruggio R, Miceli DM, Miele M, Castagnetta L, Cillari E (1999) Sex hormones modulate inflammatory mediators produced by macrophages. Ann NY Acad Sci 876:426–429

Dai R, Phillips RA, Ahmed SA (2007) Despite inhibition of nuclear localization of NF-kappa B p65, c-Rel, and RelB, 17-beta estradiol up-regulates NF-kappa B signaling in mouse splenocytes: the potential role of Bcl-3. J Immunol 179(3):1776–1783

Dalal M, Kim S, Voskuhl RR (1997) Testosterone therapy ameliorates experimental autoimmune encephalomyelitis and induces a T helper 2 bias in the autoantigen-specific T lymphocyte response. J Immunol 159(1):3–6

De M, Wood GW (1990) Influence of oestrogen and progesterone on macrophage distribution in the mouse uterus. J Endocrinol 126(3):417–424

de Oliveira AP, Domingos HV, Cavriani G, Damazo AS, Dos Santos Franco AL, Oliani SM, Oliveira-Filho RM, Vargaftig BB, de Lima WT (2007) Cellular recruitment and cytokine generation in a rat model of allergic lung inflammation are differentially modulated by progesterone and estradiol. Am J Physiol Cell Physiol 293(3):C1120–C1128

Deitch EA, Ananthakrishnan P, Cohen DB, da Xu Z, Feketeova E, Hauser CJ (2006) Neutrophil activation is modulated by sex hormones after trauma-hemorrhagic shock and burn injuries. Am J Physiol Heart Circ Physiol 291(3):H1456–H1465

Delfino FJ, Boustead JN, Fix C, Walker WH (2003) NF-kappaB and TNF-alpha stimulate androgen receptor expression in Sertoli cells. Mol Cell Endocrinol 201(1–2):1–12

Delyani JA, Murohara T, Nossuli TO, Lefer AM (1996) Protection from myocardial reperfusion injury by acute administration of 17 beta-estradiol. J Mol Cell Cardiol 28(5):1001–1008

Dinarello CA (1999) IL-18: a TH1-inducing, proinflammatory cytokine and new member of the IL-1 family. J Allergy Clin Immunol 103(1 Pt 1):11–24

Easterbrook JD, Kaplan JB, Vanasco NB, Reeves WK, Purcell RH, Kosoy MY, Glass GE, Watson J, Klein SL (2007) A survey of zoonotic pathogens carried by Norway rats in Baltimore, Maryland, USA. Epidemiol Infect 135(7):1192–1199

Elenkov IJ, Wilder RL, Bakalov VK, Link AA, Dimitrov MA, Fisher S, Crane M, Kanik KS, Chrousos GP (2001) IL-12, TNF-alpha, and hormonal changes during late pregnancy and early postpartum: implications for autoimmune disease activity during these times. J Clin Endocrinol Metab 86(10):4933–4938

Erridge C, Pridmore A, Eley A, Stewart J, Poxton IR (2004) Lipopolysaccharides of Bacteroides fragilis, Chlamydia trachomatis and Pseudomonas aeruginosa signal via toll-like receptor 2. J Med Microbiol 53(Pt 8):735–740

Faurschou M, Borregaard N (2003) Neutrophil granules and secretory vesicles in inflammation. Microbes Infect 5(14):1317–1327

Ferguson MM, McDonald FG (1985) Oestrogen as an inhibitor of human NK cell cytolysis. FEBS Lett 191(1):145–148

Fernandez-Lago L, Rodriguez-Tarazona E, Vizcaino N (1999) Differential secretion of interleukin-12 (IL-12) subunits and heterodimeric IL-12p70 protein by CD-1 mice and murine macrophages in response to intracellular infection by *Brucella abortus*. J Med Microbiol 48(12):1065–1073

Finberg RW, Re F, Popova L, Golenbock DT, Kurt-Jones EA (2004) Cell activation by Toll-like receptors: role of LBP and CD14. J Endotoxin Res 10(6):413–418

Fitzgerald KA, Rowe DC, Barnes BJ, Caffrey DR, Visintin A, Latz E, Monks B, Pitha PM, Golenbock DT (2003) LPS-TLR4 signaling to IRF-3/7 and NF-kappaB involves the toll adapters TRAM and TRIF. J Exp Med 198(7):1043–1055

Flynn MG, McFarlin BK, Phillips MD, Stewart LK, Timmerman KL (2003) Toll-like receptor 4 and CD14 mRNA expression are lower in resistive exercise-trained elderly women. J Appl Physiol 95(5):1833–1842

Fontenot JD, Gavin MA, Rudensky AY (2003) Foxp3 programs the development and function of CD4+CD25+ regulatory T cells. Nat Immunol 4(4):330–336

Fox HS, Bond BL, Parslow TG (1991) Estrogen regulates the IFN-gamma promoter. J Immunol 146(12):4362–4367

Germain SJ, Sacks GP, Sooranna SR, Sargent IL, Redman CW (2007) Systemic inflammatory priming in normal pregnancy and preeclampsia: the role of circulating syncytiotrophoblast microparticles. J Immunol 178(9):5949–5956

Ghisletti S, Meda C, Maggi A, Vegeto E (2005) 17beta-estradiol inhibits inflammatory gene expression by controlling NF-kappaB intracellular localization. Mol Cell Biol 25(8):2957–2968

Ghosh S, May MJ, Kopp EB (1998) NF-kappa B and Rel proteins: evolutionarily conserved mediators of immune responses. Annu Rev Immunol 16:225–260

Gilmore W, Weiner LP, Correale J (1997) Effect of estradiol on cytokine secretion by proteolipid protein-specific T cell clones isolated from multiple sclerosis patients and normal control subjects. J Immunol 158(1):446–451

Gomez F, Ruiz P, Lopez R, Rivera C, Romero S, Bernal JA (2000) Effects of androgen treatment on expression of macrophage Fcgamma receptors. Clin Diagn Lab Immunol 7(4):682–686

Gorelik L, Flavell RA (2002) Transforming growth factor-beta in T-cell biology. Nat Rev Immunol 2(1):46–53

Gourdy P, Araujo LM, Zhu R, Garmy-Susini B, Diem S, Laurell H, Leite-de-Moraes M, Dy M, Arnal JF, Bayard F, Herbelin A (2005) Relevance of sexual dimorphism to regulatory T cells: estradiol promotes IFN-gamma production by invariant natural killer T cells. Blood 105 (6):2415–2420

Grasso G, Muscettola M (1990) The influence of beta-estradiol and progesterone on interferon gamma production in vitro. Int J Neurosci 51(3–4):315–317

Haas C, Ryffel B, Le Hir M (1998) IFN-gamma receptor deletion prevents autoantibody production and glomerulonephritis in lupus-prone (NZB x NZW)F1 mice. J Immunol 160(8): 3713–3718

Hampton MB, Kettle AJ, Winterbourn CC (1998) Inside the neutrophil phagosome: oxidants, myeloperoxidase, and bacterial killing. Blood 92(9):3007–3017

Hao S, Zhao J, Zhou J, Zhao S, Hu Y, Hou Y (2007) Modulation of 17beta-estradiol on the number and cytotoxicity of NK cells in vivo related to MCM and activating receptors. Int Immunopharmacol 7(13):1765–1775

Hata K, Andoh A, Shimada M, Fujino S, Bamba S, Araki Y, Okuno T, Fujiyama Y, Bamba T (2002) IL-17 stimulates inflammatory responses via NF-kappaB and MAP kinase pathways in human colonic myofibroblasts. Am J Physiol Gastrointest Liver Physiol 282(6):G1035–G1044

Hayden MS, Ghosh S (2004) Signaling to NF-kappaB. Genes Dev 18(18):2195–2224

Hemmi H, Takeuchi O, Kawai T, Kaisho T, Sato S, Sanjo H, Matsumoto M, Hoshino K, Wagner H, Takeda K, Akira S (2000) A Toll-like receptor recognizes bacterial DNA. Nature 408 (6813):740–745

Hermann P, Rubio M, Nakajima T, Delespesse G, Sarfati M (1998) IFN-alpha priming of human monocytes differentially regulates gram-positive and gram-negative bacteria-induced IL-10 release and selectively enhances IL-12p70, CD80, and MHC class I expression. J Immunol 161(4):2011–2018

Hibbert L, Pflanz S, De Waal Malefyt R, Kastelein RA (2003) IL-27 and IFN-alpha signal via Stat1 and Stat3 and induce T-Bet and IL-12Rbeta2 in naive T cells. J Interferon Cytokine Res 23(9):513–522

Hildebrand F, Hubbard WJ, Choudhry MA, Thobe BM, Pape HC, Chaudry IH (2006) Are the protective effects of 17beta-estradiol on splenic macrophages and splenocytes after trauma-hemorrhage mediated via estrogen-receptor (ER)-alpha or ER-beta? J Leukoc Biol 79(6):1173–1180

Hong M, Zhu Q (2004) Macrophages are activated by 17beta-estradiol: possible permission role in endometriosis. Exp Toxicol Pathol 55(5):385–391

Hori S, Nomura T, Sakaguchi S (2003) Control of regulatory T cell development by the transcription factor Foxp3. Science 299(5609):1057–1061

Hou J, Zheng WF (1988) Effect of sex hormones on NK and ADCC activity of mice. Int J Immunopharmacol 10(1):15–22

Hsu JT, Kan WH, Hsieh CH, Choudhry MA, Schwacha MG, Bland KI, Chaudry IH (2007) Mechanism of estrogen-mediated attenuation of hepatic injury following trauma-hemorrhage: Akt-dependent HO-1 up-regulation. J Leukoc Biol 82(4):1019–1026

Huang F, Kao CY, Wachi S, Thai P, Ryu J, Wu R (2007) Requirement for both JAK-mediated PI3K signaling and ACT1/TRAF6/TAK1-dependent NF-kappaB activation by IL-17A in enhancing cytokine expression in human airway epithelial cells. J Immunol 179(10):6504–6513

Huber M, Steinwald V, Guralnik A, Brustle A, Kleemann P, Rosenplanter C, Decker T, Lohoff M (2008) IL-27 inhibits the development of regulatory T cells via STAT3. Int Immunol 20(2):223–234

Huygen K, Palfliet K (1984) Strain variation in interferon gamma production of BCG-sensitized mice challenged with PPD II. Importance of one major autosomal locus and additional sexual influences. Cell Immunol 85(1):75–81

Ikejima K, Enomoto N, Iimuro Y, Ikejima A, Fang D, Xu J, Forman DT, Brenner DA, Thurman RG (1998) Estrogen increases sensitivity of hepatic Kupffer cells to endotoxin. Am J Physiol 274(4 Pt 1):G669–G676

Ito A, Bebo BF Jr, Matejuk A, Zamora A, Silverman M, Fyfe-Johnson A, Offner H (2001) Estrogen treatment down-regulates TNF-alpha production and reduces the severity of experimental autoimmune encephalomyelitis in cytokine knockout mice. J Immunol 167(1):542–552

Ito I, Hayashi T, Yamada K, Kuzuya M, Naito M, Iguchi A (1995) Physiological concentration of estradiol inhibits polymorphonuclear leukocyte chemotaxis via a receptor mediated system. Life Sci 56(25):2247–2253

Itoh M, Takahashi T, Sakaguchi N, Kuniyasu Y, Shimizu J, Otsuka F, Sakaguchi S (1999) Thymus and autoimmunity: production of CD25+CD4+ naturally anergic and suppressive T cells as a key function of the thymus in maintaining immunologic self-tolerance. J Immunol 162(9):5317–5326

Itoh Y, Hayashi H, Miyazawa K, Kojima S, Akahoshi T, Onozaki K (2007) 17beta-estradiol induces IL-1alpha gene expression in rheumatoid fibroblast-like synovial cells through estrogen receptor alpha (ERalpha) and augmentation of transcriptional activity of Sp1 by dissociating histone deacetylase 2 from ERalpha. J Immunol 178(5):3059–3066

Janele D, Lang T, Capellino S, Cutolo M, Da Silva JA, Straub RH (2006) Effects of testosterone, 17beta-estradiol, and downstream estrogens on cytokine secretion from human leukocytes in the presence and absence of cortisol. Ann NY Acad Sci 1069:168–182

Jiang X, Ellison SJ, Alarid ET, Shapiro DJ (2007) Interplay between the levels of estrogen and estrogen receptor controls the level of the granzyme inhibitor, proteinase inhibitor 9 and susceptibility to immune surveillance by natural killer cells. Oncogene 26(28):4106–4114

Jiao Y, Fu J, Xing S, Fu B, Zhang Z, Shi M, Wang X, Zhang J, Jin L, Kang F, Wu H, Wang FS (2008) The decrease of regulatory T cells correlates with excessive activation and apoptosis of CD8(+) T cells in HIV-1-infected typical progressors, but not in long-term non-progressors. Immunology

Joosten SA, Ottenhoff TH (2008) Human CD4 and CD8 regulatory T cells in infectious diseases and vaccination. Hum Immunol 69(11):760–770

Josefsson E, Tarkowski A, Carlsten H (1992) Anti-inflammatory properties of estrogen. I. In vivo suppression of leukocyte production in bone marrow and redistribution of peripheral blood neutrophils. Cell Immunol 142(1):67–78

Kanda N, Watanabe S (2003) 17beta-estradiol inhibits the production of RANTES in human keratinocytes. J Invest Dermatol 120(3):420–427

Kanda N, Watanabe S (2005) Regulatory roles of sex hormones in cutaneous biology and immunology. J Dermatol Sci 38(1):1–7

Karpuzoglu E, Fenaux JB, Phillips RA, Lengi AJ, Elvinger F, Ansar Ahmed S (2006) Estrogen up-regulates inducible nitric oxide synthase, nitric oxide, and cyclooxygenase-2 in splenocytes activated with T cell stimulants: role of interferon-gamma. Endocrinology 147(2):662–671

Karpuzoglu E, Phillips RA, Gogal RM Jr, Ansar Ahmed S (2007) IFN-gamma-inducing transcription factor, T-bet is upregulated by estrogen in murine splenocytes: Role of IL-27 but not IL-12. Mol Immunol 44(7):1819–1825

Karpuzoglu-Sahin E, Gogal RM Jr, Hardy C, Sponenberg P, Ahmed SA (2005) Short-term administration of 17-beta estradiol to outbred male CD-1 mice induces changes in the immune system, but not in reproductive organs. Immunol Invest 34(1):1–26

Karpuzoglu-Sahin E, Hissong BD, Ansar Ahmed S (2001a) Interferon-gamma levels are upregulated by 17-beta-estradiol and diethylstilbestrol. J Reprod Immunol 52(1–2):113–127

Karpuzoglu-Sahin E, Zhi-Jun Y, Lengi A, Sriranganathan N, Ansar Ahmed S (2001b) Effects of long-term estrogen treatment on IFN-gamma, IL-2 and IL-4 gene expression and protein synthesis in spleen and thymus of normal C57BL/6 mice. Cytokine 14(4):208–217

Kawaguchi M, Kokubu F, Matsukura S, Ieki K, Odaka M, Watanabe S, Suzuki S, Adachi M, Huang SK (2003) Induction of C-X-C chemokines, growth-related oncogene alpha expression, and epithelial cell-derived neutrophil-activating protein-78 by ML-1 (interleukin-17F) involves activation of Raf1-mitogen-activated protein kinase kinase-extracellular signal-regulated kinase 1/2 pathway. J Pharmacol Exp Ther 307(3):1213–1220

Kawai T, Akira S (2007) TLR signaling. Semin Immunol 19(1):24–32

Kawasaki K, Gomi K, Kawai Y, Shiozaki M, Nishijima M (2003) Molecular basis for lipopolysaccharide mimetic action of Taxol and flavolipin. J Endotoxin Res 9(5):301–307

Kelly MN, Kolls JK, Happel K, Schwartzman JD, Schwarzenberger P, Combe C, Moretto M, Khan IA (2005) Interleukin-17/interleukin-17 receptor-mediated signaling is important for generation of an optimal polymorphonuclear response against *Toxoplasma gondii* infection. Infect Immun 73(1):617–621

Kingsley CI, Karim M, Bushell AR, Wood KJ (2002) CD25+CD4+ regulatory T cells prevent graft rejection: CTLA-4- and IL-10-dependent immunoregulation of alloresponses. J Immunol 168(3):1080–1086

Komi J, Lassila O (2000) Nonsteroidal anti-estrogens inhibit the functional differentiation of human monocyte-derived dendritic cells. Blood 95(9):2875–2882

Kopydlowski KM, Salkowski CA, Cody MJ, van Rooijen N, Major J, Hamilton TA, Vogel SN (1999) Regulation of macrophage chemokine expression by lipopolysaccharide in vitro and in vivo. J Immunol 163(3):1537–1544

Krishnan J, Selvarajoo K, Tsuchiya M, Lee G, Choi S (2007) Toll-like receptor signal transduction. Exp Mol Med 39(4):421–438

Krishnan L, Guilbert LJ, Russell AS, Wegmann TG, Mosmann TR, Belosevic M (1996) Pregnancy impairs resistance of C57BL/6 mice to Leishmania major infection and causes decreased antigen-specific IFN-gamma response and increased production of T helper 2 cytokines. J Immunol 156(2):644–652

Kukreja A, Cost G, Marker J, Zhang C, Sun Z, Lin-Su K, Ten S, Sanz M, Exley M, Wilson B, Porcelli S, Maclaren N (2002) Multiple immuno-regulatory defects in type-1 diabetes. J Clin Invest 109(1):131–140

Kusumoto K, Murakami Y, Otsuki M, Kanayama M, Takeuchi S, Takahashi S (2005) Interleukin-18 (IL-18) mRNA expression and localization of IL-18 mRNA-expressing cells in the mouse uterus. Zoolog Sci 22(9):1003–1010

Kyurkchiev D, Ivanova-Todorova E, Hayrabedyan S, Altankova I, Kyurkchiev S (2007) Female sex steroid hormones modify some regulatory properties of monocyte-derived dendritic cells. Am J Reprod Immunol 58(5):425–433

Labrie F, Belanger A, Luu-The V, Labrie C, Simard J, Cusan L, Gomez JL, Candas B (1998) DHEA and the intracrine formation of androgens and estrogens in peripheral target tissues: its role during aging. Steroids 63(5–6):322–328

Lambert KC, Curran EM, Judy BM, Milligan GN, Lubahn DB, Estes DM (2005) Estrogen receptor alpha (ERalpha) deficiency in macrophages results in increased stimulation of CD4+ T cells while 17beta-estradiol acts through ERalpha to increase IL-4 and GATA-3 expression in CD4+ T cells independent of antigen presentat. J Immunol 175(9):5716–5723

Laurence A, Tato CM, Davidson TS, Kanno Y, Chen Z, Yao Z, Blank RB, Meylan F, Siegel R, Hennighausen L, Shevach EM, O'Shea JJ (2007) Interleukin-2 signaling via STAT5 constrains T helper 17 cell generation. Immunity 26(3):371–381

Leipe J, Skapenko A, Lipsky PE, Schulze-Koops H (2005) Regulatory T cells in rheumatoid arthritis. Arthritis Res Ther 7(3):93

Lengi AJ, Phillips RA, Karpuzoglu E, Ahmed SA (2006) Estrogen selectively regulates chemokines in murine splenocytes. J Leukoc Biol 81(4):1065–1074

Lessard L, Saad F, Le Page C, Diallo JS, Peant B, Delvoye N, Mes-Masson AM (2007) NF-kappaB2 processing and p52 nuclear accumulation after androgenic stimulation of LNCaP prostate cancer cells. Cell Signal 19(5):1093–1100

Levings MK, Sangregorio R, Roncarolo MG (2001) Human cd25(+)cd4(+) t regulatory cells suppress naive and memory T cell proliferation and can be expanded in vitro without loss of function. J Exp Med 193(11):1295–1302

Lighvani AA, Frucht DM, Jankovic D, Yamane H, Aliberti J, Hissong BD, Nguyen BV, Gadina M, Sher A, Paul WE, O'Shea JJ (2001) T-bet is rapidly induced by interferon-gamma in lymphoid and myeloid cells. Proc Natl Acad Sci USA 98(26):15137–15142

Lin WJ, Yeh WC (2005) Implication of Toll-like receptor and tumor necrosis factor alpha signaling in septic shock. Shock 24(3):206–209

Liu HY, Buenafe AC, Matejuk A, Ito A, Zamora A, Dwyer J, Vandenbark AA, Offner H (2002) Estrogen inhibition of EAE involves effects on dendritic cell function. J Neurosci Res 70(2):238–248

Liu J, Guan X, Ma X (2007) Regulation of IL-27 p28 gene expression in macrophages through MyD88- and interferon-{gamma}-mediated pathways. J Exp Med 204(1):141–152

Liu L, Wang L, Zhao Y, Wang Y, Wang Z, Qiao Z (2006) Testosterone attenuates p38 MAPK pathway during Leishmania donovani infection of macrophages. Parasitol Res 99(2):189–193

Liva SM, Voskuhl RR (2001) Testosterone acts directly on CD4+ T lymphocytes to increase IL-10 production. J Immunol 167(4):2060–2067

Liyanage UK, Goedegebuure PS, Moore TT, Viehl CT, Moo-Young TA, Larson JW, Frey DM, Ehlers JP, Eberlein TJ, Linehan DC (2006) Increased prevalence of regulatory T cells (Treg) is induced by pancreas adenocarcinoma. J Immunother 29(4):416–424

Lock C, Hermans G, Pedotti R, Brendolan A, Schadt E, Garren H, Langer-Gould A, Strober S, Cannella B, Allard J, Klonowski P, Austin A, Lad N, Kaminski N, Galli SJ, Oksenberg JR, Raine CS, Heller R, Steinman L (2002) Gene-microarray analysis of multiple sclerosis lesions yields new targets validated in autoimmune encephalomyelitis. Nat Med 8(5):500–508

Ma LJ, Guzman EA, DeGuzman A, Muller HK, Walker AM, Owen LB (2007) Local cytokine levels associated with delayed-type hypersensitivity responses: modulation by gender, ovariectomy, and estrogen replacement. J Endocrinol 193(2):291–297

Mackay F, Woodcock SA, Lawton P, Ambrose C, Baetscher M, Schneider P, Tschopp J, Browning JL (1999) Mice transgenic for BAFF develop lymphocytic disorders along with autoimmune manifestations. J Exp Med 190(11):1697–1710

Maret A, Coudert JD, Garidou L, Foucras G, Gourdy P, Krust A, Dupont S, Chambon P, Druet P, Bayard F, Guery JC (2003) Estradiol enhances primary antigen-specific CD4 T cell responses and Th1 development in vivo. Essential role of estrogen receptor alpha expression in hematopoietic cells. Eur J Immunol 33(2):512–521

Marie JC, Letterio JJ, Gavin M, Rudensky AY (2005) TGF-beta1 maintains suppressor function and Foxp3 expression in CD4+CD25+ regulatory T cells. J Exp Med 201(7):1061–1067

Marzi M, Vigano A, Trabattoni D, Villa ML, Salvaggio A, Clerici E, Clerici M (1996) Characterization of type 1 and type 2 cytokine production profile in physiologic and pathologic human pregnancy. Clin Exp Immunol 106(1):127–133

Matalka KZ (2003) The effect of estradiol, but not progesterone, on the production of cytokines in stimulated whole blood, is concentration-dependent. Neuroendocrinol Lett 24(3–4):185–191

Matejuk A, Adlard K, Zamora A, Silverman M, Vandenbark AA, Offner H (2001) 17 beta-estradiol inhibits cytokine, chemokine, and chemokine receptor mRNA expression in the central nervous system of female mice with experimental autoimmune encephalomyelitis. J Neurosci Res 65(6):529–542

Matejuk A, Dwyer J, Zamora A, Vandenbark AA, Offner H (2002) Evaluation of the effects of 17beta-estradiol (17beta-e2) on gene expression in experimental autoimmune encephalomyelitis using DNA microarray. Endocrinology 143(1):313–319

Matsuoka K, Inoue N, Sato T, Okamoto S, Hisamatsu T, Kishi Y, Sakuraba A, Hitotsumatsu O, Ogata H, Koganei K, Fukushima T, Kanai T, Watanabe M, Ishii H, Hibi T (2004) T-bet upregulation and subsequent interleukin 12 stimulation are essential for induction of Th1 mediated immunopathology in Crohn's disease. Gut 53(9):1303–1308

Matthiesen L, Khademi M, Ekerfelt C, Berg G, Sharma S, Olsson T, Ernerudh J (2003) In-situ detection of both inflammatory and anti-inflammatory cytokines in resting peripheral blood mononuclear cells during pregnancy. J Reprod Immunol 58(1):49–59

Maurer M, Trajanoski Z, Frey G, Hiroi N, Galon J, Willenberg HS, Gold PW, Chrousos GP, Scherbaum WA, Bornstein SR (2001) Differential gene expression profile of glucocorticoids, testosterone, and dehydroepiandrosterone in human cells. Horm Metab Res 33(12):691–695

McMurray RW, Hoffman RW, Nelson W, Walker SE (1997) Cytokine mRNA expression in the B/W mouse model of systemic lupus erythematosus–analyses of strain, gender, and age effects. Clin Immunol Immunopathol 84(3):260–268

Miller AP, Feng W, Xing D, Weathington NM, Blalock JE, Chen YF, Oparil S (2004) Estrogen modulates inflammatory mediator expression and neutrophil chemotaxis in injured arteries. Circulation 110(12):1664–1669

Miller L, Hunt JS (1996) Sex steroid hormones and macrophage function. Life Sci 59(1):1–14

Miyake K (2004) Endotoxin recognition molecules, Toll-like receptor 4-MD-2. Semin Immunol 16(1):11–16

Miyake K (2007) Innate immune sensing of pathogens and danger signals by cell surface Toll-like receptors. Semin Immunol 19(1):3–10

Mori H, Sawairi M, Itoh N, Hanabayashi T, Tamaya T (1992) Effects of sex steroids on cell differentiation and interleukin-1 beta production in the human promyelocytic leukemia cell line HL-60. J Reprod Med 37(10):871–878

Muehlenbein MP, Alger J, Cogswell F, James M, Krogstad D (2005) The reproductive endocrine response to Plasmodium vivax infection in Hondurans. Am J Trop Med Hyg 73(1):178–187

Mullen AC, High FA, Hutchins AS, Lee HW, Villarino AV, Livingston DM, Kung AL, Cereb N, Yao TP, Yang SY, Reiner SL (2001) Role of T-bet in commitment of TH1 cells before IL-12-dependent selection. Science 292(5523):1907–1910

Murakami Y, Otsuki M, Kusumoto K, Takeuchi S, Takahashi S (2005) Estrogen inhibits interleukin-18 mRNA expression in the mouse uterus. J Reprod Dev 51(5):639–647

Murphy HS, Sun Q, Murphy BA, Mo R, Huo J, Chen J, Chensue SW, Adams M, Richardson BC, Yung R (2004) Tissue-specific effect of estradiol on endothelial cell-dependent lymphocyte recruitment. Microvasc Res 68(3):273–285

Muta T, Takeshige K (2001) Essential roles of CD14 and lipopolysaccharide-binding protein for activation of toll-like receptor (TLR)2 as well as TLR4 Reconstitution of TLR2- and TLR4-activation by distinguishable ligands in LPS preparations. Eur J Biochem 268(16):4580–4589

Nakamura K, Kitani A, Strober W (2001) Cell contact-dependent immunosuppression by CD4(+) CD25(+) regulatory T cells is mediated by cell surface-bound transforming growth factor beta. J Exp Med 194(5):629–644

Nakanishi K, Yoshimoto T, Tsutsui H, Okamura H (2001) Interleukin-18 regulates both Th1 and Th2 responses. Annu Rev Immunol 19:423–474

Nakao Y, Funami K, Kikkawa S, Taniguchi M, Nishiguchi M, Fukumori Y, Seya T, Matsumoto M (2005) Surface-expressed TLR6 participates in the recognition of diacylated lipopeptide and peptidoglycan in human cells. J Immunol 174(3):1566–1573

Nakaya M, Tachibana H, Yamada K (2006) Effect of estrogens on the interferon-gamma producing cell population of mouse splenocytes. Biosci Biotechnol Biochem 70(1):47–53

Nelius T, Filleur S, Yemelyanov A, Budunova I, Shroff E, Mirochnik Y, Aurora A, Veliceasa D, Xiao W, Wang Z, Volpert OV (2007) Androgen receptor targets NFkappaB and TSP1 to suppress prostate tumor growth in vivo. Int J Cancer 121(5):999–1008

Neufert C, Becker C, Wirtz S, Fantini MC, Weigmann B, Galle PR, Neurath MF (2007) IL-27 controls the development of inducible regulatory T cells and Th17 cells via differential effects on STAT1. Eur J Immunol 37(7):1809–1816

Ni Choileain N, MacConmara M, Zang Y, Murphy TJ, Mannick JA, Lederer JA (2006) Enhanced regulatory T cell activity is an element of the host response to injury. J Immunol 176(1):225–236

Nilsson N, Carlsten H (1994) Estrogen induces suppression of natural killer cell cytotoxicity and augmentation of polyclonal B cell activation. Cell Immunol 158(1):131–139

Nishioka T, Nishida E, Iida R, Morita A, Shimizu J (2008) In vivo expansion of CD4(+)Foxp3(+) regulatory T cells mediated by GITR molecules. Immunol Lett 121(2):97–104

Norata GD, Tibolla G, Seccomandi PM, Poletti A, Catapano AL (2006) Dihydrotestosterone decreases tumor necrosis factor-alpha and lipopolysaccharide-induced inflammatory response in human endothelial cells. J Clin Endocrinol Metab 91(2):546–554

O'Garra A, Barrat FJ (2003) In vitro generation of IL-10-producing regulatory CD4+ T cells is induced by immunosuppressive drugs and inhibited by Th1- and Th2-inducing cytokines. Immunol Lett 85(2):135–139

O'Neill LA, Bowie AG (2007) The family of five: TIR-domain-containing adaptors in Toll-like receptor signalling. Nat Rev Immunol 7(5):353–364

Oda T, Yoshie H, Yamazaki K (2003) Porphyromonas gingivalis antigen preferentially stimulates T cells to express IL-17 but not receptor activator of NF-kappaB ligand in vitro. Oral Microbiol Immunol 18(1):30–36

Offner H (2004) Neuroimmunoprotective effects of estrogen and derivatives in experimental autoimmune encephalomyelitis: therapeutic implications for multiple sclerosis. J Neurosci Res 78(5):603–624

Offner H, Polanczyk M (2006) A potential role for estrogen in experimental autoimmune encephalomyelitis and multiple sclerosis. Ann NY Acad Sci 1089:343–372

Olsen NJ, Kovacs WJ (1996) Gonadal steroids and immunity. Endocr Rev 17(4):369–384

Olsen NJ, Watson MB, Henderson GS, Kovacs WJ (1991) Androgen deprivation induces phenotypic and functional changes in the thymus of adult male mice. Endocrinology 129(5):2471–2476

Page ST, Plymate SR, Bremner WJ, Matsumoto AM, Hess DL, Lin DW, Amory JK, Nelson PS, Wu JD (2006) Effect of medical castration on CD4+ CD25+ T cells, CD8+ T cell IFN-gamma expression, and NK cells: a physiological role for testosterone and/or its metabolites. Am J Physiol Endocrinol Metab 290(5):E856–E863

Paharkova-Vatchkova V, Maldonado R, Kovats S (2004) Estrogen preferentially promotes the differentiation of CD11c+ CD11b(intermediate) dendritic cells from bone marrow precursors. J Immunol 172(3):1426–1436

Paimela T, Ryhanen T, Mannermaa E, Ojala J, Kalesnykas G, Salminen A, Kaarniranta K (2007) The effect of 17beta-estradiol on IL-6 secretion and NF-kappaB DNA-binding activity in human retinal pigment epithelial cells. Immunol Lett 110(2):139–144

Palsson-McDermott EM, O'Neill LA (2004) Signal transduction by the lipopolysaccharide receptor, Toll-like receptor-4. Immunology 113(2):153–162

Pioli PA, Jensen AL, Weaver LK, Amiel E, Shen Z, Shen L, Wira CR, Guyre PM (2007) Estradiol attenuates lipopolysaccharide-induced CXC chemokine ligand 8 production by human peripheral blood monocytes. J Immunol 179(9):6284–6290

Pirhonen J, Siren J, Julkunen I, Matikainen S (2007) IFN-{alpha} regulates Toll-like receptor-mediated IL-27 gene expression in human macrophages. J Leukoc Biol 82(5):1185–1192

Polanczyk M, Zamora A, Subramanian S, Matejuk A, Hess DL, Blankenhorn EP, Teuscher C, Vandenbark AA, Offner H (2003) The protective effect of 17beta-estradiol on experimental autoimmune encephalomyelitis is mediated through estrogen receptor-alpha. Am J Pathol 163 (4):1599–1605

Polanczyk MJ, Carson BD, Subramanian S, Afentoulis M, Vandenbark AA, Ziegler SF, Offner H (2004) Cutting edge: estrogen drives expansion of the CD4+CD25+ regulatory T cell compartment. J Immunol 173(4):2227–2230

Polanczyk MJ, Hopke C, Huan J, Vandenbark AA, Offner H (2005) Enhanced FoxP3 expression and Treg cell function in pregnant and estrogen-treated mice. J Neuroimmunol 170(1–2):85–92

Polanczyk MJ, Hopke C, Vandenbark AA, Offner H (2007) Treg suppressive activity involves estrogen-dependent expression of programmed death-1 (PD-1). Int Immunol 19(3):337–343

Puren AJ, Fantuzzi G, Dinarello CA (1999) Gene expression, synthesis, and secretion of interleukin 18 and interleukin 1beta are differentially regulated in human blood mononuclear cells and mouse spleen cells. Proc Natl Acad Sci USA 96(5):2256–2261

Ralston SH, Russell RG, Gowen M (1990) Estrogen inhibits release of tumor necrosis factor from peripheral blood mononuclear cells in postmenopausal women. J Bone Miner Res 5(9): 983–988

Rettew JA, Huet-Hudson YM, Marriott I (2008) Testosterone reduces macrophage expression in the mouse of toll-like receptor 4, a trigger for inflammation and innate immunity. Biol Reprod 78(3):432–437

Ripple MO, Henry WF, Schwarze SR, Wilding G, Weindruch R (1999) Effect of antioxidants on androgen-induced AP-1 and NF-kappaB DNA-binding activity in prostate carcinoma cells. J Natl Cancer Inst 91(14):1227–1232

Roach JC, Glusman G, Rowen L, Kaur A, Purcell MK, Smith KD, Hood LE, Aderem A (2005) The evolution of vertebrate Toll-like receptors. Proc Natl Acad Sci USA 102(27):9577–9582

Rock FL, Hardiman G, Timans JC, Kastelein RA, Bazan JF (1998) A family of human receptors structurally related to Drosophila Toll. Proc Natl Acad Sci USA 95(2):588–593

Rogers A, Clowes JA, Pereda CA, Eastell R (2007) Different effects of raloxifene and estrogen on interleukin-1beta and interleukin-1 receptor antagonist production using in vitro and ex vivo studies. Bone 40(1):105–110

Rollins BJ (1997) Chemokines. Blood 90(3):909–928

Ruggiero RJ, Likis FE (2002) Estrogen: physiology, pharmacology, and formulations for replacement therapy. J Midwifery Womens Health 47(3):130–138

Sabahi F, Rola-Plesczcynski M, O'Connell S, Frenkel LD (1995) Qualitative and quantitative analysis of T lymphocytes during normal human pregnancy. Am J Reprod Immunol 33(5):381–393

Sacks GP, Redman CW, Sargent IL (2003) Monocytes are primed to produce the Th1 type cytokine IL-12 in normal human pregnancy: an intracellular flow cytometric analysis of peripheral blood mononuclear cells. Clin Exp Immunol 131(3):490–497

Sakaguchi S, Ono M, Setoguchi R, Yagi H, Hori S, Fehervari Z, Shimizu J, Takahashi T, Nomura T (2006) Foxp3+ CD25+ CD4+ natural regulatory T cells in dominant self-tolerance and autoimmune disease. Immunol Rev 212:8–27

Sakaguchi S, Sakaguchi N, Shimizu J, Yamazaki S, Sakihama T, Itoh M, Kuniyasu Y, Nomura T, Toda M, Takahashi T (2001) Immunologic tolerance maintained by CD25+ CD4+ regulatory T cells: their common role in controlling autoimmunity, tumor immunity, and transplantation tolerance. Immunol Rev 182:18–32

Sakai M, Tsuda H, Tanebe K, Sasaki Y, Saito S (2002) Interleukin-12 secretion by peripheral blood mononuclear cells is decreased in normal pregnant subjects and increased in preeclamptic patients. Am J Reprod Immunol 47(2):91–97

Sakazaki F, Ueno H, Nakamuro K (2008) 17beta-Estradiol enhances expression of inflammatory cytokines and inducible nitric oxide synthase in mouse contact hypersensitivity. Int Immunopharmacol 8(5):654–660

Salomon B, Lenschow DJ, Rhee L, Ashourian N, Singh B, Sharpe A, Bluestone JA (2000) B7/CD28 costimulation is essential for the homeostasis of the CD4+CD25+ immunoregulatory T cells that control autoimmune diabetes. Immunity 12(4):431–440

Satoskar A, Alexander J (1995) Sex-determined susceptibility and differential IFN-gamma and TNF-alpha mRNA expression in DBA/2 mice infected with Leishmania mexicana. Immunology 84(1):1–4

Schuurs AH, Verheul HA (1990) Effects of gender and sex steroids on the immune response. J Steroid Biochem 35(2):157–172

Screpanti I, Santoni A, Gulino A, Herberman RB, Frati L (1987) Estrogen and antiestrogen modulation of the levels of mouse natural killer activity and large granular lymphocytes. Cell Immunol 106(2):191–202

Shanker G, Sorci-Thomas M, Adams MR (1994) Estrogen modulates the expression of tumor necrosis factor alpha mRNA in phorbol ester-stimulated human monocytic THP-1 cells. Lymphokine Cytokine Res 13(6):377–382

Shevach EM (2002) CD4+ CD25+ suppressor T cells: more questions than answers. Nat Rev Immunol 2(6):389–400

Shimaoka Y, Hidaka Y, Tada H, Nakamura T, Mitsuda N, Morimoto Y, Murata Y, Amino N (2000) Changes in cytokine production during and after normal pregnancy. Am J Reprod Immunol 44(3):143–147

Shimizu J, Yamazaki S, Takahashi T, Ishida Y, Sakaguchi S (2002) Stimulation of CD25(+)CD4(+) regulatory T cells through GITR breaks immunological self-tolerance. Nat Immunol 3 (2):135–142

Shin HJ, Lee H, Park JD, Hyun HC, Sohn HO, Lee DW, Kim YS (2007) Kinetics of binding of LPS to recombinant CD14, TLR4, and MD-2 proteins. Mol Cells 24(1):119–124

Simpson ER (2003) Sources of estrogen and their importance. J Steroid Biochem Mol Biol 86 (3–5):225–230

Siracusa MC, Overstreet MG, Housseau F, Scott AL, Klein SL (2008) 17beta-estradiol alters the activity of conventional and IFN-producing killer dendritic cells. J Immunol 180(3): 1423–1431

Soto AM, Sonnenschein C, Chung KL, Fernandez MF, Olea N, Serrano FO (1995) The E-SCREEN assay as a tool to identify estrogens: an update on estrogenic environmental pollutants. Environ Health Perspect 103(Suppl 7):113–122

Squadrito F, Altavilla D, Squadrito G, Campo GM, Arlotta M, Arcoraci V, Minutoli L, Serrano M, Saitta A, Caputi AP (1997) 17Beta-oestradiol reduces cardiac leukocyte accumulation in myocardial ischaemia reperfusion injury in rat. Eur J Pharmacol 335(2–3):185–192

Stratton MS, Greenstein B, Udayakumar TS, Nagle RB, Bowden GT (2002) Androgens block interleukin-1 beta-induced promatrilysin expression in prostate carcinoma cells. Prostate 53(1):1–8

Straub RH (2007) The complex role of estrogens in inflammation. Endocr Rev 28(5):521–574

Strengell M, Matikainen S, Siren J, Lehtonen A, Foster D, Julkunen I, Sareneva T (2003) IL-21 in synergy with IL-15 or IL-18 enhances IFN-gamma production in human NK and T cells. J Immunol 170(11):5464–5469

Szabo SJ, Kim ST, Costa GL, Zhang X, Fathman CG, Glimcher LH (2000) A novel transcription factor, T-bet, directs Th1 lineage commitment. Cell 100(6):655–669

Szabo SJ, Sullivan BM, Stemmann C, Satoskar AR, Sleckman BP, Glimcher LH (2002) Distinct effects of T-bet in TH1 lineage commitment and IFN-gamma production in CD4 and CD8 T cells. Science 295(5553):338–342

Tai P, Wang J, Jin H, Song X, Yan J, Kang Y, Zhao L, An X, Du X, Chen X, Wang S, Xia G, Wang B (2008) Induction of regulatory T cells by physiological level estrogen. J Cell Physiol 214(2):456–464

Takeda A, Hamano S, Yamanaka A, Hanada T, Ishibashi T, Mak TW, Yoshimura A, Yoshida H (2003) Cutting edge: role of IL-27/WSX-1 signaling for induction of T-bet through activation of STAT1 during initial Th1 commitment. J Immunol 170(10):4886–4890

Takeshita F, Gursel I, Ishii KJ, Suzuki K, Gursel M, Klinman DM (2004) Signal transduction pathways mediated by the interaction of CpG DNA with Toll-like receptor 9. Semin Immunol 16(1):17–22

Tallant T, Deb A, Kar N, Lupica J, de Veer MJ, DiDonato JA (2004) Flagellin acting via TLR5 is the major activator of key signaling pathways leading to NF-kappa B and proinflammatory gene program activation in intestinal epithelial cells. BMC Microbiol 4:33

Tanimura N, Saitoh S, Matsumoto F, Akashi-Takamura S, Miyake K (2008) Roles for LPS-dependent interaction and relocation of TLR4 and TRAM in TRIF-signaling. Biochem Biophys Res Commun 368(1):94–99

Tapping RI, Tobias PS (2003) Mycobacterial lipoarabinomannan mediates physical interactions between TLR1 and TLR2 to induce signaling. J Endotoxin Res 9(4):264–268

Tchorzewski H, Krasomski G, Biesiada L, Glowacka E, Banasik M, Lewkowicz P (2000) IL-12, IL-6 and IFN-gamma production by lymphocytes of pregnant women with rheumatoid arthritis remission during pregnancy. Mediators Inflamm 9(6):289–293

Teranishi CT, Flynn MG, Phillips MD, Liu J, Horrall M (2001) The effects of resistance training on peripheral blood cytokine production in women aged 18–39 and 65–79 yr (Abstract). Med Sci Sports Exerc 33:S78

Tiidus PM, Zajchowski S, Enns D, Holden D, Bombardier E, Belcastro AN (2002) Differential effect of oestrogen on post-exercise cardiac muscle myeloperoxidase and calpain activities in female rats. Acta Physiol Scand 174(2):131–136

Tournier C, Dong C, Turner TK, Jones SN, Flavell RA, Davis RJ (2001) MKK7 is an essential component of the JNK signal transduction pathway activated by proinflammatory cytokines. Genes Dev 15(11):1419–1426

Tracey KJ, Cerami A (1993) Tumor necrosis factor, other cytokines and disease. Annu Rev Cell Biol 9:317–343

Trinchieri G (2003) Interleukin-12 and the regulation of innate resistance and adaptive immunity. Nat Rev Immunol 3(2):133–146

Vabulas RM, Ahmad-Nejad P, Ghose S, Kirschning CJ, Issels RD, Wagner H (2002) HSP70 as endogenous stimulus of the Toll/interleukin-1 receptor signal pathway. J Biol Chem 277 (17):15107–15112

Vegeto E, Ghisletti S, Meda C, Etteri S, Belcredito S, Maggi A (2004) Regulation of the lipopolysaccharide signal transduction pathway by 17beta-estradiol in macrophage cells. J Steroid Biochem Mol Biol 91(1–2):59–66

Verthelyi D, Klinman DM (2000) Sex hormone levels correlate with the activity of cytokine-secreting cells in vivo. Immunology 100(3):384–390

Wang C, Deng L, Hong M, Akkaraju GR, Inoue J, Chen ZJ (2001a) TAK1 is a ubiquitin-dependent kinase of MKK and IKK. Nature 412(6844):346–351

Wang Q, Dziarski R, Kirschning CJ, Muzio M, Gupta D (2001b) Micrococci and peptidoglycan activate TLR2–>MyD88–>IRAK–>TRAF–>NIK–>IKK–>NF-kappaB signal transduction pathway that induces transcription of interleukin-8. Infect Immun 69 (4):2270–2276

Watanabe N, Iwamura T, Shinoda T, Fujita T (1997) Regulation of NFKB1 proteins by the candidate oncoprotein BCL-3: generation of NF-kappaB homodimers from the cytoplasmic pool of p50–p105 and nuclear translocation. Embo J 16(12):3609–3620

Wessendorf G, Scheibl P, Zerbe PS (1998) Effect of estrogens on the immune system with regard to bovine placental retention. Dtsch Tierarztl Wochenschr 105(1):32–34

Wilcoxen SC, Kirkman E, Dowdell KC, Stohlman SA (2000) Gender-dependent IL-12 secretion by APC is regulated by IL-10. J Immunol 164(12):6237–6243

Williams RH, Foster DW, Kronenberg HM, Larsen PR, Wilson J (1998) Williams textbook of Endocrinology, 9th edn. WB Saunders, Philadelphia

Wong KH, Negishi H, Adashi EY (2002) Expression, hormonal regulation, and cyclic variation of chemokines in the rat ovary: key determinants of the intraovarian residence of representatives of the white blood cell series. Endocrinology 143(3):784–791

Wood GW, Hausmann E, Choudhuri R (1997) Relative role of CSF-1, MCP-1/JE, and RANTES in macrophage recruitment during successful pregnancy. Mol Reprod Dev 46(1): 62-9; discussion 69-70.

Wunderlich F, Benten WP, Lieberherr M, Guo Z, Stamm O, Wrehlke C, Sekeris CE, Mossmann H (2002) Testosterone signaling in T cells and macrophages. Steroids 67(6):535–538

Yamada H, Nakashima Y, Okazaki K, Mawatari T, Fukushi JI, Kaibara N, Hori A, Iwamoto Y, Yoshikai Y (2007) Th1 but not Th17 cells predominate in the joints of patients with rheumatoid arthritis. Ann Rheum Dis 67(9):1299–1304

Yamamoto Y, Gaynor RB (2004) IkappaB kinases: key regulators of the NF-kappaB pathway. Trends Biochem Sci 29(2):72–79

Yang L, Hu Y, Hou Y (2006) Effects of 17beta-estradiol on the maturation, nuclear factor kappa B p65 and functions of murine spleen CD11c-positive dendritic cells. Mol Immunol 43(4):357–366

Ye P, Garvey PB, Zhang P, Nelson S, Bagby G, Summer WR, Schwarzenberger P, Shellito JE, Kolls JK (2001) Interleukin-17 and lung host defense against Klebsiella pneumoniae infection. Am J Respir Cell Mol Biol 25(3):335–340

Yin Z, Chen C, Szabo SJ, Glimcher LH, Ray A, Craft J (2002) T-Bet expression and failure of GATA-3 cross-regulation lead to default production of IFN-gamma by gammadelta T cells. J Immunol 168(4):1566–1571

Yokoyama Y, Schwacha MG, Samy TS, Bland KI, Chaudry IH (2002) Gender dimorphism in immune responses following trauma and hemorrhage. Immunol Res 26(1–3):63–76

You HJ, Kim JY, Jeong HG (2003) 17 beta-estradiol increases inducible nitric oxide synthase expression in macrophages. Biochem Biophys Res Commun 303(4):1129–1134

Yu HP, Hsieh YC, Suzuki T, Choudhry MA, Schwacha MG, Blanc KI, Chaudry IH (2007) Mechanism of the nongenomic effects of estrogen on intestinal myeloperoxidase activity following trauma-hemorrhage: up-regulation of the PI-3K/Akt pathway. J Leukoc Biol 82(3):774–780

Yu JJ, Gaffen SL (2008) Interleukin-17: a novel inflammatory cytokine that bridges innate and adaptive immunity. Front Biosci 13:170–177

Yun KI, Chae CH, Lee CW (2008) Effect of estrogen on the expression of cytokines of the temporomandibular joint cartilage cells of the mouse. J Oral Maxillofac Surg 66(5):882–887

Zhang J, Wong CH, Xia W, Mruk DD, Lee NP, Lee WM, Cheng CY (2005) Regulation of Sertoli-germ cell adherens junction dynamics via changes in protein-protein interactions of the N-cadherin-beta-catenin protein complex which are possibly mediated by c-Src and myotubularin-related protein 2: an in vivo study using an androgen suppression model. Endocrinology 146(3):1268–1284

Zhang QH, Hu YZ, Cao J, Zhong YQ, Zhao YF, Mei QB (2004) Estrogen influences the differentiation, maturation and function of dendritic cells in rats with experimental autoimmune encephalomyelitis. Acta Pharmacol Sin 25(4):508–513

Zhang X, Wang L, Zhang H, Guo D, Qiao Z, Qiao J (2001) Estrogen inhibits lipopolysaccharide-induced tumor necrosis factor-alpha release from murine macrophages. Methods Find Exp Clin Pharmacol 23(4):169–173

Chapter 3
Sex Steroid Receptors in Immune Cells

Susan Kovats, Esther Carreras, and Hemant Agrawal

Abstract Lymphocytes and myeloid cells express estrogen, progesterone, and androgen receptors, and studies show that sex steroid hormones directly modulate their activation, lifespan, and functional response during innate and adaptive immunity. Hematopoietic progenitors also express estrogen and androgen receptors, and profound effects of sex hormones on development of lymphoid and myeloid cells have been reported. The sex steroid receptors act as nuclear transcription factors, via multiple ligand-dependent or ligand-independent mechanisms. Sex steroid receptors also mediate rapid signaling events that synergize with membrane receptor signaling. The basis of sex-based differences in immunity will be clarified by determination of the potentially diverse molecular mechanisms by which sex steroid receptor signaling regulates immune cell development and function.

3.1 Introduction

Immune cells express estrogens receptors (ER), androgen receptors (AR) and progesterone receptors (PR), suggesting that endogenous sex hormones and ectopic ER, AR, or PR ligands directly modulate their function (reviewed in Whitacre et al. 1999). In nonimmune cell types, sex steroid receptor signaling regulates many normal cellular processes including proliferation, survival, and fate determining gene expression. Thus, sex steroid receptor mediated responses to circulating sex hormones or exogenous ligands may have an important role in the normal immunology of both males and females (Heldring et al. 2007). Studies of mammalian immune responses in the normal state, during autoimmunity, or after infection or trauma provide evidence that ER, AR, and PR ligands also modulate innate and

S. Kovats (✉)
Arthritis and Immunology Research Program, Oklahoma Medical Research Foundation, 825 NE 13th Street, Oklahoma City, OK 73104, USA
e-mail: Susan-Kovats@omrf.org

S.L. Klein and C.W. Roberts (eds.), *Sex Hormones and Immunity to Infection,*
DOI 10.1007/978-3-642-02155-8_3, © Springer-Verlag Berlin Heidelberg 2010

adaptive immunity and hematopoiesis (Dorner et al. 1980; Medina et al. 2001; Olsen and Kovacs 2001; Klein 2004; Kovacs 2005; Nalbandian and Kovats 2005; Grimaldi et al. 2006; Marriott and Huet-Hudson 2006). Although there are profound differences in immunity to infection and autoimmunity between males and females, during pregnancy or upon systemic treatment with sex hormones (Ahmed et al. 1985; Grossman 1985; Schuurs and Verheul 1990; Cutolo et al. 2004; McClain et al. 2007), studies are just beginning to link many of these differences to specific immune cell types that express sex steroid receptors.

3.2 Estrogen Receptors

3.2.1 Estrogen Receptor Expression and Ligands

3.2.1.1 ER Expression in Immune Cells

Two subtypes of ER, α and β, form homodimers and heterodimers. In many cell types, ERα and ERβ have nonredundant functions, and their relative roles in the immune system are beginning to be defined. ER expression in immune cells has been determined primarily by polymerase chain reaction (PCR) or by use of specific antibodies with immunoblotting or intracellular flow cytometry. Immune cells in human and murine blood, lymphoid organs, and bone marrow all express ERα, and in some cases ERβ (Whitacre et al. 1999). ERα expression by B and T lymphocytes, dendritic cells (DC), macrophages, monocytes, natural killer (NK) cells, and mast cells has been reported (Komi and Lassila 2000; Curran et al. 2001; Grimaldi et al. 2002; Mor et al. 2003; Paharkova-Vatchkova et al. 2004; Phiel et al. 2005; Harkonen and Vaananen 2006). ERβ is not as ubiquitously expressed by immune cells. Within human peripheral blood mononuclear cells (PBMCs), CD4$^+$ T cells express higher amounts of ERα than ERβ, CD8$^+$ T cells and monocytes express low amounts of both ER, and B cells express more ERβ than ERα (Phiel et al. 2005). ERα and ERβ are also expressed in murine splenic B lymphocytes (Grimaldi et al. 2002). Murine splenic DC and peritoneal macrophages express ERα, but not ERβ (Lambert et al. 2004) although murine DC generated during GM-CSF-dependent differentiation of bone marrow precursors express both ERα and ERβ (Paharkova-Vatchkova et al. 2004; Mao et al. 2005). ER expression is autoregulated (Castles et al. 1997); however, whether changes in concentration of ER ligands alter ER expression in immune cells has not been well studied.

ERs are expressed on hematopoietic progenitors. CD34$^+$ hematopoietic progenitors in human adult bone marrow express both ERα and ERβ, whereas CD34$^+$ hematopoietic progenitors in cord blood do not express ER (Igarashi et al. 2001). Murine fetal liver progenitors express no detectable ER, whereas adult murine bone marrow progenitors express ERα by 3 weeks of age and ERβ by 8 weeks (Igarashi et al. 2001). ERα but not ERβ mRNA can be detected in bone marrow myeloid progenitors that give rise to DC (Carreras et al. 2008). This suggests that ER expression by hematopoietic progenitors is developmentally regulated, with expression being suppressed during fetal life and activated during neonatal life, synchronous with ontogenetic formation of the immune system.

Most studies assessing estrogen modulation of immunity have not separated the relative expression from the functional roles of ERα and ERβ in immune cells. However, selective agonists for ERα and ERβ (Harrington et al. 2003; Cvoro et al. 2008; Kawasaki et al. 2008), and mice lacking ERα or ERβ (Islander et al. 2003; Shim et al. 2003; Polanczyk et al. 2004; Shim et al. 2004), are now being used to delineate the roles of the individual ERs in immunity. ERα and ERβ deficient mice develop immune systems that appear normal. However, their immune systems may become dysregulated as they age (>1 year). Thus ERα$^{-/-}$ mice develop systemic autoimmunity (Shim et al. 2004) and aged ERβ$^{-/-}$ mice develop myeloprolifera-tive disease resembling chronic myeloid leukemia (Shim et al. 2003). Studies of immunity in ERα or ERβ deficient mice indicate that each contributes distinctly to immune responses, sometimes by expression in nonimmune cells; most studies, however, indicate that ERα plays a more pronounced role in immunity than does ERβ (Liu et al. 2003; Maret et al. 2003; Polanczyk et al. 2003). For example, 17β-estradiol (E2) acting via ERα is required for an appropriate innate immune response to bacterial lipopolysaccharide (LPS) or viral infections in brains of female mice (Soucy et al. 2005). ERα, but not ERβ, expression is required in hematopoietic cells to achieve enhanced primary helper T cell type 1 (Th1) responses to low doses of E2 in vivo (Maret et al. 2003)

In sum, hematopoietic progenitors and terminally differentiated immune cells express ER, implying that ER ligands may regulate both the development (see below) and mature function of lymphocytes and myeloid cells.

3.2.1.2 Endogenous, Pharmacological or Environmental ER Ligands

Throughout life, the body is exposed to variable levels of endogenous and exoge-nous ER ligands, which influence both differentiation and function of immune cells. Endogenous estrogens include estrone (E1), 17β-estradiol (E2), and estriol (E3) with E2 being the major form in adults. E3 may also be detected at high levels, but only during pregnancy. E2 levels, generally, occur within or near the K_D of the E2-receptor interaction (0.1–1.0 nM equivalent to 27–272 pg ml^{-1}) (Askanase and Buyon 2002). An exception occurs in humans at term of pregnancy, when levels reach 16,000–30,000 pg ml^{-1}. Serum E2 in humans peaks at 200–500 pg ml^{-1} during the menstrual cycle; estrogen replacement therapy produces serum levels of ~100 pg ml^{-1} (Askanase and Buyon 2002). Reported serum levels of E2 in female mice cycle between ~25–35 pg ml^{-1} during diestrus and ~70–200 pg ml^{-1} during estrus/proestrus/ovulation; levels in male mice are ~8–15 pg ml^{-1} (Foster et al. 1983; Walmer et al. 1992; Couse and Korach 1999). Differences in serum E2 levels between inbred mouse strains also have been reported (Foster et al. 1983).

Sex steroids can enter cells by at least two mechanisms. Unbound hormones may traverse the plasma membrane by virtue of their lipid solubility and small size by the "free" hormone mode of entry. Most hormones, however, are present in serum as part of a bound complex with sex hormone binding globulin (SHBG). Recently, it was reported that sex steroids bound to SHBG may be internalized via megalin, a member of the low-density lipoprotein receptor-related protein family (Hammes et al. 2005). Megalin-mediated endocytosis is followed by release of hormone inside endocytic vesicles; how the steroid then enters the cytosol and contacts cognate receptors is unknown. Both pathways may operate in vivo because the phenotype of megalin-deficient mice is not identical to that of AR- or ER-deficient mice (Hammes et al. 2005). The current practice for measuring serum concentrations of sex hormones is to measure the free steroids; however, the finding that SHBG-bound steroids also are biologically active has caused a reconsideration of this method (Adams 2005).

Pharmacological ER ligands termed selective ER modulators (SERMs), such as tamoxifen, toremifene, and raloxifene, are used to prevent and treat breast cancer and osteoporosis (Dutertre and Smith 2000). Women taking daily oral tamoxifen have plasma levels of 200 nM (Physicians' Desk Reference). Studies in murine models show that systemic tamoxifen exposure modulates immune cell function (Dayan et al. 1997; Sthoeger et al. 2003). Other exogenous ER ligands include dietary phytoestrogens such as soy isoflavones (Ren et al. 2001), and environmental estrogens termed "endocrine disruptors," such as organochlorine pesticides and the industrial chemical bisphenol A used in plastics (Ahmed 2000). Murine models of "endocrine disruptor" exposure are beginning to be used to study the effects of these molecules on immune function (Ahmed 2000; Klein et al. 2002; Sobel et al. 2005). The phytoestrogen genistein provides an example. Studies based on data derived from murine models indicate that genistein has a ninefold greater affinity for ERβ than ERα (Barkhem et al. 1998) and suggest that its effect may be based on different affinities for the two receptors.

3.2.1.3 Choosing Appropriate Model Systems for ER Studies

Current models used to study estrogenic effects on immune cells, generally, involve in vitro or in vivo exposure to physiological, supraphysiological, or pregnancy levels of E2. In addition to imposing a distinct elevated hormone environment (in most cases), it also abolishes cycling of E2 levels that are potentially important to immune cell biology. A second approach uses genetically engineered mice that lack ERα or ERβ in all tissues. Global disruption of ERα, however, creates gross hormonal imbalances. For example, in female ER$\alpha^{-/-}$ mice, serum E2 levels are ten times higher than in wild-type mice, progesterone levels are decreased, and androgen synthesis and serum testosterone levels are increased (Couse et al. 1995; Taniguchi et al. 2007). These changes in hormone levels might alter the physiological role of ERβ, AR, or PR. Furthermore, lack of cross talk between the ERα and ERβ signaling pathways in single ER knockouts could be anticipated to influence the experimental outcome and the derived conclusions regarding normal function of the two ER.

One additional confounding issue with these two widely used approaches is that elevated systemic levels of E2 profoundly deplete populations of hematopoietic progenitors (see below), and this in turn leads to alterations in number and phenotype of lymphocytes and DC in bone marrow and spleen (Medina et al. 2001; Harman et al. 2006; Welner et al. 2007; Carreras et al. 2008). Elevated androgens also modulate lymphocyte development (Smithson et al. 1998; Olsen and Kovacs 2001). A second issue is that studies in vitro have demonstrated low concentrations of E2, as is found in diestrus, ovariectomized, or postmenopausal females, are proinflammatory, and high concentrations of E2, such as during an ovulatory/proestrus phase or pregnancy, are antiinflammatory (Straub 2007). Therefore, a normal female has both diestrus and ovulatory levels of E2 within a narrow time window, and a uniform in vitro or in vivo exposure to E2 may not yield physiologically representative information. New approaches are, therefore, needed to distinguish the role of ER signaling in specific immune cell types and their progenitors in a physiologically relevant hormone environment characteristic of adult mice.

The wide range of E2 or other ER ligand concentrations used in published studies has made it difficult to reach a consensus regarding the effects of these ligands on immune responses (Straub 2007). In future studies, interpretation of the biological significance of data will be greatly aided by documentation of several relevant factors. These include the duration of ER ligand exposure, the concentration of the ER ligands achieved in vivo or in vitro, ERα or ERβ expression by specific cell types under study, the age and sex of the animal, and whether it has been ovariectomized or placed on a phytoestrogen-free diet.

3.2.2 Mechanisms of Estrogen Receptor Signaling

3.2.2.1 ER Structure and Functional Domains

ER proteins are members of the nuclear receptor super family. These proteins share modular functional domains that control transcription (reviewed in Heldring et al. 2007) (Fig. 3.1). The C region contains two zinc fingers that allow ER to dimerize and bind DNA at specific estrogen response elements (ERE) (see below). The D region is a hinge domain that facilitates ER dimerization and binding to heat shock proteins (HSP) that act as chaperones when ER is not bound to ligands. The C-terminal E/F region contains the ligand-binding domain. The N-terminal A/B region contains the activation function (AF) 1 domain, and the E/F region contains the AF-2 domain. These regions are both important for ligand-dependent transcriptional activity. The AF domains interact with transcriptional coactivators and corepressors. The AF-1 domain may also be activated in a ligand-independent manner through phosphorylation by MAPK at specific serine residues; however, it must be noted that human ERβ lacks this region. Splice variants such as ERα46,

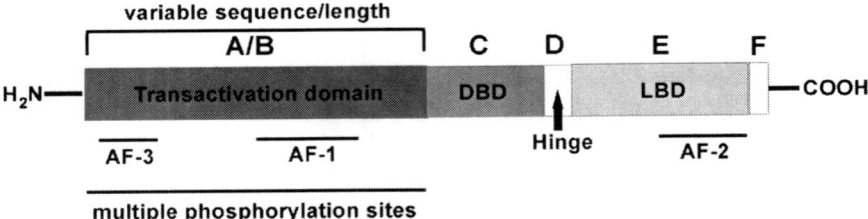

Fig. 3.1 Domain structure of nuclear hormone receptors. (**a, b**) The amino (NH$_2$)-terminal transactivation domain is of variable length, has multiple phosphorylation sites, and contains the activation function 1 (AF-1) region required for complete ligand-activated transcriptional activity. PR also has a unique AF-3 region. (**c**) The DNA-binding domain (DBD) is composed of two zinc fingers that bind specifically to short DNA sequences in the major groove, e.g., hormone response elements. (**d**) The hinge region is a small flexible region located between the DBD and LBD that contains a nuclear localization signal (NLS). (**e, f**) The carboxy (C)-terminal ligand-binding domain (LBD) also contains the AF-2 region, important for ligand-dependent transcriptional activity

which lacks the A/B (AF-1 bearing) domain, are often coexpressed with full-length ERα mRNA and may perform independent functions (Marino et al. 2006b). ERα46 has been postulated to act as a dominant negative inhibitor of ERα function when expressed with full-length ERα.

3.2.2.2 ERs Function as Nuclear Transcription Factors

Prior to ligand binding, ERs are retained in the cytosol by complexing with HSP. Ligand binding releases HSP and the ligand-bound ER dimers move to the nucleus. Nuclear ER control transcription by directly binding estrogen response sequences or by forming complexes with other transcription factors that bind to DNA (O'Lone et al. 2004) (Fig. 3.2). Each structurally distinct ligand imparts a unique conformation to ER dimers, which then dictates recruitment of distinct profiles of coregulators into multiprotein transcription complexes (Heldring et al. 2007). Ligation of ER may lead to various patterns of gene expression in different cell types, depending on the ligand form and concentration, the relative expression of the two ERs, and the availability of cell-specific coactivators or corepressors (Frasor et al. 2003; McDonnell 2004).

The development of SERMs as therapeutics is based on the idea that synthetic ligands can be used to specifically modulate ER conformation, and therefore transcriptional specificity, to yield the desired combination of tissue-specific agonist or antagonist activity (Jordan et al. 2001). For example, the SERM tamoxifen is an antagonist for mammary cells, but an agonist for endometrial cells, while raloxifene is an antagonist for both. In mammary cells, tamoxifen and raloxifene induce the recruitment of corepressors to target promoters; however, in endometrial cells, tamoxifen, but not raloxifene induces the recruitment of coactivators to target promoters (Shang and Brown 2002). Cellular responses to SERMs are, therefore,

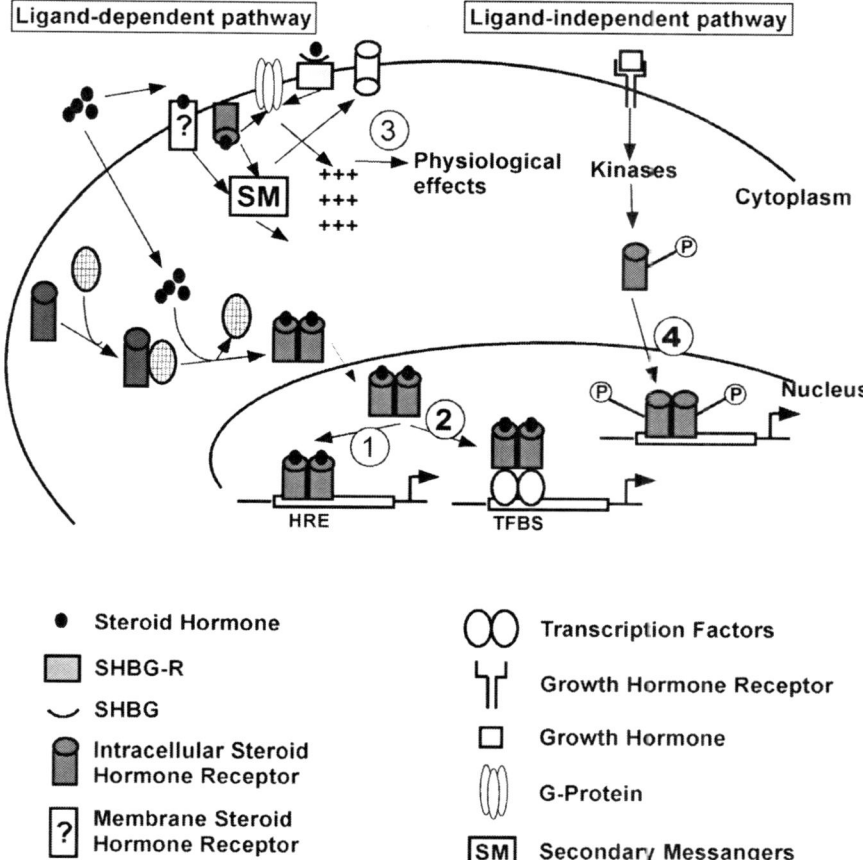

Fig. 3.2 Sex steroid receptor signaling pathways. AR, ER, and PR mediate their effects in ligand dependent (1–3) and ligand-independent (4) pathways. (1) Upon ligand binding, cytosolic hormone receptors move to the nucleus where they complex with coregulatory proteins and bind to specific hormone response elements (HRE) proximal to genes. (2) Ligand-bound hormone receptors may complex with and enhance binding of other transcription factors to their cognate sequence elements near promoters. (3) Hormone receptors, often in conjunction with membrane receptors, transduce rapid signals by activation of second messengers, leading to phosphorylation of kinases and changes in ion flux across membranes. This pathway may be mediated by a membrane associated hormone receptor or via ion channels, G-protein-coupled receptors, or a SHBG-R. (4) In the ligand-independent pathway, hormone receptors are phosphorylated after membrane receptor signaling, after which they translocate to the nucleus and bind to HRE

determined by the cell type and promoter-specific differences in coregulator recruitment.

Estrogen response elements consist of palindrome sequences proximal to genes and function in an orientation- and position-independent manner similar to a transcriptional enhancer (Marino et al. 2006b). ERE do not always contain a perfect consensus sequence and slight variations may alter ER affinity for DNA, with consequent effects on binding of coactivators. A recent study mapped approximately 1,000 ERα binding sites in MCF-7 breast cancer cells (Lin et al. 2007). Interestingly, only 5% of the binding sites were located within 5 kb upstream of transcription initiation sites, while the majority mapped to introns or locations far from the gene coding region, with transcriptional regulatory mechanisms operative over regions of up to 100 kb. Of the ERα binding sites identified, 71% contained a full consensus ERE, 25% contained an ERE half site and 4% had no recognizable ERE sequence. The ERα binding sites were often adjacent to Specificity protein 1 (Sp-1) binding sites, suggesting cooperative regulation of these promoters by ERα and Sp-1.

Other studies estimate that roughly 35% of human E2-responsive genes do not contain consensus ERE (O'Lone et al. 2004). The ER regulates these genes by complexing with and enhancing the DNA binding of other transcription factors. Sp-1 is the predominant mediator of indirect ER–DNA binding, but the ER also binds ATF-2/c-jun, ATF-1/CREB, NF-Y, and activator protein 1 (AP-1) complexes (reviewed in Marino et al. 2006b).

Transcriptional coregulators bind to multiple members of the nuclear receptor super family as coactivators or corepressors (or both) (reviewed in Marino et al. 2006b; Heldring et al. 2007). Coregulators function through multiple mechanisms, including chromatin remodeling, histone modification (acetylation or methylation), recruitment of RNA polymerase II, regulation of splicing, and coordinated degradation. One study using chromatin immunoprecipitation assays showed that ER binding to a target promoter stimulated the assembly of coactivators in a specific functional order (Shang et al. 2000). In the continuous presence of estrogens, the ER transcription complex cycles on and off the promoter, providing a mechanism to sense the changes in estrogen or coregulator levels. One mechanism to silence transcription exploits covalent posttranslational modifications (lysine acetylation and arginine methylation) of ERs and coactivators to inhibit their interactions (Smith and O'Malley 2004). Although cell type-specific responses to ER ligands are dictated by the cellular complement of coregulators, little is known about coregulator expression in immune cells.

Tissue-specific responses to ER ligands depend on ER-mediated regulation of discrete sets of genes, and on the relative expression of ERαα or ERββ homodimers and ERαβ heterodimers. It will, therefore, be important to define which ER subtype is involved in an immune response. ERα and ERβ have distinct as well as shared target genes (Kian Tee et al. 2004; Stossi et al. 2004). Exclusive ligation of ERα or ERβ causes distinct molecular outcomes in several model systems, with ERβ often opposing gene expression networks regulated by ERα, (Chang et al. 2006; Williams et al. 2008). For example, ERα and ERβ work in opposition at AP-1 sites; ERα–E2 dimers increase transcription of the cyclin D1 gene whereas ERβ–E2 dimers decrease it (Liu et al. 2002). ERβ may antagonize ERα-dependent transcription by altering the recruitment of regulatory factors or increasing ERα degradation (Matthews et al. 2006).

In a ligand-independent mechanism, ERs can act as transcription factors after being phosphorylated by kinases (such as MAPK) (Heldring et al. 2007) (Fig. 3.2). One instructive example is in the autoregulation of the *Tnfα* gene in osteoblasts. Here, ligand-free and E2-bound ERα oppose each other in transcription of the *Tnfα* gene (Cvoro et al. 2006). Using chromatin immunoprecipitation assays of cells stimulated with TNF-α, ligand-free phosphorylated ER was found in complex with c-jun, p50-NF-κB, p65-NF-κB, CBP, and HSP90 on the *Tnfα* promoter. Binding of E2 by ER in this complex stimulated recruitment of the coregulator GRIP1 and led to the release of ER and repression of *Tnfα* transcription. This mechanism is specific to the *Tnfα* gene because E2–ER–GRIP1 complexes activated transcription of a keratin gene in the same cells. These data highlight the complexity of ER-mediated gene regulation. Because the cellular concentration of E2 directly affects the balance of ligand-bound and ligand-free ER, it may account for reported biphasic effects of E2 on specific cellular responses. For example, human T cell production of TNF-α is stimulated by low amounts of E2 and inhibited by higher concentrations of E2 (Gilmore et al. 1997).

Ligand-independent ERα signaling may be an important regulator of cell function in vivo. Using novel mice bearing a mutant ERα allele that cannot bind endogenous estrogens, but retains the capacity for ligand-independent activation, it was shown that ligand-free ERα could regulate growth-factor-mediated uterine epithelial cell proliferation (Sinkevicius et al. 2008). These reports of distinct functions of ligand-free and ligand-bound ER suggest that studies may yield different conclusions based on the experimental approach, such as using ER-deficient cells versus manipulating ER ligand concentrations.

3.2.2.3 ERs are Signal Transduction Molecules

Estrogens also elicit rapid (within seconds or minutes) changes in cells such as Ca^{++} fluxes across membranes, activation of G proteins, or generation of inositol phosphate (Fig. 3.2) (reviewed in Levin 2002; Marino et al. 2006b). In most cases, ligand-bound ER stimulates surface receptors such as IGF-1 or EGFR, leading to activation of ERK/MAPK, phospholipase C/protein kinase C, PI3K/AKT, or cAMP/protein kinase A signaling pathways. Palmitoylation of ER at a specific motif in the E domain facilitates its association with caveolin-1 on the cytoplasmic surface of the plasma membrane (PM) (Acconcia et al. 2005; Marino et al. 2006a). This motif is highly conserved in human and murine ERα, ERβ, AR, and PR A and PR B (Pedram et al. 2007), but despite various lines of evidence there is no consensus on whether these rapid signaling responses involve the classical ER proteins or whether they act via an undefined PM-associated estrogen binding protein (Warner and Gustafsson 2006; Heldring et al. 2007).

A G-protein-coupled receptor located in the endoplasmic reticulum, GPR30, binds E2 at an EC50 of 0.5 nM, an affinity similar to that of ER for E2 (Prossnitz et al. 2008). GPR30 binding by E2 leads to intracellular Ca^{++} mobilization, PI3K activation, and synthesis of phosphatidylinositol 3,4,5-triphosphate in COS7 cells. Thus, some estrogen responses may not be mediated by classical ERs, reinforcing the need to consider effects of estrogens and the classical ER independently.

3.2.2.4   ER Signaling Regulates Pathways Important for Immune Responses

Known genomic targets of ligated ER include a variety of genes involved in cell differentiation, growth, and survival (O'Lone et al. 2004). ER regulation in signaling pathways and its control of transcription factors is derived largely from studies of nonimmune cell types. Some of these mechanisms operate in immune cells and are discussed here.

NF-κB

Many signaling pathways in immune cells culminate in activation of NF-κB, a transcription factor necessary to express genes important for innate and adaptive immunity and immune cell development. NF-κB activation is an important consequence of toll-like receptor signaling after ligation with pathogen-associated molecular patterns and leads to production of proinflammatory cytokines. E2 is often considered to have antiinflammatory effects, because in many cell types, including monocytes and macrophages, ER signaling inhibits some proinflammatory functions of NF-κB; reduction of IL-6 transcription is a frequently cited example (McKay and Cidlowski 1999; Ghisletti et al. 2005; Harkonen and Vaananen 2006). However, E2 can also increase expression of other NF-κB-regulated proteins including IFN-γ, MCP-1, IL-12p40, and inducible nitric oxide (NO) synthase (Gourdy et al. 2005; Karpuzoglu et al. 2006; Calippe et al. 2008). Therefore, it is important to realize that ERα and β proteins can either antagonize or synergize with NF-κB to regulate genes during inflammation (reviewed in Kalaitzidis and Gilmore 2005; De Bosscher et al. 2006).

Mechanistically, ER proteins can inhibit the NF-κB pathway in multiple ways, including (1) inhibition of IKK activity important for phosphorylation and subsequent degradation of the inhibitory subunit IκB, (2) inhibition of IκB degradation, (3) blockade of DNA binding to NF-κB, (4) competitive binding of transcriptional coactivators shared with NF-κB, and (5) direct inhibitory binding to DNA-bound NF-κB (Kalaitzidis and Gilmore 2005). However, recent work shows that in murine splenocytes, E2 inhibits the nuclear translocation of a subset of NF-κB subunits, yet increases NF-κB signaling via a novel mechanism involving the NF-κB p50 subunit and the coactivator Bcl-3 (Dai et al. 2007). Inhibition of NF-κB activity at the *Il6* gene by ERα requires the ligand- and DNA-binding domains, but not AF-1. Interestingly, the synergistic positive effect of ERα on NF-κB activity at the serotonin 1-A receptor gene requires the AF-1 region, indicating that differential interaction of the ER AF domains with transcriptional coregulators may determine the effects of ER on NF-κB function at specific promoters (Kalaitzidis and Gilmore 2005). The complexity of this regulation is also exemplified by the fact that ERα can selectivity inhibit NF-κB target genes in the same cell. For example, ERα signaling decreases IL-6, but not TRAF1 transcription in breast cancer cells (Bhat-Nakshatri et al. 2004).

STAT Proteins

ER ligands have been reported to regulate phosphorylation, nuclear translocation, and transcriptional activity of STAT3 and STAT5 in endothelial and epithelial cells, implying rapid ER-mediated signaling events. Depending on the cell type and costimulus, ER ligands can enhance or decrease STAT activity, in some cases through a physical interaction between ER and STAT5 or STAT3 (Yamamoto et al. 2000; Faulds et al. 2001; Bjornstrom and Sjoberg 2002). In other models, E2 was shown to increase expression of inhibitors of cytokine signaling including suppressor of cytokine signaling (SOCS) 2 (Leung et al. 2003) and protein inhibitor of activated STAT (PIAS3) (Wang et al. 2001). E2 acting via ER can suppress IL-6 and regulate myeloma growth by inducing PIAS3 mRNA and increasing the physical association of PIAS3 with STAT3, which results in repression of STAT3 DNA binding (Wang et al. 2001). As STAT3 and STAT5 are involved in cytokine receptor signaling in immune cells, it is likely that ER signaling regulates cytokines in multiple immune cell types.

TGF-β

A connection between ER and the TGF-β pathway has been made in nonimmune cell types, raising the possibility that ER signaling modulates intracellular pathways involving immune cell production of, or responses to, TGF-β. ER ligands induce production of active TGF-β in several model systems (Takahashi et al. 1994; Gao et al. 2004; Dhandapani et al. 2005). ER signaling also modulates expression of genes in the TGF-β receptor signaling pathway in human breast cancer cells (Chang et al. 2006), and ER physically interacts with SMAD3, a downstream mediator of TGF-β receptor signaling (Matsuda et al. 2001a). It is notable that PIAS family members interact with SMAD family members to positively or negatively regulate their transcriptional activity, suggesting that ER regulation of PIAS might influence TGF-β/SMAD-regulated pathways (Shuai and Liu 2005).

Survival and Proliferation Pathways

Estradiol can upregulate expression of the antiapoptotic *Bcl2* gene and downregulate the proapoptotic *Bim* gene (Bynoe et al. 2000; Subramanian and Shaha 2007; Wang et al. 2007; Yao et al. 2007). Indeed, ER signaling regulates apoptosis and survival of immune cells, but a uniform effect of estrogens on survival is not evident based on published studies. Estrogen can increase FasL on monocytes, and depending on the differentiation state of the monocyte/macrophage, it can induce apoptosis (Mor et al. 2003). Estrogen acting via ERα also induces FasL and apoptosis in bone marrow-derived osteoclasts, providing an explanation for the prevention of bone loss by estrogen (Nakamura et al. 2007). E2 administration immediately after trauma-hemorrhage prevents splenic DC apoptosis and the decrease in other DC functions that normally occur in this model, suggesting that E2 promotes DC survival (Kawasaki et al. 2008). Ectopic exposure to E2 in vivo regulates survival pathways in B lymphocytes by increasing the levels of Bcl-2 (Bynoe et al. 2000). Constitutive Bcl-2 expression, however, prevents depletion of hematopoietic progenitors imposed by ectopic E2 (see below) (Medina et al. 2000; Grimaldi et al. 2002).

Estrogens promote cell proliferation by regulating DNA synthesis and cell cycle progression and survival. Mechanisms by which estrogens regulate these processes may involve rapid signaling events. ERα–E2-mediated activation of ERK/MAPK and PI3K/AKT causes increased expression of the antiapoptotic Bcl-2, decreased activation of the proapoptotic caspase-3, and increased expression of cyclin D1, thus promoting G1 to S phase transition (reviewed in Marino et al. 2006b). ERα binds in a ligand-dependent manner to the p85α regulatory subunit of PI3K, which stimulates PI3K-dependent activation of Akt, an important mediator of cell survival (Simoncini et al. 2000).

3.2.3 ER Signaling and ER Ligands in Immune Cell Development

3.2.3.1 Hematopoietic Progenitors

Variations in systemic estrogen levels have led to profound effects on hematopoietic progenitor populations in pregnant, ovariectomized, and E2-treated mice (Kincade et al. 2000). Hematopoietic progenitors are found within a population of lineage marker negative cells that lack surface proteins normally associated with myeloid, lymphoid, or erythroid cells (Pelayo et al. 2005). Progenitor populations, defined by specific surface proteins, include common myeloid progenitors, common lymphoid progenitors, and early lymphoid progenitors. These progenitor populations express ERs, and their numbers are regulated by estrogens in vivo.

Early lymphoid progenitors and lymphoid-restricted progenitors are selectively depleted by in vivo treatment with supraphysiological amounts of E2 (Medina et al. 2001). Elevated systemic E2 may decrease progenitor survival, since B cell progenitors in transgenic mice bearing the antiapoptotic gene *Bcl2* are resistant to E2-mediated depletion (Medina et al. 2000). Reduction in early lymphoid progenitors in E2-treated mice also reduces numbers of B lineage precursors and plasmacytoid DC in bone marrow (Medina et al. 2000; Welner et al. 2007). Plasmacytoid DC derived from myeloid progenitors also have been shown to be reduced, but not depleted by E2 treatment in vivo (Harman et al. 2006). The studies of mice subjected to a constant supraphysiological level of E2 suggest differential sensitivity of myeloid and lymphoid progenitors to negative regulation by ER agonists. Studies in a Flt3 ligand (Flt3L)-driven DC culture model show that E2 and ERα signaling decreases DC differentiation, probably by reducing the survival of myeloid progenitors (Carreras et al. 2008). In sum, ER signaling regulates hematopoietic progenitor homeostasis, with agonist ER ligands such as E2 serving to limit the number of myeloid and lymphoid progenitors in the steady state.

3.2.3.2 Dendritic Cells

Bone marrow progenitor cells can be driven to differentiate into DC by GM-CSF or Flt3 ligand, the two cytokines that direct DC differentiation in vivo (Shortman and Naik 2007). Several groups have demonstrated that E2 acting via ERα promotes the GM-CSF mediated differentiation of langerin-expressing DC (Paharkova-Vatchkova et al. 2004; Mao et al. 2005; Douin-Echinard et al. 2008; Siracusa et al. 2008). The SERMs tamoxifen, raloxifene, and toremifene also regulate GM-CSF-mediated human and murine DC differentiation from human monocytes and synovial macrophages or murine bone marrow precursors, respectively (Komi and Lassila 2000; Komi et al. 2001; Nalbandian et al. 2005).

Contrasting with results from the GM-CSF model, E2 acting via ERα inhibits Flt3L-mediated DC differentiation in part by decreasing myeloid progenitor survival (Carreras et al. 2008). E2 acts via ERα on the same highly purified myeloid progenitors to promote GM-CSF-mediated DC differentiation and inhibit Flt3L-mediated DC differentiation (Carreras et al. 2008). Thus, the same myeloid progenitor population responds differentially to ER ligands depending on the external cytokine milieu (Carreras et al. 2008). This suggests that ER and cytokine receptor signaling pathways interact to regulate cellular differentiation.

The relatively short lifespan of DCs in lymphoid organs is 3–12 days (Kamath et al. 2002), which means that in vivo exposures to agonist or antagonist ER ligands even for limited periods of time might alter de novo DC differentiation mediated by GM-CSF or Flt3L, with a downstream effect on DC numbers and intrinsic functional capacity. Thus, in studies of DC function after modulation of ER ligands or ER expression in vivo, it will be important to take into account effects of sex steroids on DC functional responses as well as potentially distinct and separate effects on new DC differentiation.

3.2.3.3 Lymphocytes

Profound depletion of lymphoid progenitors due to elevated systemic E2, as mentioned above, leads to decreased numbers of selected B cell subsets in bone marrow and spleen (Kincade et al. 2000). This has physiological correlates as well; B lymphopoiesis is significantly reduced in pregnant mice (Medina and Kincade 1994), but elevated in hypogonadal mice (Smithson et al. 1994). Studies in which bone marrow was transplanted from ERα-deficient into wild-type mice to make bone marrow "chimeras" concluded that ERα expression in hematopoietic cells is required for the effects of estrogen on hematopoiesis and B cell numbers (Thurmond et al. 2000).

Estrogens also affect T cell development (reviewed in Pernis 2007). Supraphysiological or pregnancy levels of estrogen induce thymic atrophy by multiple mechanisms. Depletion of bone marrow thymic homing progenitors and early thymic progenitors in the thymus, and reduced proliferation of developing $CD4^-CD8^-$ thymocytes to pre-TCR signals both lead to a paucity of $CD4^+CD8^+$ thymocytes (Zoller and Kersh 2006; Zoller et al. 2007). Thymic involution requires both ERα and ERβ, with ERα required in both hematopoietic and stromal cells (Staples et al. 1999; Erlandsson et al. 2001).

Another aspect of development influenced by estrogens is the acquisition of lymphocyte self-tolerance. Estrogen-mediated breaches in transitional B cell tolerance were observed in a murine lupus model using mice transgenic for an anti-DNA B cell receptor heavy chain (Bynoe et al. 2000). Exogenous estrogen-mediated loss of tolerance is associated with changes in the proportions of immature transitional B cells (recent bone marrow emigrants) and marginal zone B cells (Grimaldi et al. 2001). B cell autoreactivity measurements revealed increased fractions of high-affinity anti-DNA antibody clones in estrogen-treated animals compared with controls (Grimaldi et al. 2006). Elevated expression of *Cd22*, *Shp1*, and *Bcl2* in B cells isolated from estrogen-treated BALB/c mice suggests that protection from apoptosis or downregulation of B cell receptor signaling could explain estrogen-mediated B cell effects (Grimaldi et al. 2002). Increased expression of these same genes in B cell transfectants expressing constitutively active ERα provides evidence that estrogen acts directly on B cells to alter transitional B cell tolerance (Grimaldi et al. 2002).

3.3 Androgen Receptors

3.3.1 *Androgen Receptor Expression and Ligands*

3.3.1.1 AR Expression in Immune Cells

Receptors for androgens have been detected in immune cells (Angele et al. 2000; Nalbandian and Kovats 2005; Sader et al. 2005). This suggests a direct effect of androgens on development or function; however, the mechanisms by which this might occur are not well understood. Expression of AR on immune cells has been well documented using a variety of methods including ligand binding assays, flow cytometry, immunoblotting, fluorescent microscopy, and reverse transcription (RT)-PCR (Kovacs and Olsen 1987; Takeda et al. 1990; Viselli et al. 1995; Benten et al. 2002a; Benten et al. 2004; Sader et al. 2005). Variable expression of classical intracellular AR (iAR) has been documented in different immune cell types. PM testosterone-binding sites, suggestive of surface receptors for androgens, have also been reported. It is currently unclear whether surface AR is encoded by the gene that encodes iAR because to date a specific surface AR has not been cloned or purified.

CD45$^+$ rat leukocytes from spleen, liver, and thymus express iAR as determined by intracellular flow cytometry analyzes (Butts et al. 2007b). Functionally, active iAR, but not surface AR, was detected by flow cytometry analysis of thymic T cells. CD4$^-$CD8$^-$ and CD8$^+$ thymocytes have been shown to express the highest levels of iAR, with CD4$^+$ thymocytes expressing intermediate levels and CD4$^+$CD8$^+$ thymocytes expressing the lowest levels (Viselli et al. 1995). In peripheral lymphoid organs, classical iARs are present in CD4$^+$ and CD8$^+$ splenic T cells and testosterone-binding sites can be detected on T cell surfaces (Benten et al. 1999a; Liva and Voskuhl 2001). This suggests that during development the presence or absence of iARs and surface ARs can vary within a single cell type (Benten et al. 2002a).

ARs have also been found in macrophages and B cells. Mouse macrophage cell lines IC-21 and RAW264.7 and bone marrow-derived macrophages display functional surface AR. Macrophages respond to PM-impermeable testosterone by increasing intracellular Ca^{2+} mobilization. These macrophages are devoid of classical iARs as measured by RT-PCR, flow cytometry, and immunoblotting techniques (Benten et al. 1999b; Guo et al. 2002; Liu et al. 2005), and respond to testosterone via G-protein-coupled receptors for testosterone (Wunderlich et al. 2002). Recently, it was determined that macrophages express the iAR and a surface AR (Wunderlich et al. 2002; Benten et al. 2004; Ahmadi and McCruden 2006). By contrast, B cells express iAR, but not surface AR, as measured by RT-PCR, flow cytometry, and binding of plasma membrane-impermeable testosterone (Benten et al. 2002b). DC expression of the AR is not well studied. In our studies with bone marrow-derived DCs (BMDCs) in GM-CSF supplemented cultures, BMDCs express ERs but not ARs (Paharkova-Vatchkova et al. 2004).

In addition to fully differentiated immune cells, ARs are also present in hematopoietic progenitors of humans and mice (Igarashi et al. 2001). They are expressed in mice at an earlier age than ERs, suggesting differential effects of sex hormones on hematopoietic progenitors during postnatal development of the animal (Igarashi et al. 2001).

3.3.1.2 AR Ligands

AR function is regulated by ligand binding. AR ligands are classified according to their structure as steroidal or nonsteroidal, and are further pharmacologically classified as androgens (agonists) or antiandrogens (antagonists), depending on whether they activate or downregulate the expression of AR target genes.

The endogenous steroidal androgen testosterone is synthesized primarily from cholesterol in interstitial (Leydig) cells of testes in men and in the adrenal cortex, liver, and ovary in women. Testosterone is irreversibly converted in target tissues into a physiologically active metabolite dihydrotestosterone (DHT) by 5α-reductase in the presence of NADPH (Russell et al. 1994). DHT is a more potent AR ligand and binds to ARs with higher affinity compared with testosterone (Oettel 2003). Testosterone production is estimated to be 3–10 mg day^{-1} in healthy young men and 100–300 day^{-1} in young women. It is normally present in serum at 5–15 ng ml^{-1} in males and 0.1–0.8 ng ml^{-1} in females (Burger 2002; Davison and Bell 2006). Ninety-eight to ninety-nine percent of testosterone in the blood is bound with high affinity to sex hormone-binding globulin (SHBG) and low affinity to albumin, leaving 1–2% circulating in a free, unbound state (Dunn et al. 1981). DHT levels are even lower as DHT has a higher binding affinity to SHBG. Endogenous testosterone levels decrease in aging men and women (Schulman and Lunenfeld 2002; Davison et al. 2005).

Exogenous AR ligands function as agonists or antagonists. These are either steroidal (e.g., nandrolone, fluoxymestrone, oxendolone, cyproterone), in which the structure of naturally occurring hormones is modified, or nonsteroidal (e.g., flutamide, R-bicalutamide LG120907) (Gao et al. 2005). Selective androgen receptor modulators (SARMs) can act as full agonists in anabolic tissues such as muscle and bone, or antagonists or weak agonists in androgenic tissues such as prostate and seminal vesicles. Examples of SARM include S-1, S-4, THQ-1, THQ-60, and BMS 564929 (Gao and Dalton, 2007).

3.3.2 Mechanisms of Androgen Receptor Signaling

3.3.2.1 AR Structure and Functional Domains

AR is a member of the steroid hormone receptor family of nuclear transcription factors. The gene was initially cloned, in 1988, by Chang et al. (1988a, 1988b and Lubahn et al. (1988), and soon after by others (Trapman et al. 1988; Brinkmann et al. 1989; Tilley et al. 1989). The AR gene is located at Xq11–12 (Kuiper et al. 1989) and encodes a protein with a molecular mass of about 110 kDa (Brinkmann and Trapman 2000). In humans, two isoforms of AR encoded by a single gene have been identified and are referred to as forms A and B. AR-A has a mass of 87 kDa and was identified as an amino (NH$_2$)-terminally truncated form of the 110 kDa AR-B protein. AR-A constitutes approximately 20% of the immunoreactive receptor population and AR-B approximately 80% (Wilson and McPhaul 1994). AR-A is thought to be produced by translation initiation at the first internal methionine codon (Met-188) of AR-B (Gao and McPhaul 1998). AR isoforms are expressed in many human tissues (especially, in reproductive tissues), and AR isoforms have been found in other species (Ikeuchi et al. 1999; Sperry and Thomas 1999; Takeo and Yamashita 1999). Whether AR isoforms may have distinct functions is unclear.

For example, AR-A in cultured human bone cells and skin fibroblasts can antagonize AR-B function (Liegibel et al. 2003). The AR gene has eight exons, which code for functionally distinct regions of the protein. Like other nuclear hormone receptors, AR has a modular structure, and the four functional domains of AR are referred to as A through F (Fig. 3.1) (Zhou et al. 1994; Jenster et al. 1995; McEwan 2004).

3.3.2.2 Nuclear AR Signaling Pathways

Androgens are hydrophobic and are found in blood bound to albumin or SHBG. According to the classical model of androgen action, unbound androgens freely cross the PM of target cells and activate iARs by binding (Mendel 1989). It was recently shown that SHBG-bound androgen can be internalized by endocytosis after binding to the cell surface low-density lipoprotein receptor megalin (Hammes et al. 2005). The androgen–SHBG–megalin complex then releases the androgen within endocytic vesicles in a process that may be dependent on vesicle acidification. The iAR that is not bound to androgen is present in the cytoplasm in a complex with HSP (Pratt and Toft 1997). Binding of cytoplasmic androgens to iARs induces a series of conformational changes that includes dissociation of HSPs, dimerization, phosphorylation, and translocation of the iAR–androgen complex to the nucleus. The receptor–hormone complex then binds to androgen response elements (ARE) in the promoter/enhancer of AR target genes and recruits coregulators and the transcriptional machinery (Heinlein and Chang 2002a; Shang et al. 2002). This is known as the "genomic pathway" of androgen function (Fig. 3.2) and it elicits a response within hours or days.

There is also evidence for a ligand-independent activity for AR (Fig. 3.2). Ligand-free ARs function in this pathway as transcription factors after their phosphorylation by kinases that are activated during growth factor receptor signaling (Cenni and Picard, 1999). In the absence of androgens, ARs can be activated during cellular responses to growth factors such as epidermal growth factor (EGF), keratinocyte growth factor (KGF), insulin-like growth factor I (IGF-I) (Culig et al. 1994) or via activation of PKA or protein kinase C (PKC) signaling pathways (Nazareth and Weigel 1996; Darne et al. 1998). This type of receptor activation is also documented for other steroid hormone receptors such as ER and PR (Cenni and Picard, 1999). However, the precise mechanism for ligand-independent activation of the AR is not understood.

3.3.2.3 Functions of AR in Nongenomic Signal Transducing Pathways

As is the case for ERs and PRs, the nongenomic or nonclassical pathway for androgen action has been documented in the reproductive, cardiovascular, immune, and musculoskeletal systems (Benten et al. 1997; Benten et al. 1999a; Guo et al. 2002; Estrada et al. 2003; Walker and Cheng 2005; Vicencio et al. 2006).

Compared to genomic effects, nongenomic effects of androgens occur very rapidly within seconds or minutes and incorporate components of some well-studied PM-associated signaling pathways (Fig. 3.2) (Heinlein and Chang 2002b). The nongenomic pathway of androgen action is associated with rapid changes in the concentration of membrane ion-channels, activation or repression of cytoplasmic kinase-signaling cascades, or changes in cytoplasmic Ca^{2+} levels. Rapid membrane-associated changes may be due to direct effects of androgens on ion-channels or transporters, e.g., Ca^{2+} or Na^{+}-K^{+} ATPase, or arise from alteration of signaling pathways due to stimulation of kinases or phosphatases (Michels and Hoppe 2008). Initiation of these indirect changes not only modulates ion-channels but can also mediate transcription (Peterziel et al. 1999; Er et al. 2004). However, the biological relevance of these observations is not clear, because in some experiments nonphysiological amounts of androgens have been used.

The nongenomic action of androgens can be explained by three different mechanisms. Firstly, androgen–AR complexes may interact with the tyrosine kinase c-Src (Src) and may stimulate c-Src kinase activity and activate the mitogen-activated protein kinase pathway (Kousteni et al. 2001). Secondly, androgens may bind to SHBG attached to the SHBG-receptor (SHBG-R) and increase intracellular levels of cAMP and protein kinase A (PKA) activity, which would then activate iAR transcriptional activity (Rosner et al. 1999; Nakhla et al. 1999). Thirdly, androgens may bind to a membrane-associated iAR or an as yet unidentified surface AR (Foradori et al. 2008).

Studies with different cell types document the binding of androgens to the PM (Benten et al. 1999a, b; Liva and Voskuhl 2001; Kampa et al. 2002; Ahmadi and McCruden 2006), but to date a specific membrane receptor used by androgens to mediate nongenomic actions has not been identified. Classical AR association with plasma membranes has been shown in studies of cells such as *Xenopus* oocytes, Steroli cells, and T cells (Konoplya and Popoff 1992; Benten et al. 1999a; Lutz et al. 2003; Cheng et al. 2007), and this association may be facilitated by other membrane proteins since the classical AR lacks a transmembrane or hydrophobic domain. Recently, an association of ARs with lipid rafts was shown in the human prostate cancer cell line LNCaP. Androgens in these cells signal via the plasma membrane protein caveolin-1 and activate the PI3K/AKT signaling pathway (Li et al. 2003; Freeman et al. 2005). Studies with Chinese Hamster Ovary (CHO) cells revealed a highly conserved nine amino acid membrane localization motif in the ligand-binding domain (LBD) of steroid receptors including ARs. This conserved motif is palmitoylated, which facilitates membrane association of the iAR and enables interaction with caveolin-1. Both of these events are required for rapid signaling (Pedram et al. 2007).

3.3.2.4 AR Cross-Talk with Other Cellular Signaling Pathways

Ligated ARs regulate transcription of genes necessary for cell differentiation, growth, and survival. AR signaling also overlaps with a number of other signaling pathways important for immune function (Matsuda et al. 2001b; Ueda et al. 2002; Culig 2004). These include the NF-κB and JAK/STAT signaling pathways (De Bosscher et al. 2006; Kaarbo et al. 2007). Cross-talk of ARs with these pathways is not well studied in immune cells, but information from experiments with prostate cancer cells is available (Matsuda et al. 2001b; Kaarbo et al. 2007). Because NF-κB and JAK/STAT are involved in cytokine receptor signaling in immune cells, it is likely that AR signaling regulates these molecules in different immune cell types.

Many signaling pathways in immune cells lead to the activation of NF-κB, which in turn is required for increased expression of many crucial inflammatory and immune response genes (Gilmore 2006). There is a reciprocal negative cross-talk between NF-κB and AR signaling. Specifically, activated ARs keep NF-κB in an inactive form by suppressing the phosphorylation and subsequent degradation of inhibitor of NF-κB (IκB) (Palvimo et al. 1996). NF-κB can also repress the AR gene promoter (Supakar et al. 1995; Nakajima et al. 1996). Activation of NF-κB blocks proliferation of AR-dependent prostate cancer cells with no effect on cancer cells that lack AR (Nakajima et al. 1996). In contrast, constitutive activation of NF-κB is observed in AR-negative prostate cancer cell lines (Suh et al. 2002). TNF-α signaling can also induce growth arrest and apoptosis of androgen-responsive human prostate cancer LNCaP cells by inhibiting the phosphorylation of IκB (Chopra et al. 2004).

Direct interactions between members of the JAK/STAT signaling pathway family and ARs have been reported (Matsuda et al. 2001b; Ueda et al. 2002; Yamamoto et al. 2003). Several studies have shown that IL-6 activates AR-mediated gene expression via STAT3 (Hobisch et al. 1998; Chen et al. 2000). In prostate cancer cells, IL-6-induced activation of STAT3 is enhanced upon AR signaling induced by DHT (Matsuda et al. 2001b). This activation is inhibited by flutamide, an AR antagonist, suggesting direct physical interactions between STAT3 and ARs.

3.3.3 Effects of Androgens and ARs on Immune Function

Androgens exert suppressive effects on both humoral and cellular immune responses (Cutolo et al. 2004; Klein 2004). Their role in immune system function has been extensively studied in various species using four experimental approaches: (a) comparison of males and females (Da Silva 1995; Olsen and Kovacs 2001; Moore and Wilson 2002), (b) finding relationships between immune function and endogenous testosterone levels in males (Saino et al. 1997; Granger et al. 2000; Kurtis et al. 2001; Roubenoff et al. 2002), (c) assessment of immune function following castration or testosterone supplementation (Olsen et al. 1991; Benten et al. 1993; Angele et al. 1998; Bhasin et al. 2000; Singh et al. 2002), and (d) in vitro analyzes of testosterone effects on immune function (Chao et al. 1995; Smithson et al. 1998; Wunderlich et al. 2002). Studies using these approaches are discussed in Chaps. 2, 4–6. How androgens regulate immunity in vivo can be addressed in mouse models with defective AR signaling such as testicular feminization (*Tfm*) male mice or conditional AR knockout mice (Yeh et al. 2002; Matsumoto et al. 2005; Matsumoto et al. 2008).

3.3.4 Androgens and AR in Immune Cell Development

Several lines of evidence suggest that androgens can exert suppressive effects on the development of B and T lymphocytes (Frey-Wettstein and Craddock 1970; Viselli et al. 1997; Olsen and Kovacs 2001; Heng et al. 2005). Castrated mice or mice lacking a functional AR (*Tfm* male mice), exhibit increased bone marrow B cell and thymic T cell populations as well as increased numbers of splenic lymphocytes (Aboudkhil et al. 1991; Olsen et al. 1991; Wilson et al. 1995; Smithson et al. 1998; Ellis et al. 2001; Sutherland et al. 2005; Goldberg et al. 2007; Radojevic et al. 2007). In addition, castration of male mice results in expansion of pre-B cells and newly formed B cells in bone marrow and a dramatic increase in surface IgM expressing naive B cells in spleen (Wilson et al. 1995). Androgen replacement with either testosterone or DHT reverses this effect (Windmill et al. 1993; Viselli et al. 1997). One mechanism for androgen-induced suppression of lymphocyte development is suggested by the finding that testosterone administered to intact female and castrated male mice accelerates thymocyte apoptosis and decreases thymopoiesis (Olsen et al. 1998).

Expression of ARs has been documented in hematopoietic progenitors and lymphoid and nonlymphoid cells in the thymus and bone marrow of mice and humans (Kovacs and Olsen 1987; Takeda et al. 1990; Mantalaris et al. 2001). This suggests that androgens and ARs regulate B and T cell development. Bone marrow chimeric mice bearing AR-positive hematopoietic and AR-negative stromal/epithelial cellular compartments have been used to address whether androgens regulate lymphocyte development by acting directly on developing lymphoid precursors or by influencing supporting cells, such as thymic epithelial cells and bone marrow stromal cells. Androgenic hormone manipulation in these mice suggests that AR expressing thymic epithelial cells and bone marrow stromal cells are mediators of androgenic effects on immature lymphocytes (Olsen et al. 2001a; Olsen et al. 2001b). Furthermore, bone marrow stromal cells can produce TGF-β in response to DHT in vitro, and B cell precursors do not respond to suppressive effects of androgens unless they are cocultured with TGF-β producing stromal cells or grown in media that has been conditioned by androgen-treated stromal cells (Olsen et al. 2001a). This suggests that stromal cells mediate some of the observed effects of androgen on B cell development in bone marrow.

Androgens also influence DC development. For example, the density of epidermal Langerhan cells (LC) varies by sex, with male mice having fewer LC than do female mice. Gonadectomy, but not ovariectomy, elevates LC density, suggesting that androgens negatively regulate the density of these cells in male mice (Koyama et al. 1987). Systemic and topical application of testosterone or its metabolite to gonadectomized males and intact females results in a significant decrease in LC density (Koyama et al. 1989). However, in GM-CSF-driven DC differentiation cultures, addition of testosterone or DHT does not promote DC differentiation (Paharkova-Vatchkova et al. 2004). Taken together with studies showing hematopoietic progenitor expression of ARs, the studies outlined above indicate that androgens can exert profound effects on immune cell development.

3.4 Progesterone Receptors

3.4.1 Progesterone Receptor Expression and Receptor Ligands

3.4.1.1 PR Expression in Immune Cells

There is clear evidence that the immune system is modulated by progesterone (see Chaps. 9 and 10); however, the immune cell types that express PR, and the conditions under which they express PR remain under investigation. Blood lymphocytes of pregnant women have been shown to react with a specific monoclonal antibody to the nuclear PR (nPR), which does not react with lymphocytes in nonpregnant women (Szekeres-Bartho et al. 1990). Other studies, however, find very little or no expression of nPR in lymphocytes from pregnant or nonpregnant women (Mansour et al. 1994). RT-PCR and western blot have not detected any nPR expression in normal human blood lymphocytes (Bamberger et al. 1999). Examination of receptors by flow cytometry, western blotting, and confocal microcopy indicate that isolated peripheral blood NK cells from males and females express nPR but purified T and B cells do not express nPR (Arruvito et al. 2008). Interestingly, T lymphocytes and immortalized T cells (Jurkat cells) express membrane progesterone receptors (mPRα) and (mPRβ) (Dosiou et al. 2008). Rat leukocytes from spleen, thymus, and liver, and rat BMDCs express nPRs as measured by flow cytometry (Butts et al. 2007a,b). Macrophages also express PR (Khan et al. 2005; Jones et al. 2008).

3.4.1.2 PR Ligands

Progesterone (P4) is a steroid hormone secreted primarily by the gonads and adrenal cortex in nonpregnant women. In many species, including humans and rodents, the placenta assumes the role of synthesizing P4 during pregnancy (Stites and Siiteri 1983). Human P4 levels rise from 1–2 nM during the follicular phase of the menstrual cycle to 20–40 nM in the late luteal phase. P4 levels during pregnancy increase to 100–500 nM in serum and can reach 1–10 μM in placenta (Stites and Siiteri 1983). Although progesterone is considered a female sex hormone, P4 serum levels in males can reach 1–2 nM and are known to play a role in male reproduction

(Oettel and Mukhopadhyay 2004). Progesterone in blood is generally found bound to transcortin (corticosteroid-binding globulin).

P4 analogs have evolved in the past 50 years because of their clinical application. PR ligands are currently used in contraception and in postmenopausal hormone therapy. They are also undergoing clinical evaluation as treatments for a variety of gynecologic disorders (Cadepond et al. 1997; Spitz 2003). PR ligands are characterized as PR agonists, PR antagonists, or selective PR modulators with mixed agonistic and antagonistic effect depending on the cell type (Winneker et al. 2005; Spitz, 2006). At physiological concentrations, P4 and P4 analogs can also bind to the glutocorticoid receptor and induce effects similar to glucocorticoids (Kontula et al. 1983; Selman et al. 1994; Bamberger et al. 1999).

3.4.2   Mechanism of Progesterone Receptor Signaling

3.4.2.1   PR Structure and Functional Domains

The biological actions of human progesterone P4 are mediated by nPRs, members of the superfamily of nuclear receptors that includes steroid receptors, vitamin D, and thyroid hormone (Evans 1988; Carson-Jurica et al. 1990), Two isoforms of nPR, nPR-A and nPR-B, are transcribed from different promoters on the same gene and are responsible for the majority of nPR function (Graham and Clarke 1997). PR-A (94 kDa) is a truncated form of PR-B (116 kDa) and contains a 164 amino acid N-terminal deletion.

As are all the members of the nuclear receptor superfamily, PR is organized into structural–functional domains that are conserved at different degrees between species (Fig. 3.1) (Carson-Jurica et al. 1990). Progestins bind to the moderately conserved carboxy-terminal LBD, which also is necessary for receptor dimerization and gene transactivation (AF-2). The centrally-located DNA-binding domain (DBD) is the most conserved region and encodes two type II zinc fingers (Evans 1988). The hinge region between DBD and LBD contains other transactivation domains. The amino-terminal domain is the most variable and also contains transactivation functions (AF-1) (Meyer et al. 1990). PR-B has a unique activation function site (AF-3) in the first 164 amino acids, which is absent in PR-A (Sartorius et al. 1994). Most progesterone target tissues express both PR isoforms in varying ratios. In most cells, hPR-A inhibits transcription at progesterone-responsive promoters while hPR-B activates transcription at these promoters (Edwards 2005).

3.4.2.2   PRs Function as Nuclear Transcription Factors

In the absence of P4, nPR is stabilized in the cytoplasm by complexing to HSPs. Following ligation with P4, PRs dissociate from HSPs and form homo- or heterodimers that act as transcription factors by binding to P4 responsive elements (PRE) in target promoters (Fig. 3.2). Ligated PR can also regulate genes that do not contain PRE by interacting with other transcription factors that directly bind DNA (Tseng et al. 2003).

PR expression is induced primarily by estrogen via the ER although the PR promoter does not contain consensus palindromic EREs. Therefore, it is not always clear which physiological responses are mediated by P4, estrogen, or the combination of the two hormones. Mouse models with null mutations for nPR-A, nPR-B, or null mutations in both isoforms have been useful for discriminating physiological functions specifically attributable to P4 (Lydon et al. 1995; Mulac-Jericevic et al. 2000; Mulac-Jericevic et al. 2003).

Mice lacking both nPR-A and nPR-B isoforms (PRKO) have pleiotropic defects in the reproductive system (Lydon et al. 1995). In addition, treatment of PRKO mice with E2 and P4 produces an acute inflammatory response in the uterus that is not seen in wild-type (WT) mice (Lydon et al. 1995). Treatment of WT mice with E2 but not E2 in combination with P4 results in the recruitment of macrophages and neutrophils to the uterus. P4 does not antagonize the inflammatory effect of E2 in PRKO mice (Tibbetts et al. 1999). These studies indicate that P4, acting via PRs, ablates the inflammatory response induced by E2 treatment.

Thymic involution during pregnancy depends on expression of nPR in thymic stromal cells (Tibbetts et al. 1999). Mice deficient in nPR-A show thymic involution after treatment with E2 and P4 (Mulac-Jericevic et al. 2000) suggesting that involution is mediated by nPR-B.

3.4.2.3 PR Ligand-Independent Signaling

PRs are substrates for kinase phosphorylation, adding an additional level of regulation to PR action (Rao et al. 1987). Phosphorylation of nPR-A or nPR-B at specific sites affects nuclear translocation/transcriptional activity and stability of the PR (Lange et al. 2000; Qiu and Lange, 2003). Ligation of nPR-A or nPR-B with P4 increases the overall phosphorylation of the receptor (Zhang et al. 1995).

MAPK can mediate cross-talk between nPRs and growth-factor-signaling pathways. Phosphorylation activates PRs for binding to PRE (Denner et al. 1990; Vegeto et al. 1999; Gonzalez-Flores et al. 2004). Treatment with EGF, which stimulates MAPK, results in phosphorylation of the PR and nuclear translocation in the absence of P4 (Qiu et al. 2003).

3.4.2.4 PR as Signal Transducing Molecules

Rapid "nongenomic" P4 actions have been described in cancer cells, reproductive tissues, and immune cells. The identity of the PR responsible for nongenomic effects and the biological significance of these mechanisms of P4 action, however, are controversial. The nongenomic mechanism starts at or near the cell surface by binding of P4 to membrane-associated receptors and activation of intracellular signaling pathways. Diverse proteins, including the classical nPRs, other steroid nuclear receptors, and a new membrane PR unrelated to nuclear receptors, have been proposed as mediators of rapid P4 actions (reviewed in Losel et al. 2003). Both isoforms of the nPR contain motifs that can interact with the SH3 domain of Src, but only PR-B, which is distributed in the nucleus and cytoplasm, mediates rapid activation of Src/MAPK (Erk-1/2) in breast cancer cells. This contrasts with PR-A, which is found primarily in the nucleus (Boonyaratanakornkit et al. 2007).

Progesterone-specific membrane-binding sites were first detected in amphibian oocytes (Kostellow et al. 1982; Blondeau and Baulieu, 1984). The specific receptor was later identified and characterized in sea trout (Zhu et al. 2003). This mPR (Mw ~40 kDa), which has three subtypes mPR α, β, and γ, belongs to the seven-transmembrane P4 adiponectin Q receptor (PAQR) family that is closely related to G proteins. In sea trout, mPR are expressed at high levels in reproductive tissue, brain, and pituitary (Zhu et al. 2003; Kazeto et al. 2005). Low expression in a variety of nonreproductive organs, including spleen, has been detected by real-time RT-PCR in channel catfish (Kazeto et al. 2005). In humans, mPRα is expressed mainly in reproductive tissues, kidney, and adrenals (Zhu et al. 2003), but it is also found in lymphocytes and other cells (Dosiou et al. 2008).

P4 membrane receptor component 1 (PGMRC1) (Mw ~26 kDa) is a single transmembrane protein expressed in reproductive and nonreproductive organs, and has a high affinity for P4 with low affinity for other steroids (Meyer et al. 1996; Thomas, 2008). No information is available regarding expression of this protein in immune cells.

Some of the physiological effects of P4 on immune cells are mediated by nongenomic mechanisms. P4 directly blocks K^+ channels in activated T lymphocytes, which produces a depolarization of the membrane potential and subsequent reduction in Ca^{2+} entry. Low intracellular Ca^{2+} concentration interferes with IL-2 transcription in response to T cell receptor ligation and leads to immunosuppression (Ehring et al. 1998). Progesterone also suppresses T cell proliferation by inducing acidification of the cytosol through inhibition of Na^+/H^+ exchange (Chien et al. 2007).

3.4.3   PR Signaling and PR Ligands in Immune Cell Development

The immunomodulatory function of P4 has been studied primarily within the context of pregnancy (see Chap. 9) and in susceptibility to parasites (see Chaps. 6 and 10). Nevertheless, there are some studies on the role of low (nonpregnancy) P4 levels in immune cell differentiation and function. Low concentrations (1–10 nM) of P4 suppress rat BMDC secretion of proinflammatory cytokines in response to LPS stimulation (Butts et al. 2007a) and induce apoptosis of peripheral blood NK cells (Arruvito et al. 2008). In response to viral infection, plasmacytoid DCs secrete

IFN-α, which has potent antiviral effects. P4 and the hormonal contraceptive depot medroxyprogesterone acetate (DMPA) inhibit IFN-α production by plasmacytoid DCs (Hughes et al. 2008). This may explain the high risk of viral infection in women using DMPA (Kaushic et al. 2003). P4 prevents TNF-α-induced apoptosis of undifferentiated and PMA-differentiated monoblastoid U937 cells (Vegeto et al. 1999). In mice, B-lymphopoiesis is reduced by treatment with E2 but not by treatment with P4. The treatment of P4 together with E2 has a synergistic effect, reducing the dose of E2 needed to inhibit B-lymphopoiesis (Medina and Kincade, 1994). These data suggest that dampening of inflammatory responses is an important role of P4.

3.5 Conclusions

Immune cells express ERs, PRs, and ARs, implying that endogenous sex hormones and other ligands for these receptors directly regulate their activation, lifespan, or functional responses during innate and adaptive immunity. Immune function is being studied after manipulation of sex hormone receptor ligands or receptor expression in vitro and in vivo, and due to the ligand-independent activity of these receptors, different conclusions may be reached using these two approaches. Because hematopoietic progenitors express ERs and ARs and possibly PRs, it will be important to distinguish between the potentially distinct effects of sex steroids on lymphoid and myeloid cell differentiation versus pathways of immune cell activation.

Acknowledgements The authors thank Dr. John Knight for expert editorial assistance.

References

Aboudkhil S, Bureau JP, Garrelly L, Vago P (1991) Effects of castration, Depo-testosterone and cyproterone acetate on lymphocyte T subsets in mouse thymus and spleen. Scand J Immunol 34:647–653

Acconcia F, Ascenzi P, Bocedi A, Spisni E, Tomasi V, Trentalance A, Visca P, Marino M (2005) Palmitoylation-dependent estrogen receptor alpha membrane localization: regulation by 17beta-estradiol. Mol Biol Cell 16:231–237

Adams JS (2005) "Bound" to work: the free hormone hypothesis revisited. Cell 122:647–649

Ahmadi K, McCruden AB (2006) Macrophage may responses to androgen via its receptor. Med Sci Monit 12:BR15–BR20

Ahmed SA (2000) The immune system as a potential target for environmental estrogens (endocrine disrupters): a new emerging field. Toxicology 150:191–206

Ahmed SA, Penhale WJ, Talal N (1985) Sex hormones, immune responses, and autoimmune diseases. Mechanisms of sex hormone action. Am J Pathol 121:531–551

Angele MK, Ayala A, Cioffi WG, Bland KI, Chaudry IH (1998) Testosterone: the culprit for producing splenocyte immune depression after trauma hemorrhage. Am J Physiol 274:C1530–C1536

Angele MK, Schwacha MG, Ayala A, Chaudry IH (2000) Effect of gender and sex hormones on immune responses following shock. Shock 14:81–90

Arruvito L, Giulianelli S, Flores AC, Paladino N, Barboza M, Lanari C, Fainboim L (2008) NK cells expressing a progesterone receptor are susceptible to progesterone-induced apoptosis. J Immunol 180:5746–5753

Askanase AD, Buyon JP (2002) Reproductive health in SLE. Best Pract Res Clin Rheumatol 16:265–280

Bamberger CM, Else T, Bamberger AM, Beil FU, Schulte HM (1999) Dissociative glucocorticoid activity of medroxyprogesterone acetate in normal human lymphocytes. J Clin Endocrinol Metab 84:4055–4061

Barkhem T, Carlsson B, Nilsson Y, Enmark E, Gustafsson J, Nilsson S (1998) Differential response of estrogen receptor alpha and estrogen receptor beta to partial estrogen agonists/antagonists. Mol Pharmacol 54:105–112

Benten WP, Wunderlich F, Herrmann R, Kuhn-Velten WN (1993) Testosterone-induced compared with oestradiol-induced immunosuppression against Plasmodium chabaudi malaria. J Endocrinol 139:487–494

Benten WP, Lieberherr M, Sekeris CE, Wunderlich F (1997) Testosterone induces Ca2+ influx via non-genomic surface receptors in activated T cells. FEBS Lett 407:211–214

Benten WP, Lieberherr M, Giese G, Wrehlke C, Stamm O, Sekeris CE, Mossmann H, Wunderlich F (1999a) Functional testosterone receptors in plasma membranes of T cells. Faseb J 13:123–133

Benten WP, Lieberherr M, Stamm O, Wrehlke C, Guo Z, Wunderlich F (1999b) Testosterone signaling through internalizable surface receptors in androgen receptor-free macrophages. Mol Biol Cell 10:3113–3123

Benten WP, Becker A, Schmitt-Wrede HP, Wunderlich F (2002a) Developmental regulation of intracellular and surface androgen receptors in T cells. Steroids 67:925–931

Benten WP, Stephan C, Wunderlich F (2002b) B cells express intracellular but not surface receptors for testosterone and estradiol. Steroids 67:647–654

Benten WP, Guo Z, Krucken J, Wunderlich F (2004) Rapid effects of androgens in macrophages. Steroids 69:585–590

Bhasin S, Storer TW, Javanbakht M, Berman N, Yarasheski KE, Phillips J, Dike M, Sinha-Hikim I, Shen R, Hays RD, Beall G (2000) Testosterone replacement and resistance exercise in HIV-infected men with weight loss and low testosterone levels. Jama 283:763–770

Bhat-Nakshatri P, Campbell RA, Patel NM, Newton TR, King AJ, Marshall MS, Ali S, Nakshatri H (2004) Tumour necrosis factor and PI3-kinase control oestrogen receptor alpha protein level and its transrepression function. Br J Cancer 90:853–859

Bjornstrom L, Sjoberg M (2002) Signal transducers and activators of transcription as downstream targets of nongenomic estrogen receptor actions. Mol Endocrinol 16:2202–2214

Blondeau JP, Baulieu EE (1984) Progesterone receptor characterized by photoaffinity labelling in the plasma membrane of Xenopus laevis oocytes. Biochem J 219:785–792

Boonyaratanakornkit V, McGowan E, Sherman L, Mancini MA, Cheskis BJ, Edwards DP (2007) The role of extranuclear signaling actions of progesterone receptor in mediating progesterone regulation of gene expression and the cell cycle. Mol Endocrinol 21:359–375

Brinkmann AO, Trapman J (2000) Genetic analysis of androgen receptors in development and disease. Adv Pharmacol 47:317–341

Brinkmann AO, Faber PW, van Rooij HC, Kuiper GG, Ris C, Klaassen P, van der Korput JA, Voorhorst MM, van Laar JH, Mulder E et al (1989) The human androgen receptor: domain structure, genomic organization and regulation of expression. J Steroid Biochem 34:307–310

Burger HG (2002) Androgen production in women. Fertil Steril 77(Suppl 4):S3–S5

Butts CL, Shukair SA, Duncan KM, Bowers E, Horn C, Belyavskaya E, Tonelli L, Sternberg EM (2007a) Progesterone inhibits mature rat dendritic cells in a receptor-mediated fashion. Int Immunol 19:287–296

Butts CL, Shukair SA, Duncan KM, Harris CW, Belyavskaya E, Sternberg EM (2007b) Evaluation of steroid hormone receptor protein expression in intact cells using flow cytometry. Nucl Recept Signal 5:e007

Bynoe MS, Grimaldi CM, Diamond B (2000) Estrogen up-regulates Bcl-2 and blocks tolerance induction of naive B cells. Proc Natl Acad Sci U S A 97:2703–278

Cadepond F, Ulmann A, Baulieu EE (1997) RU486 (mifepristone): mechanisms of action and clinical uses. Annu Rev Med 48:129–156

Calippe B, Douin-Echinard V, Laffargue M, Laurell H, Rana-Poussine V, Pipy B, Guery JC, Bayard F, Arnal JF, Gourdy P (2008) Chronic estradiol administration in vivo promotes the proinflammatory response of macrophages to tlr4 activation: involvement of the phosphatidy-linositol 3-kinase pathway. J Immunol 180:7980–7988

Carreras E, Turner S, Paharkova-Vatchkova V, Mao A, Dascher C, Kovats S (2008) Estradiol acts directly on bone marrow myeloid progenitors to differentially regulate GM-CSF or Flt3 Ligand mediated dendritic cell differentiation. J. Immunol. 180:727–738

Carson-Jurica MA, Schrader WT, O'Malley BW (1990) Steroid receptor family: structure and functions. Endocr Rev 11:201–220

Castles CG, Oesterreich S, Hansen R, Fuqua SA (1997) Auto-regulation of the estrogen receptor promoter. J Steroid Biochem Mol Biol 62:155–163

Cenni B, Picard D (1999) Ligand-independent Activation of Steroid Receptors: New Roles for Old Players. Trends Endocrinol Metab 10:41–46

Chang CS, Kokontis J, Liao ST (1988a) Molecular cloning of human and rat complementary DNA encoding androgen receptors. Science 240:324–326

Chang CS, Kokontis J, Liao ST (1988b) Structural analysis of complementary DNA and amino acid sequences of human and rat androgen receptors. Proc Natl Acad Sci USA 85:7211–7215

Chang EC, Frasor J, Komm B, Katzenellenbogen BS (2006) Impact of estrogen receptor beta on gene networks regulated by estrogen receptor alpha in breast cancer cells. Endocrinology 147:4831–4842

Chao TC, Van Alten PJ, Greager JA, Walter RJ (1995) Steroid sex hormones regulate the release of tumor necrosis factor by macrophages. Cell Immunol 160:43–49

Chen T, Wang LH, Farrar WL (2000) Interleukin 6 activates androgen receptor-mediated gene expression through a signal transducer and activator of transcription 3-dependent pathway in LNCaP prostate cancer cells. Cancer Res 60:2132–2135

Cheng J, Watkins SC, Walker WH (2007) Testosterone activates mitogen-activated protein kinase via Src kinase and the epidermal growth factor receptor in sertoli cells. Endocrinology 148:2066–2074

Chien EJ, Liao CF, Chang CP, Pu HF, Lu LM, Shie MC, Hsieh DJ, Hsu MT (2007) The non-genomic effects on Na+/H+-exchange 1 by progesterone and 20alpha-hydroxyprogesterone in human T cells. J Cell Physiol 211:544–550

Chopra DP, Menard RE, Januszewski J, Mattingly RR (2004) TNF-alpha-mediated apoptosis in normal human prostate epithelial cells and tumor cell lines. Cancer Lett 203:145–154

Couse JF, Korach KS (1999) Estrogen receptor null mice: what have we learned and where will they lead us? Endocr Rev 20:358–417

Couse JF, Curtis SW, Washburn TF, Lindzey J, Golding TS, Lubahn DB, Smithies O, Korach KS (1995) Analysis of transcription and estrogen insensitivity in the female mouse after targeted disruption of the estrogen receptor gene. Mol Endocrinol 9:1441–1454

Culig Z (2004) Androgen receptor cross-talk with cell signalling pathways. Growth Factors 22:179–184

Culig Z, Hobisch A, Cronauer MV, Radmayr C, Trapman J, Hittmair A, Bartsch G, Klocker H (1994) Androgen receptor activation in prostatic tumor cell lines by insulin-like growth factor-I, keratinocyte growth factor, and epidermal growth factor. Cancer Res 54:5474–5478

Curran EM, Berghaus LJ, Vernetti NJ, Saporita AJ, Lubahn DB, Estes DM (2001) Natural killer cells express estrogen receptor-alpha and estrogen receptor-beta and can respond to estrogen via a non-estrogen receptor-alpha-mediated pathway. Cell Immunol 214:12–20

Cutolo M, Sulli A, Capellino S, Villaggio B, Montagna P, Seriolo B, Straub RH (2004) Sex hormones influence on the immune system: basic and clinical aspects in autoimmunity. Lupus Lupus 13:635–638

Cvoro A, Tzagarakis-Foster C, Tatomer D, Paruthiyil S, Fox MS, Leitman DC (2006) Distinct roles of unliganded and liganded estrogen receptors in transcriptional repression. Mol Cell 21:555–564

Cvoro A, Tatomer D, Tee MK, Zogovic T, Harris HA, Leitman DC (2008) Selective estrogen receptor-beta agonists repress transcription of proinflammatory genes. J Immunol 180:630–636

Da Silva JA (1995) Sex hormones, glucocorticoids and autoimmunity: facts and hypotheses. Ann Rheum Dis 54:6–16

Dai R, Phillips RA, Ahmed SA (2007) Despite inhibition of nuclear localization of NF-kappa B p65, c-Rel, and RelB, 17-beta estradiol up-regulates NF-kappa B signaling in mouse spleno-cytes: the potential role of Bcl-3. J Immunol 179:1776–1783

Darne C, Veyssiere G, Jean C (1998) Phorbol ester causes ligand-independent activation of the androgen receptor. Eur J Biochem 256:541–549

Davison SL, Bell R (2006) Androgen physiology. Semin Reprod Med 24:71–77

Davison SL, Bell R, Donath S, Montalto JG, Davis SR (2005) Androgen levels in adult females: changes with age, menopause, and oophorectomy. J Clin Endocrinol Metab 90:3847–3853

Dayan M, Zinger H, Kalush F, Mor G, Amir-Zaltzman Y, Kohen F, Sthoeger Z, Mozes E (1997) The beneficial effects of treatment with tamoxifen and anti-oestradiol antibody on experimental systemic lupus erythematosus are associated with cytokine modulations. Immunology 90:101–108

De Bosscher K, Vanden Berghe W, Haegeman G (2006) Cross-talk between nuclear receptors and nuclear factor kappaB. Oncogene 25:6868–6886

Denner LA, Weigel NL, Maxwell BL, Schrader WT, O'Malley BW (1990) Regulation of progesterone receptor-mediated transcription by phosphorylation. Science 250:1740–1743

Dhandapani KM, Wade FM, Mahesh VB, Brann DW (2005) Astrocyte-derived transforming growth factor-{beta} mediates the neuroprotective effects of 17{beta}-estradiol: involvement of nonclassical genomic signaling pathways. Endocrinology 146:2749–2759

Dorner G, Eckert R, Hinz G (1980) Androgen-dependent sexual dimorphism of the immune system. Endokrinologie 76:112–114

Dosiou C, Hamilton AE, Pang Y, Overgaard MT, Tulac S, Dong J, Thomas P, Giudice LC (2008) Expression of membrane progesterone receptors on human T lymphocytes and Jurkat cells and activation of G-proteins by progesterone. J Endocrinol 196:67–77

Douin-Echinard V, Laffont S, Seillet C, Delpy L, Krust A, Chambon P, Gourdy P, Arnal JF, Guery JC (2008) Estrogen receptor alpha, but not beta, is required for optimal dendritic cell differentiation and CD40-induced cytokine production. J Immunol 180:3661–3669

Dunn JF, Nisula BC, Rodbard D (1981) Transport of steroid hormones: binding of 21 endogenous steroids to both testosterone-binding globulin and corticosteroid-binding globulin in human plasma. J Clin Endocrinol Metab 53:58–68

Dutertre M, Smith CL (2000) Molecular mechanisms of selective estrogen receptor modulator (SERM) action. J Pharmacol Exp Ther 295:431–47

Edwards DP (2005) Regulation of signal transduction pathways by estrogen and progesterone. Annu Rev Physiol 67:335–376

Ehring GR, Kerschbaum HH, Eder C, Neben AL, Fanger CM, Khoury RM, Negulescu PA, Cahalan MD (1998) A nongenomic mechanism for progesterone-mediated immunosuppression: inhibition of K+ channels, Ca2+ signaling, and gene expression in T lymphocytes. J Exp Med 188:1593–1602

Ellis TM, Moser MT, Le PT, Flanigan RC, Kwon ED (2001) Alterations in peripheral B cells and B cell progenitors following androgen ablation in mice. Int Immunol 13:553–558

Er F, Michels G, Gassanov N, Rivero F, Hoppe UC (2004) Testosterone induces cytoprotection by activating ATP-sensitive K+channels in the cardiac mitochondrial inner membrane. Circulation 110:3100–3107

Erlandsson MC, Ohlsson C, Gustafsson JA, Carlsten H (2001) Role of oestrogen receptors alpha and beta in immune organ development and in oestrogen-mediated effects on thymus. Immunology 103:17–25

Estrada M, Espinosa A, Muller M, Jaimovich E (2003) Testosterone stimulates intracellular calcium release and mitogen-activated protein kinases via a G protein-coupled receptor in skeletal muscle cells. Endocrinology 144:3586–3597

Evans RM (1988) The steroid and thyroid hormone receptor superfamily. Science 240:889–895

Faulds MH, Pettersson K, Gustafsson JA, Haldosen LA (2001) Cross-talk between ERs and signal transducer and activator of transcription 5 is E2 dependent and involves two functionally separate mechanisms. Mol Endocrinol 15:1929–1940

Foradori CD, Weiser MJ, Handa RJ (2008) Non-genomic actions of androgens. Front Neuroendocrinol 29:169–181

Foster HL, Small JD, Fox JG (1983) Normative biology, immunology and husbandry. Academic Press Inc., Orlando, FL

Frasor J, Danes JM, Komm B, Chang KC, Lyttle CR, Katzenellenbogen BS (2003) Profiling of estrogen up- and down-regulated gene expression in human breast cancer cells: insights into gene networks and pathways underlying estrogenic control of proliferation and cell phenotype. Endocrinology 144:4562–4574

Freeman MR, Cinar B, Lu ML (2005) Membrane rafts as potential sites of nongenomic hormonal signaling in prostate cancer. Trends Endocrinol Metab 16:273–279

Frey-Wettstein M, Craddock CG (1970) Testosterone-induced depletion of thymus and marrow lymphocytes as related to lymphopoiesis and hematopoiesis. Blood 35:257–271

Gao W, Dalton JT (2007) Expanding the therapeutic use of androgens via selective androgen receptor modulators (SARMs). Drug Discov Today 12:241–248

Gao T, McPhaul MJ (1998) Functional activities of the A and B forms of the human androgen receptor in response to androgen receptor agonists and antagonists. Mol Endocrinol 12:654–663

Gao Y, Qian WP, Dark K, Toraldo G, Lin AS, Guldberg RE, Flavell RA, Weitzmann MN, Pacifici R (2004) Estrogen prevents bone loss through transforming growth factor beta signaling in T cells. Proc Natl Acad Sci USA 101:16618–16623

Gao W, Bohl CE, Dalton JT (2005) Chemistry and structural biology of androgen receptor. Chem Rev 105:3352–3370

Ghisletti S, Meda C, Maggi A, Vegeto E (2005) 17beta-estradiol inhibits inflammatory gene expression by controlling NF-kappaB intracellular localization. Mol Cell Biol 25:2957–2968

Gilmore TD (2006) Introduction to NF-kappaB: players, pathways, perspectives. Oncogene 25:6680–6684

Gilmore W, Weiner LP, Correale J (1997) Effect of estradiol on cytokine secretion by proteolipid protein-specific T cell clones isolated from multiple sclerosis patients and normal control subjects. J Immunol 158:446–451

Goldberg GL, Alpdogan O, Muriglan SJ, Hammett MV, Milton MK, Eng JM, Hubbard VM, Kochman A, Willis LM, Greenberg AS, Tjoe KH, Sutherland JS, Chidgey A, van den Brink MR, Boyd RL (2007) Enhanced immune reconstitution by sex steroid ablation following allogeneic hemopoietic stem cell transplantation. J Immunol 178:7473–7484

Gonzalez-Flores O, Shu J, Camacho-Arroyo I, Etgen AM (2004) Regulation of lordosis by cyclic 3', 5'-guanosine monophosphate, progesterone, and its 5alpha-reduced metabolites involves mitogen-activated protein kinase. Endocrinology 145:5560–5567

Gourdy P, Araujo LM, Zhu R, Garmy-Susini B, Diem S, Laurell H, Leite-de-Moraes M, Dy M, Arnal JF, Bayard F, Herbelin A (2005) Relevance of sexual dimorphism to regulatory T cells: estradiol promotes IFN-gamma production by invariant natural killer T cells. Blood 105:2415–2420

Graham JD, Clarke CL (1997) Physiological action of progesterone in target tissues. Endocr Rev 18:502–519

Granger DA, Booth A, Johnson DR (2000) Human aggression and enumerative measures of immunity. Psychosom Med 62:583–590

Grimaldi CM, Michael DJ, Diamond B (2001) Cutting edge: expansion and activation of a population of autoreactive marginal zone B cells in a model of estrogen-induced lupus. J Immunol 167:1886–190

Grimaldi CM, Cleary J, Dagtas AS, Moussai D, Diamond B (2002) Estrogen alters thresholds for B cell apoptosis and activation. J Clin Invest 109:1625–133

Grimaldi CM, Jeganathan V, Diamond B (2006) Hormonal regulation of B cell development: 17 beta-estradiol impairs negative selection of high-affinity DNA-reactive B cells at more than one developmental checkpoint. J Immunol 176:2703–2710

Grossman CJ (1985) Interactions between the gonadal steroids and the immune system. Science 227:257–261

Guo Z, Benten WP, Krucken J, Wunderlich F (2002) Nongenomic testosterone calcium signaling. Genotropic actions in androgen receptor-free macrophages. J Biol Chem 277:29600–29607

Hammes A, Andreassen TK, Spoelgen R, Raila J, Hubner N, Schulz H, Metzger J, Schweigert FJ, Luppa PB, Nykjaer A, Willnow TE (2005) Role of endocytosis in cellular uptake of sex steroids. Cell 122:751–762

Harkonen PL, Vaananen HK (2006) Monocyte-macrophage system as a target for estrogen and selective estrogen receptor modulators. Ann N Y Acad Sci 1089:218–227

Harman BC, Miller JP, Nikbakht N, Gerstein R, Allman D (2006) Mouse plasmacytoid dendritic cells derive exclusively from estrogen-resistant myeloid progenitors. Blood 108:878–885

Harrington WR, Sheng S, Barnett DH, Petz LN, Katzenellenbogen JA, Katzenellenbogen BS (2003) Activities of estrogen receptor alpha- and beta-selective ligands at diverse estrogen responsive gene sites mediating transactivation or transrepression. Mol Cell Endocrinol 206:13–22

Heinlein CA, Chang C (2002a) Androgen receptor (AR) coregulators: an overview. Endocr Rev 23:175–200

Heinlein CA, Chang C (2002b) The roles of androgen receptors and androgen-binding proteins in nongenomic androgen actions. Mol Endocrinol 16:2181–2187

Heldring N, Pike A, Andersson S, Matthews J, Cheng G, Hartman J, Tujague M, Strom A, Treuter E, Warner M, Gustafsson JA (2007) Estrogen receptors: how do they signal and what are their targets. Physiol Rev 87:905–931

Heng TS, Goldberg GL, Gray DH, Sutherland JS, Chidgey AP, Boyd RL (2005) Effects of castration on thymocyte development in two different models of thymic involution. J Immunol 175:2982–2993

Hobisch A, Eder IE, Putz T, Horninger W, Bartsch G, Klocker H, Culig Z (1998) Interleukin-6 regulates prostate-specific protein expression in prostate carcinoma cells by activation of the androgen receptor. Cancer Res 58:4640–4645

Hughes GC, Thomas S, Li C, Kaja MK, Clark EA (2008) Cutting edge: progesterone regulates IFN-alpha production by plasmacytoid dendritic cells. J Immunol 180:2029–2033

Igarashi H, Kouro T, Yokota T, Comp PC, Kincade PW (2001) Age and stage dependency of estrogen receptor expression by lymphocyte precursors. Proc Natl Acad Sci U S A 98:15131–15136

Ikeuchi T, Todo T, Kobayashi T, Nagahama Y (1999) cDNA cloning of a novel androgen receptor subtype. J Biol Chem 274:25205–25209

Islander U, Erlandsson MC, Hasseus B, Jonsson CA, Ohlsson C, Gustafsson JA, Dahlgren U, Carlsten H (2003) Influence of oestrogen receptor alpha and beta on the immune system in aged female mice. Immunology 110:149–157

Jenster G, van der Korput HA, Trapman J, Brinkmann AO (1995) Identification of two transcription activation units in the N-terminal domain of the human androgen receptor. J Biol Chem 270:7341–7346

Jones LA, Anthony JP, Henriquez FL, Lyons RE, Nickdel MB, Carter KC, Alexander J, Roberts CW (2008) Toll-like receptor-4-mediated macrophage activation is differentially regulated by progesterone via the glucocorticoid and progesterone receptors. Immunology 125(1):59–69

Jordan VC, Gapstur S, Morrow M (2001) Selective estrogen receptor modulation and reduction in risk of breast cancer, osteoporosis, and coronary heart disease. J Natl Cancer Inst 93:1449–1457

Kaarbo M, Klokk TI, Saatcioglu F (2007) Androgen signaling and its interactions with other signaling pathways in prostate cancer. Bioessays 29:1227–1238

Kalaitzidis D, Gilmore TD (2005) Transcription factor cross-talk: the estrogen receptor and NF-kappaB. Trends Endocrinol Metab 16:46–52

Kamath AT, Henri S, Battye F, Tough DF, Shortman K (2002) Developmental kinetics and lifespan of dendritic cells in mouse lymphoid organs. Blood 100:1734–1741

Kampa M, Papakonstanti EA, Hatzoglou A, Stathopoulos EN, Stournaras C, Castanas E (2002) The human prostate cancer cell line LNCaP bears functional membrane testosterone receptors that increase PSA secretion and modify actin cytoskeleton. Faseb J 16:1429–1431

Karpuzoglu E, Fenaux JB, Phillips RA, Lengi AJ, Elvinger F, Ansar Ahmed S (2006) Estrogen up-regulates inducible nitric oxide synthase, nitric oxide, and cyclooxygenase-2 in splenocytes activated with T cell stimulants: role of interferon-gamma. Endocrinology 147:662–671

Kaushic C, Ashkar AA, Reid LA, Rosenthal KL (2003) Progesterone increases susceptibility and decreases immune responses to genital herpes infection. J Virol 77:4558–4565

Kawasaki T, Choudhry MA, Suzuki T, Schwacha MG, Bland KI, Chaudry IH (2008) 17beta-Estradiol's salutary effects on splenic dendritic cell functions following trauma-hemorrhage are mediated via estrogen receptor-alpha. Mol Immunol 45:376–385

Kazeto Y, Goto-Kazeto R, Thomas P, Trant JM (2005) Molecular characterization of three forms of putative membrane-bound progestin receptors and their tissue-distribution in channel catfish, Ictalurus punctatus. J Mol Endocrinol 34:781–791

Khan KN, Masuzaki H, Fujishita A, Kitajima M, Sekine I, Matsuyama T, Ishimaru T (2005) Estrogen and progesterone receptor expression in macrophages and regulation of hepatocyte growth factor by ovarian steroids in women with endometriosis. Hum Reprod 20:2004–2013

Kian Tee M, Rogatsky I, Tzagarakis-Foster C, Cvoro A, An J, Christy RJ, Yamamoto KR, Leitman DC (2004) Estradiol and selective estrogen receptor modulators differentially regulate target genes with estrogen receptors alpha and beta. Mol Biol Cell 15:1262–1272

Kincade PW, Medina KL, Payne KJ, Rossi MI, Tudor KS, Yamashita Y, Kouro T (2000) Early B-lymphocyte precursors and their regulation by sex steroids. Immunol Rev 175:128–37

Klein SL (2004) Hormonal and immunological mechanisms mediating sex differences in parasite infection. Parasite Immunol 26:247–264

Klein SL, Wisniewski AB, Marson AL, Glass GE, Gearhart JP (2002) Early exposure to genistein exerts long-lasting effects on the endocrine and immune systems in rats. Mol Med 8:742–749

Komi J, Lassila O (2000) Nonsteroidal anti-estrogens inhibit the functional differentiation of human monocyte-derived dendritic cells. Blood 95:2875–282

Komi J, Mottonen M, Luukkainen R, Lassila O (2001) Non-steroidal anti-oestrogens inhibit the differentiation of synovial macrophages into dendritic cells. Rheumatology (Oxford) 40:185–191

Konoplya EF, Popoff EH (1992) Identification of the classical androgen receptor in male rat liver and prostate cell plasma membranes. Int J Biochem 24:1979–1983

Kontula K, Paavonen T, Luukkainen T, Andersson LC (1983) Binding of progestins to the glucocorticoid receptor. Correlation to their glucocorticoid-like effects on in vitro functions of human mononuclear leukocytes. Biochem Pharmacol 32:1511–1518

Kostellow AB, Weinstein SP, Morrill GA (1982) Specific binding of progesterone to the cell surface and its role in the meiotic divisions in Rana oocytes. Biochim Biophys Acta 720:356–363

Kousteni S, Bellido T, Plotkin LI, O'Brien CA, Bodenner DL, Han L, Han K, DiGregorio GB, Katzenellenbogen JA, Katzenellenbogen BS, Roberson PK, Weinstein RS, Jilka RL, Manolagas SC (2001) Nongenotropic, sex-nonspecific signaling through the estrogen or androgen receptors: dissociation from transcriptional activity. Cell 104:719–730

Kovacs EJ (2005) Aging, traumatic injury, and estrogen treatment. Exp Gerontol 40:549–555

Kovacs WJ, Olsen NJ (1987) Androgen receptors in human thymocytes. J Immunol 139:490–493

Koyama Y, Nagao S, Ohashi K, Takahashi H, Marunouchi T (1987) Sex differences in the densities of epidermal Langerhans cells of the mouse. J Invest Dermatol 88:541–544

Koyama Y, Nagao S, Ohashi K, Takahashi H, Marunouchi T (1989) Effect of systemic and topical application of testosterone propionate on the density of epidermal Langerhans cells in the mouse. J Invest Dermatol 92:86–90

Kuiper GG, Faber PW, van Rooij HC, van der Korput JA, Ris-Stalpers C, Klaassen P, Trapman J, Brinkmann AO (1989) Structural organization of the human androgen receptor gene. J Mol Endocrinol 2:R1–4

Kurtis JD, Mtalib R, Onyango FK, Duffy PE (2001) Human resistance to Plasmodium falciparum increases during puberty and is predicted by dehydroepiandrosterone sulfate levels. Infect Immun 69:123–128

Lambert KC, Curran EM, Judy BM, Lubahn DB, Estes DM (2004) Estrogen receptor-alpha deficiency promotes increased TNF-alpha secretion and bacterial killing by murine macrophages in response to microbial stimuli in vitro. J Leukoc Biol 75:1166–1172

Lange CA, Shen T, Horwitz KB (2000) Phosphorylation of human progesterone receptors at serine-294 by mitogen-activated protein kinase signals their degradation by the 26 S proteasome. Proc Natl Acad Sci USA 97:1032–1037

Leung KC, Doyle N, Ballesteros M, Sjogren K, Watts CK, Low TH, Leong GM, Ross RJ, Ho KK (2003) Estrogen inhibits GH signaling by suppressing GH-induced JAK2 phosphorylation, an effect mediated by SOCS-2. Proc Natl Acad Sci USA 100:1016–1021

Levin ER (2002) Cellular functions of plasma membrane estrogen receptors. Steroids 67:471–475

Li L, Ren CH, Tahir SA, Ren C, Thompson TC (2003) Caveolin-1 maintains activated Akt in prostate cancer cells through scaffolding domain binding site interactions with and inhibition of serine/threonine protein phosphatases PP1 and PP2A. Mol Cell Biol 23:9389–9404

Liegibel UM, Sommer U, Boercsoek I, Hilscher U, Bierhaus A, Schweikert HU, Nawroth P, Kasperk C (2003) Androgen receptor isoforms AR-A and AR-B display functional differences in cultured human bone cells and genital skin fibroblasts. Steroids 68:1179–1187

Lin CY, Vega VB, Thomsen JS, Zhang T, Kong SL, Xie M, Chiu KP, Lipovich L, Barnett DH, Stossi F, Yeo A, George J, Kuznetsov VA, Lee YK, Charn TH, Palanisamy N, Miller LD, Cheung E, Katzenellenbogen BS, Ruan Y, Bourque G, Wei CL, Liu ET (2007) Whole-genome cartography of estrogen receptor alpha binding sites. PLoS Genet 3:e87

Liu MM, Albanese C, Anderson CM, Hilty K, Webb P, Uht RM, Price RHJ, Pestell RG, Kushner PJ (2002) Opposing action of estrogen receptors alpha and beta on cyclin D1 gene expression. J Biol Chem 277:24353–24360

Liu HB, Loo KK, Palaszynski K, Ashouri J, Lubahn DB, Voskuhl RR (2003) Estrogen receptor alpha mediates estrogen's immune protection in autoimmune disease. J Immunol 171:6936–6940

Liu L, Benten WP, Wang L, Hao X, Li Q, Zhang H, Guo D, Wang Y, Wunderlich F, Qiao Z (2005) Modulation of Leishmania donovani infection and cell viability by testosterone in bone marrow-derived macrophages: signaling via surface binding sites. Steroids 70:604–614

Liva SM, Voskuhl RR (2001) Testosterone acts directly on CD4+ T lymphocytes to increase IL-10 production. J Immunol 167:2060–2067

Losel RM, Falkenstein E, Feuring M, Schultz A, Tillmann HC, Rossol-Haseroth K, Wehling M (2003) Nongenomic steroid action: controversies, questions, and answers. Physiol Rev 83:965–1016

Lubahn DB, Joseph DR, Sullivan PM, Willard HF, French FS, Wilson EM (1988) Cloning of human androgen receptor complementary DNA and localization to the X chromosome. Science 240:327–330

Lutz LB, Jamnongjit M, Yang WH, Jahani D, Gill A, Hammes SR (2003) Selective modulation of genomic and nongenomic androgen responses by androgen receptor ligands. Mol Endocrinol 17:1106–1116

Lydon JP, DeMayo FJ, Funk CR, Mani SK, Hughes AR, Montgomery CA Jr, Shyamala G, Conneely OM, O'Malley BW (1995) Mice lacking progesterone receptor exhibit pleiotropic reproductive abnormalities. Genes Dev 9:2266–2278

Mansour I, Reznikoff-Etievant MF, Netter A (1994) No evidence for the expression of the progesterone receptor on peripheral blood lymphocytes during pregnancy. Hum Reprod 9:1546–1549

Mantalaris A, Panoskaltsis N, Sakai Y, Bourne P, Chang C, Messing EM, Wu JH (2001) Localization of androgen receptor expression in human bone marrow. J Pathol 193:361–366

Mao A, Paharkova-Vatchkova V, Hardy J, Miller MM, Kovats S (2005) Estrogen selectively promotes the differentiation of dendritic cells with characteristics of Langerhans cells. J Immunol 175:5146–5151

Maret A, Coudert JD, Garidou L, Foucras G, Gourdy P, Krust A, Dupont S, Chambon P, Druet P, Bayard F, Guery JC (2003) Estradiol enhances primary antigen-specific CD4 T cell responses and Th1 development in vivo. Essential role of estrogen receptor alpha expression in hematopoietic cells. Eur J Immunol 33:512–521

Marino M, Ascenzi P, Acconcia F (2006a) S-palmitoylation modulates estrogen receptor alpha localization and functions. Steroids 71:298–303

Marino M, Galluzzo P, Ascenzi P (2006b) Estrogen signaling multiple pathways to impact gene transcription. Curr Genomics 7:497–508

Marriott I, Huet-Hudson YM (2006) Sexual dimorphism in innate immune responses to infectious organisms. Immunol Res 34:177–192

Matsuda T, Yamamoto T, Muraguchi A, Saatcioglu F (2001a) Cross-talk between transforming growth factor-beta and estrogen receptor signaling through Smad3. J Biol Chem 276:42908–42914

Matsuda T, Junicho A, Yamamoto T, Kishi H, Korkmaz K, Saatcioglu F, Fuse H, Muraguchi A (2001b) Cross-talk between signal transducer and activator of transcription 3 and androgen receptor signaling in prostate carcinoma cells. Biochem Biophys Res Commun 283:179–187

Matsumoto T, Takeyama K, Sato T, Kato S (2005) Study of androgen receptor functions by genetic models. J Biochem 138:105–110

Matsumoto T, Shiina H, Kawano H, Sato T, Kato S (2008) Androgen receptor functions in male and female physiology. J Steroid Biochem Mol Biol 109:236–241

Matthews J, Wihlen B, Tujague M, Wan J, Strom A, Gustafsson JA (2006) Estrogen receptor (ER) beta modulates ERalpha-mediated transcriptional activation by altering the recruitment of c-Fos and c-Jun to estrogen-responsive promoters. Mol Endocrinol 20:534–543

McClain MA, Gatson NN, Powell ND, Papenfuss TL, Gienapp IE, Song F, Shawler TM, Kithcart A, Whitacre CC (2007) Pregnancy suppresses experimental autoimmune encephalomyelitis through immunoregulatory cytokine production. J Immunol 179:8146–8152

McDonnell DP (2004) The molecular determinants of estrogen receptor pharmacology. Maturitas 48(Suppl 1):S7–S12

McEwan IJ (2004) Molecular mechanisms of androgen receptor-mediated gene regulation: structure-function analysis of the AF-1 domain. Endocr Relat Cancer 11 281–293

McKay LI, Cidlowski JA (1999) Molecular control of immune/inflammatory responses: interactions between nuclear factor-kappa B and steroid receptor-signaling pathways. Endocr Rev 20:435–59

Medina KL, Kincade PW (1994) Pregnancy-related steroids are potential negative regulators of B lymphopoiesis. Proc Natl Acad Sci USA 91:5382–536

Medina KL, Strasser A, Kincade PW (2000) Estrogen influences the differentiation, proliferation, and survival of early B-lineage precursors. Blood 95:2059–267

Medina KL, Garrett KP, Thompson LF, Rossi MI, Payne KJ, Kincade PW (2001) Identification of very early lymphoid precursors in bone marrow and their regulation by estrogen. Nat Immunol 2:718–24

Mendel CM (1989) The free hormone hypothesis: a physiologically based mathematical model. Endocr Rev 10:232–274

Meyer ME, Pornon A, Ji JW, Bocquel MT, Chambon P, Gronemeyer H (1990) Agonistic and antagonistic activities of RU486 on the functions of the human progesterone receptor. EMBO J 9:3923–3932

Meyer C, Schmid R, Scriba PC, Wehling M (1996) Purification and partial sequencing of high-affinity progesterone-binding site(s) from porcine liver membranes. Eur J Biochem 239:726–731

Michels G, Hoppe UC (2008) Rapid actions of androgens. Front Neuroendocrinol 29:182–198

Moore SL, Wilson K (2002) Parasites as a viability cost of sexual selection in natural populations of mammals. Science 297:2015–2018

Mor G, Sapi E, Abrahams VM, Rutherford T, Song J, Hao XY, Muzaffar S, Kohen F (2003) Interaction of the estrogen receptors with the Fas ligand promoter in human monocytes. J Immunol 170:114–122

Mulac-Jericevic B, Mullinax RA, DeMayo FJ, Lydon JP, Conneely OM (2000) Subgroup of reproductive functions of progesterone mediated by progesterone receptor-B isoform. Science 289:1751–1754

Mulac-Jericevic B, Lydon JP, DeMayo FJ, Conneely OM (2003) Defective mammary gland morphogenesis in mice lacking the progesterone receptor B isoform. Proc Natl Acad Sci USA 100:9744–9749

Nakajima Y, DelliPizzi AM, Mallouh C, Ferreri NR (1996) TNF-mediated cytotoxicity and resistance in human prostate cancer cell lines. Prostate 29:296–302

Nakamura T, Imai Y, Matsumoto T, Sato S, Takeuchi K, Igarashi K, Harada Y, Azuma Y, Krust A, Yamamoto Y, Nishina H, Takeda S, Takayanagi H, Metzger D, Kanno J, Takaoka K, Martin TJ, Chambon P, Kato S (2007) Estrogen Prevents Bone Loss via Estrogen Receptor alpha and Induction of Fas Ligand in Osteoclasts. Cell 130:811–823

Nakhla AM, Leonard J, Hryb DJ, Rosner W (1999) Sex hormone-binding globulin receptor signal transduction proceeds via a G protein. Steroids 64:213–216

Nalbandian G, Kovats S (2005) Understanding sex biases in immunity: effects of estrogen on the differentiation and function of antigen-presenting cells. Immunol Res 31:91–106

Nalbandian G, Paharkova-Vatchkova V, Mao A, Nale S, Kovats S (2005) The selective estrogen receptor modulators, tamoxifen and raloxifene, impair dendritic cell differentiation and activation. J Immunol 175:2666–2675

Nazareth LV, Weigel NL (1996) Activation of the human androgen receptor through a protein kinase A signaling pathway. J Biol Chem 271:19900–19907

Oettel M (2003) Testosterone metabolism, dose-response relationships and receptor polymorphisms: selected pharmacological/toxicological considerations on benefits versus risks of testosterone therapy in men. Aging Male 6:230–256

Oettel M, Mukhopadhyay AK (2004) Progesterone: the forgotten hormone in men? Aging Male 7:236–257

O'Lone R, Frith MC, Karlsson EK, Hansen U (2004) Genomic targets of nuclear estrogen receptors. Mol Endocrinol 18:1859–1875

Olsen NJ, Kovacs WJ (2001) Effects of androgens on T and B lymphocyte development. Immunol Res 23:281–288

Olsen NJ, Watson MB, Henderson GS, Kovacs WJ (1991) Androgen deprivation induces phenotypic and functional changes in the thymus of adult male mice. Endocrinology 129:2471–2476

Olsen NJ, Viselli SM, Fan J, Kovacs WJ (1998) Androgens accelerate thymocyte apoptosis. Endocrinology 139:748–752

Olsen NJ, Gu X, Kovacs WJ (2001a) Bone marrow stromal cells mediate androgenic suppression of B lymphocyte development. J Clin Invest 108:1697–1704

Olsen NJ, Olson G, Viselli SM, Gu X, Kovacs WJ (2001b) Androgen receptors in thymic epithelium modulate thymus size and thymocyte development. Endocrinology 142:1278–1283

Paharkova-Vatchkova V, Maldonado R, Kovats S (2004) Estrogen preferentially promotes the differentiation of CD11c+CD11b(intermediate) dendritic cells from bone marrow precursors. J Immunol 172:1426–1436

Palvimo JJ, Reinikainen P, Ikonen T, Kallio PJ, Moilanen A, Janne OA (1996) Mutual transcriptional interference between RelA and androgen receptor. J Biol Chem 271:24151–24156

Pedram A, Razandi M, Sainson RC, Kim JK, Hughes CC, Levin ER (2007) A conserved mechanism for steroid receptor translocation to the plasma membrane. J Biol Chem 282:22278–22288

Pelayo R, Welner R, Perry SS, Huang J, Baba Y, Yokota T, Kincade PW (2005) Lymphoid progenitors and primary routes to becoming cells of the immune system. Curr Opin Immunol 17:100–107

Pernis AB (2007) Estrogen and CD4+ T cells. Curr Opin Rheumatol 19:414–420

Peterziel H, Mink S, Schonert A, Becker M, Klocker H, Cato AC (1999) Rapid signalling by androgen receptor in prostate cancer cells. Oncogene 18:6322–6329

Phiel KL, Henderson RA, Adelman SJ, Elloso MM (2005) Differential estrogen receptor gene expression in human peripheral blood mononuclear cell populations. Immunol Lett 97:107–113

Polanczyk M, Zamora A, Subramanian S, Matejuk A, Hess DL, Blankenhorn EP, Teuscher C, Vandenbark AA, Offner H (2003) The protective effect of 17beta-estradiol on experimental autoimmune encephalomyelitis is mediated through estrogen receptor-alpha. Am J Pathol 163:1599–1605

Polanczyk M, Yellayi S, Zamora A, Subramanian S, Tovey M, Vandenbark AA, Offner H, Zachary JF, Fillmore PD, Blankenhorn EP, Gustafsson JA, Teuscher C (2004) Estrogen receptor-1 (Esr1) and -2 (Esr2) regulate the severity of clinical experimental allergic encephalomyelitis in male mice. Am J Pathol 164:1915–1924

Pratt WB, Toft DO (1997) Steroid receptor interactions with heat shock protein and immunophilin chaperones. Endocr Rev 18:306–360

Prossnitz ER, Arterburn JB, Smith HO, Oprea TI, Sklar LA, Hathaway HJ (2008) Estrogen signaling through the transmembrane G protein-coupled receptor GPR30. Annu Rev Physiol 70:165–190

Qiu M, Lange CA (2003) MAP kinases couple multiple functions of human progesterone receptors: degradation, transcriptional synergy, and nuclear association. J Steroid Biochem Mol Biol 85:147–157

Qiu M, Olsen A, Faivre E, Horwitz KB, Lange CA (2003) Mitogen-activated protein kinase regulates nuclear association of human progesterone receptors. Mol Endocrinol 17:628–642

Radojevic K, Arsenovic-Ranin N, Kosec D, Pesic V, Pilipovic I, Perisic M, Plecas-Solarovic B, Leposavic G (2007) Neonatal castration affects intrathymic kinetics of T-cell differentiation and the spleen T-cell level. J Endocrinol 192:669–682

Rao KV, Peralta WD, Greene GL, Fox CF (1987) Cellular progesterone receptor phosphorylation in response to ligands activating protein kinases. Biochem Biophys Res Commun 146:1357–1365

Ren MQ, Kuhn G, Wegner J, Chen J (2001) Isoflavones, substances with multi-biological and clinical properties. Eur J Nutr 40:135–146

Rosner W, Hryb DJ, Khan MS, Nakhla AM, Romas NA (1999) Androgen and estrogen signaling at the cell membrane via G-proteins and cyclic adenosine monophosphate. Steroids 64:100–106

Roubenoff R, Grinspoon S, Skolnik PR, Tchetgen E, Abad L, Spiegelman D, Knox T, Gorbach S (2002) Role of cytokines and testosterone in regulating lean body mass and resting energy expenditure in HIV-infected men. Am J Physiol Endocrinol Metab 283:E138–45

Russell DW, Berman DM, Bryant JT, Cala KM, Davis DL, Landrum CP, Prihoda JS, Silver RI, Thigpen AE, Wigley WC (1994) The molecular genetics of steroid 5 alpha-reductases. Recent Prog Horm Res 49:275–284

Sader MA, McGrath KC, Hill MD, Bradstock KF, Jimenez M, Handelsman DJ, Celermajer DS, Death AK (2005) Androgen receptor gene expression in leucocytes is hormonally regulated: implications for gender differences in disease pathogenesis. Clin Endocrinol (Oxf) 62:56–63

Saino N, Bolzern AM, Moller AP (1997) Immunocompetence, ornamentation, and viability of male barn swallows (Hirundo rustica). Proc Natl Acad Sci USA 94:549–552

Sartorius CA, Melville MY, Hovland AR, Tung L, Takimoto GS, Horwitz KB (1994) A third transactivation function (AF3) of human progesterone receptors located in the unique N-terminal segment of the B-isoform. Mol Endocrinol 8:1347–1360

Schulman C, Lunenfeld B (2002) The ageing male. World J Urol 20:4–10

Schuurs AH, Verheul HA (1990) Effects of gender and sex steroids on the immune response. J Steroid Biochem 35:157–172

Selman PJ, Mol JA, Rutteman GR, van Garderen E, Rijnberk A (1994) Progestin-induced growth hormone excess in the dog originates in the mammary gland. Endocrinology 134:287–292

Shang Y, Brown M (2002) Molecular determinants for the tissue specificity of SERMs. Science 295:2465–248

Shang Y, Hu X, DiRenzo J, Lazar MA, Brown M (2000) Cofactor dynamics and sufficiency in estrogen receptor-regulated transcription. Cell 103:843–852

Shang Y, Myers M, Brown M (2002) Formation of the androgen receptor transcription complex. Mol Cell 9:601–610

Shim GJ, Wang L, Andersson S, Nagy N, Kis LL, Zhang Q, Makela S, Warner M, Gustafsson JA (2003) Disruption of the estrogen receptor beta gene in mice causes myeloproliferative disease resembling chronic myeloid leukemia with lymphoid blast crisis. Proc Natl Acad Sci USA 100:6694–6699

Shim GJ, Kis LL, Warner M, Gustafsson JA (2004) Autoimmune glomerulonephritis with spontaneous formation of splenic germinal centers in mice lacking the estrogen receptor alpha gene. Proc Natl Acad Sci USA 101:1720–1724

Shortman K, Naik SH (2007) Steady-state and inflammatory dendritic-cell development. Nat Rev Immunol 7:19–30

Shuai K, Liu B (2005) Regulation of gene-activation pathways by PIAS proteins in the immune system. Nat Rev Immunol 5:593–605

Simoncini T, Hafezi-Moghadam A, Brazil DP, Ley K, Chin WW, Liao JK (2000) Interaction of oestrogen receptor with the regulatory subunit of phosphatidylinositol-3-OH kinase. Nature 407:538–541

Singh AB, Hsia S, Alaupovic P, Sinha-Hikim I, Woodhouse L, Buchanan TA, Shen R, Bross R, Berman N, Bhasin S (2002) The effects of varying doses of T on insulin sensitivity, plasma lipids, apolipoproteins, and C-reactive protein in healthy young men. J Clin Endocrinol Metab 87:136–143

Sinkevicius KW, Burdette JE, Woloszyn K, Hewitt SC, Hamilton K, Sugg SL, Temple KA, Wondisford FE, Korach KS, Woodruff TK, Greene GL (2008) An estrogen receptor-alpha knock-in mutation provides evidence of ligand-independent signaling and allows modulation of ligand-induced pathways in vivo. Endocrinology 149:2970–2979

Siracusa MC, Overstreet MG, Housseau F, Scott AL, Klein SL (2008) 17{beta}-Estradiol Alters the Activity of Conventional and IFN-Producing Killer Dendritic Cells. J Immunol 180:1423–1431

Smith CL, O'Malley BW (2004) Coregulator function: a key to understanding tissue specificity of selective receptor modulators. Endocr Rev 25:45–71

Smithson G, Beamer WG, Shultz KL, Christianson SW, Shultz LD, Kincade PW (1994) Increased B lymphopoiesis in genetically sex steroid-deficient hypogonadal (hpg) mice. J Exp Med 180:717–20

Smithson G, Couse JF, Lubahn DB, Korach KS, Kincade PW (1998) The role of estrogen receptors and androgen receptors in sex steroid regulation of B lymphopoiesis. J Immunol 161:27–34

Sobel ES, Gianini J, Butfiloski EJ, Croker BP, Schiffenbauer J, Roberts SM (2005) Acceleration of autoimmunity by organochlorine pesticides in (NZB x NZW)F1 mice. Environ Health Perspect 113:323–328

Soucy G, Boivin G, Labrie F, Rivest S (2005) Estradiol is required for a proper immune response to bacterial and viral pathogens in the female brain. J Immunol 174:6391–6398

Sperry TS, Thomas P (1999) Characterization of two nuclear androgen receptors in Atlantic croaker: comparison of their biochemical properties and binding specificities. Endocrinology 140:1602–1611

Spitz IM (2003) Progesterone antagonists and progesterone receptor modulators: an overview. Steroids 68:981–993

Spitz IM (2006) Progesterone receptor antagonists. Curr Opin Investig Drugs 7:882–890

Staples JE, Gasiewicz TA, Fiore NC, Lubahn DB, Korach KS, Silverstone AE (1999) Estrogen receptor alpha is necessary in thymic development and estradiol-induced thymic alterations. J Immunol 163:4168–4174

Sthoeger ZM, Zinger H, Mozes E (2003) Beneficial effects of the anti-oestrogen tamoxifen on systemic lupus erythematosus of (NZBxNZW)F1 female mice are associated with specific reduction of IgG3 autoantibodies. Ann Rheum Dis 62:341–346

Stites DP, Siiteri PK (1983) Steroids as immunosuppressants in pregnancy. Immunol Rev 75:117–138

Stossi F, Barnett DH, Frasor J, Komm B, Lyttle CR, Katzenellenbogen BS (2004) Transcriptional profiling of estrogen-regulated gene expression via estrogen recepto⁻ (ER) alpha or ERbeta in human osteosarcoma cells: distinct and common target genes for these receptors. Endocrinology 145:3473–3486

Straub RH (2007) The complex role of estrogens in inflammation. Endocr Rev 28:521–574

Subramanian M, Shaha C (2007) Up-regulation of Bcl-2 through ERK phosphorylation is associated with human macrophage survival in an estrogen microenvironment. J Immunol 179:2330–2338

Suh J, Payvandi F, Edelstein LC, Amenta PS, Zong WX, Gelinas C, Rabson AB (2002) Mechanisms of constitutive NF-kappaB activation in human prostate cancer cells. Prostate 52:183–200

Supakar PC, Jung MH, Song CS, Chatterjee B, Roy AK (1995) Nuclear factor kappa B functions as a negative regulator for the rat androgen receptor gene and NF-kappa B activity increases during the age-dependent desensitization of the liver. J Biol Chem The Journal of biological chemistry 270:837–842

Sutherland JS, Goldberg GL, Hammett MV, Uldrich AP, Berzins SP, Heng TS, Blazar BR, Millar JL, Malin MA, Chidgey AP, Boyd RL (2005) Activation of thymic regeneration in mice and humans following androgen blockade. J Immunol 175:2741–2753

Szekeres-Bartho J, Szekeres G, Debre P, Autran B, Chaouat G (1990) Reactivity of lymphocytes to a progesterone receptor-specific monoclonal antibody. Cell Immunol 125:273–283

Takahashi T, Eitzman B, Bossert NL, Walmer D, Sparrow K, Flanders KC, McLachlan J, Nelson KG (1994) Transforming growth factors beta 1, beta 2, and beta 3 messenger RNA and protein expression in mouse uterus and vagina during estrogen-induced growth: a comparison to other estrogen-regulated genes. Cell Growth Differ 5:919–935

Takeda H, Chodak G, Mutchnik S, Nakamoto T, Chang C (1990) Immunohistochemical localization of androgen receptors with mono- and polyclonal antibodies to androgen receptor. J Endocrinol 126:17–25

Takeo J, Yamashita S (1999) Two distinct isoforms of cDNA encoding rainbow trout androgen receptors. J Biol Chem 274:5674–5680

Taniguchi F, Couse JF, Rodriguez KF, Emmen JM, Poirier D, Korach KS (2007) Estrogen receptor-alpha mediates an intraovarian negative feedback loop on thecal cell steroidogenesis via modulation of Cyp17a1 (cytochrome P450, steroid 17alpha-hydroxylase/17, 20 lyase) expression. FASEB J 21:586–595

Thomas P (2008) Characteristics of membrane progestin receptor alpha (mPRalpha) and progesterone membrane receptor component 1 (PGMRC1) and their roles in mediating rapid progestin actions. Front Neuroendocrinol 29:292–312

Thurmond TS, Murante FG, Staples JE, Silverstone AE, Korach KS, Gasiewicz TA (2000) Role of estrogen receptor alpha in hematopoietic stem cell development and B lymphocyte maturation in the male mouse. Endocrinology 141:2309–2318

Tibbetts TA, Conneely OM, O'Malley BW (1999) Progesterone via its receptor antagonizes the pro-inflammatory activity of estrogen in the mouse uterus. Biol Reprod 60:1158–1165

Tilley WD, Marcelli M, Wilson JD, McPhaul MJ (1989) Characterization and expression of a cDNA encoding the human androgen receptor. Proc Natl Acad Sci USA 86:327–331

Trapman J, Klaassen P, Kuiper GG, van der Korput JA, Faber PW, van Rooij HC, Geurts van Kessel A, Voorhorst MM, Mulder E, Brinkmann AO (1988) Cloning, structure and expression of a cDNA encoding the human androgen receptor. Biochem Biophys Res Commun 153:241–248

Tseng L, Tang M, Wang Z, Mazella J (2003) Progesterone receptor (hPR) upregulates the fibronectin promoter activity in human decidual fibroblasts. DNA Cell Biol 22:633–640

Ueda T, Bruchovsky N, Sadar MD (2002) Activation of the androgen receptor N-terminal domain by interleukin-6 via MAPK and STAT3 signal transduction pathways. J Biol Chem 277:7076–7085

Vegeto E, Pollio G, Pellicciari C, Maggi A (1999) Estrogen and progesterone induction of survival of monoblastoid cells undergoing TNF-alpha-induced apoptosis. FASEB J 13:793–803

Vicencio JM, Ibarra C, Estrada M, Chiong M, Soto D, Parra V, Diaz-Araya G, Jaimovich E, Lavandero S (2006) Testosterone induces an intracellular calcium increase by a nongenomic mechanism in cultured rat cardiac myocytes. Endocrinology 147:1386–1395

Viselli SM, Olsen NJ, Shults K, Steizer G, Kovacs WJ (1995) Immunochemical and flow cytometric analysis of androgen receptor expression in thymocytes. Mol Cell Endocrinol 109:19–26

Viselli SM, Reese KR, Fan J, Kovacs WJ, Olsen NJ (1997) Androgens alter B cell development in normal male mice. Cell Immunol 182:99–104

Walker WH, Cheng J (2005) FSH and testosterone signaling in Sertoli cells. Reproduction 130:15–28

Walmer DK, Wrona MA, Hughes CL, Nelson KG (1992) Lactoferrin expression in the mouse reproductive tract during the natural estrous cycle: correlation with circulating estradiol and progesterone. Endocrinology 131:1458–1466

Wang LH, Yang XY, Mihalic K, Xiao W, Li D, Farrar WL (2001) Activation of estrogen receptor blocks interleukin-6-inducible cell growth of human multiple myeloma involving molecular cross-talk between estrogen receptor and STAT3 mediated by co-regulator PIAS3. J Biol Chem 276:31839–31844

Wang F, Roberts SM, Butfiloski EJ, Morel L, Sobel ES (2007) Acceleration of autoimmunity by organochlorine pesticides: a comparison of splenic B-cell effects of chlordecone and estradiol in (NZBxNZW)F1 mice. Toxicol Sci 99:141–152

Warner M, Gustafsson JA (2006) Nongenomic effects of estrogen: why all the uncertainty? Steroids 71:91–95

Welner RS, Pelayo R, Garrett KP, Chen X, Perry SS, Sun XH, Kee BL, Kincade PW (2007) Interferon-producing killer dendritic cells (IKDCs) arise via a unique differentiation pathway from primitive c-kitHiCD62L+lymphoid progenitors. Blood 109:4825–4931

Whitacre CC, Reingold SC, O'Looney PA (1999) A gender gap in autoimmunity. Science 283:1277–1278

Williams C, Edvardsson K, Lewandowski SA, Strom A, Gustafsson JA (2008) A genome-wide study of the repressive effects of estrogen receptor beta on estrogen receptor alpha signaling in breast cancer cells. Oncogene 27:1019–1032

Wilson CM, McPhaul MJ (1994) A and B forms of the androgen receptor are present in human genital skin fibroblasts. Proc Natl Acad Sci USA 91:1234–1238

Wilson CA, Mrose SA, Thomas DW (1995) Enhanced production of B lymphocytes after castration. Blood Blood 85:1535–1539

Windmill KF, Meade BJ, Lee VW (1993) Effect of prepubertal gonadectomy and sex steroid treatment on the growth and lymphocyte populations of the rat thymus. Reprod Fertil Dev 5:73–81

Winneker RC, Fensome A, Wrobel JE, Zhang Z, Zhang P (2005) Nonsteroidal progesterone receptor modulators: structure activity relationships. Semin Reprod Med 23:46–57

Wunderlich F, Benten WP, Lieberherr M, Guo Z, Stamm O, Wrehlke C, Sekeris CE, Mossmann H (2002) Testosterone signaling in T cells and macrophages. Steroids 67:535–538

Yamamoto T, Matsuda T, Junicho A, Kishi H, Saatcioglu F, Muraguchi A (2000) Cross-talk between signal transducer and activator of transcription 3 and estrogen receptor signaling. FEBS Lett 486:143–148

Yamamoto T, Sato N, Sekine Y, Yumioka T, Imoto S, Junicho A, Fuse H, Matsuda T (2003) Molecular interactions between STAT3 and protein inhibitor of activated STAT3, and androgen receptor. Biochem Biophys Res Commun 306:610–615

Yao M, Nguyen TV, Pike CJ (2007) Estrogen regulates Bcl-w and Bim expression: role in protection against beta-amyloid peptide-induced neuronal death. J Neurosci 27:1422–1433

Yeh S, Tsai MY, Xu Q, Mu XM, Lardy H, Huang KE, Lin H, Yeh SD, Altuwaijri S, Zhou X, Xing L, Boyce BF, Hung MC, Zhang S, Gan L, Chang C (2002) Generation and characterization of androgen receptor knockout (ARKO) mice: an in vivo model for the study of androgen functions in selective tissues. Proc Natl Acad Sci USA 99:13498–13503

Zhang Y, Beck CA, Poletti A, Edwards DP, Weigel NL (1995) Identification of a group of Ser-Pro motif hormone-inducible phosphorylation sites in the human progesterone receptor. Mol Endocrinol 9:1029–1040

Zhou ZX, Sar M, Simental JA, Lane MV, Wilson EM (1994) A ligand-dependent bipartite nuclear targeting signal in the human androgen receptor. Requirement for the DNA-binding domain and modulation by NH2-terminal and carboxyl-terminal sequences. J Biol Chem 269:13115–13123

Zhu Y, Rice CD, Pang Y, Pace M, Thomas P (2003) Cloning, expression, and characterization of a membrane progestin receptor and evidence it is an intermediary in meiotic maturation of fish oocytes. Proc Natl Acad Sci USA 100:2231–2236

Zoller AL, Kersh GJ (2006) Estrogen induces thymic atrophy by eliminating early thymic progenitors and inhibiting proliferation of beta-selected thymocytes. J Immunol 176:7371–7378

Zoller AL, Schnell FJ, Kersh GJ (2007) Murine pregnancy leads to reduced proliferation of maternal thymocytes and decreased thymic emigration. Immunology 121:207–215

Chapter 4
Sex Differences in Susceptibility to Viral Infection

Sabra L. Klein and Sally Huber

Abstract Males and females differ in their susceptibility to a variety of viral pathogens. Although behavioral factors can influence exposure to viruses, several studies illustrate that physiological differences between males and females cause dimorphic responses to infection. Females often exhibit reduced susceptibility to viral infections because they typically mount stronger immune responses than males. Innate recognition and response to viruses as well as downstream adaptive immune responses differ between males and females during viral infections. This often results in sex differences in cytokine responses to infection that play a critical role in determining susceptibility to viruses. Immune responses to viruses can vary with changes in hormone concentrations naturally observed over the menstrual or estrous cycle, from contraception use, and during pregnancy.

4.1 Basic Virology

4.1.1 What are Viruses?

Viruses are obligate intracellular parasites. They contain little more than nucleic acid and virus-coded proteins with or without a lipid bilayer (envelope) and are inert except when present in cells. Plants, animals, and bacteria are all capable of acting as hosts to viruses. Viruses are categorized by the type of nucleic acid used as its genome, whether the virus particles are "enveloped" or "nonenveloped," and by the symmetry of the virus particle. The envelope is always of host cellular origin. A virus, which can infect different cell types, such as the mouse coronavirus that can infect hepatocytes and fibroblasts, will have identical proteins as these are coded by

S.L. Klein
John Hopkins Bloomberg, School of Public Health, Baltimore MD USA
e-mail: saklein@jhsph.edu

S.L. Klein and C.W. Roberts (eds.), *Sex Hormones and Immunity to Infection*,
DOI 10.1007/978-3-642-02155-8_4, © Springer-Verlag Berlin Heidelberg 2010

the viral genes, but the lipid composition may differ according to the content of the plasma membrane from which the virus buds. The composition of the lipid might, in some instances, affect the ability of the virus to infect new cells (discussed below). Nearly all viruses have a capsid, consisting of a protective shell of viral structural proteins, which prevents degradation of the viral genome by RNases or DNases in the environment. The most common structural styles for the structural proteins are helical, spherical, and icosahedral.

4.1.2 DNA and RNA Viruses

Viruses have evolved either RNA or DNA genomes. Those with RNA genomes, referred to as "RNA viruses" are designated as having either "positive" or "negative" sense. The RNA genome of positive sense viruses can directly bind to ribosomes and translate viral proteins whereas the RNA genome of negative sense viruses must first be copied using a virus RNA-dependent-RNA polymerase before viral proteins can be synthesized. Replication of the genomic RNA results in a transient double-stranded RNA (dsRNA) or replicative intermediate which can activate cellular defense mechanisms through toll-like receptor (TLR) 3 or RNA helicases (Gitlin et al. 2006; Hardarson et al. 2007).

Viral RNA-dependent-RNA polymerases lack adequate proof-reading and RNA viruses have high mutation rates (Domingo and Holland 1997; Domingo 2000). High mutation rates permit rapid antigenic shift of viral proteins and hamper the ability of the host immune response to clear the infection, but also limits the potential size of the viral genome (Eigen 1993; Tolou et al. 2002). Production of new progeny genomes requires complementary RNA viral infection which usually shuts down host cell RNA and protein synthesis resulting in the infected cell becoming a "viral factory" with nearly all protein and nucleic acid synthesis directed to making new virus particles.

Viruses with DNA genomes, referred to as "DNA viruses" usually require cell replication for successful completion of their life cycle. Most DNA viruses directly cause the cell to enter the replicative cycle. This is accomplished by one of several different mechanisms. Viral proteins can bind and inhibit cellular tumor suppressor proteins of the Rb family such as p53, p105^{Rb}, p107, and p130 (Chellappan et al. 1992) leading to release of E2F transcription factors needed for S-phase DNA synthesis; act as cyclin or growth factor homologues; bind to and neutralize cellular proteins inhibiting cyclin-dependent kinases (Jung et al. 1994; Dobner et al. 1996); or activate signal transduction pathways such as NF-kB, RAS, and c-src (DiMaio et al. 1998). Initiation of viral DNA replication occurs at specific origin sites and, unlike cellular origin sites, multiple DNA replicative initiations can occur from the origin. The retroviruses (HIV) and hepadnaviruses (Hepatitis B) have distinctive replication methods.

Retroviruses are RNA viruses which have a reverse transcriptase that makes a DNA copy of the RNA viral genome. The DNA copy is integrated into the nucleus of the host cell. In contrast, hepadnaviruses are DNA viruses, but a linear RNA copy is made and the progeny (termed "virions") DNA genomes are produced using a reverse transcriptase from this RNA copy (Nassal 2008). Why hepadnaviruses have evolved this method for replication of their genomes is unclear.

4.1.3 Replication and Utilization of Host Cell Machinery for Survival

The basic replication cycle for all viruses involves virus attachment, penetration into the cell, uncoating of the viral genome, production of virus proteins and progeny genomes, packaging of the progeny virions, and release of the progeny viruses to begin the replication cycle all over again. Viruses bind to one or more cellular molecules called virus receptors to initiate the virus replication cycle (Maddon et al. 1986; Choe et al. 1996; Feng et al. 1996). Virus receptors, generally, are either proteins or carbohydrates and are major determinants of the tissue tropism and species specificity of the virus (Munk et al. 2007). In addition to entry through virus receptor binding, viruses can gain entry into host cells as virus–antibody complexes that infect cells through Fc immunoglobulin receptors (Daughaday et al. 1981). Subsequent to attachment, virus entry proceeds through either receptor-mediated endocytosis and trafficking of virus–receptor complexes to vesicles or direct virus entry through the plasma membrane by fusion of viral envelop and cellular membrane (Nieva and Agirre 2003). Generally, RNA viruses replicate in the cell cytoplasm while DNA viruses replicate in the host cell nucleus. Most DNA viruses uncoat their genome prior to trafficking along microtubules, binding to the nuclear envelop and inserting the viral genome (Leopold and Crystal 2007). Nonenveloped viruses usually store progeny virions in crystalline structures in the cytoplasm until virus-directed lysis of the cell releases them. Enveloped viruses bud from cellular membranes. Viral glycoproteins insert into the plasma membrane and their cytoplasmic tails associate with viral matrix proteins and the viral nucleocapsid consisting of the viral genome and associated nucleoproteins. This association both excludes cellular proteins from the lipid bilayer and causes the budding of the progeny virus from the cell surface. As viruses are dependent on host cell machinery for survival and replication, the host hormonal and immunological milieu may directly affect the virus life-cycle. Concurrently, replication of viruses inside host cells may directly modify host hormonal and immunological responses; thus, this represents a coevolved arms race between viruses and hosts.

4.1.4 Virus Receptors

Most virus receptors belong to the immunoglobulin, integrin, or other superfamilies with important biological functions. Virus cross-linking the receptor may in many cases mimic the natural ligation of these molecules and result in highly effective transduction of signals in cells. Examples include: (1) HIV, which uses CD4 and coreceptors CCR5 and CXCR4, and activates MAPK pathways leading to increased CCL2 (MCP-1), CCL3 (MIP-1α) and MMP9 induction (Collman et al. 2000; Liu et al. 2000b; Del Corno et al. 2001; Misse et al. 2001); (2) Epstein–Barr virus gp350 which cross-links CD21 and results in B cell proliferation and NF-kB, protein kinase C and IL-6 activation (D'Addario et al. 1999; D'Addario et al. 2000; D'Addario et al. 2001); (3) herpesvirus which uses herpesvirus entry mediator protein, a member of the TNF receptor superfamily, which can suppress T cell proliferation and activation (Cheung et al. 2005); and (4) coxsackieviruses which use decay accelerating factor (DAF) (Bergelson et al. 1995), a complement regulatory molecule which also suppresses T cell responses (Sun et al. 1999; Liu et al. 2005). Signaling through the viral receptor can be beneficial for the virus by stimulating cell replication and making cells more susceptible to virus replication (Liu et al. 2000a), but also initiates antigen-independent, polyclonal inflammatory responses which promote virus clearance. Sex-associated hormones can affect the expression of host cell surface proteins that function as receptors for viral entry. Most notable are the chemokine receptors which act as receptors/coreceptors for HIV-1 (Mo et al. 2005; Kubarek and Jagodzinski 2007). Estrogen also enhances expression of αvβ3, an integrin that acts as a viral receptor/coreceptor for adenovirus, coxsackievirus A9, and hantaviruses (Wickham et al. 1993; Roivainen et al. 1994; Wickham et al. 1994; Li et al. 1995; Gavrilovskaya et al. 1998; Woodward et al. 2001), and DAF, a receptor for coxsackieviruses (Song et al. 1996). Increased expression of DAF on human B lymphocytes correlates with increased susceptibility of cells to virus replication (Sartini et al. 2004).

4.2   Host Immune Responses Against Viruses

4.2.1   Detection of Viruses by Host Cells

The innate immune system recognizes viruses using a limited number pattern recognition receptors (PRRs) that distinguish conserved molecular patterns (called pathogen-associated molecular patterns (PAMPs)), including nucleic acids (Medzhitov and Janeway 2002). Viral nucleic acids are recognized by host cells via TLRs and RNA helicases in a cell-specific manner (Melchjorsen et al. 2005). TLR3 recognizes dsRNA, which is a common feature of both DNA and RNA viruses, and is expressed in several cells, including macrophages and dendritic cells (DCs). TLR7 recognizes single-stranded RNA (ssRNA) and viruses that contain ssRNA genomes and is expressed in plasmacytoid DCs (pDCs) and B cells. TLR9 is primarily expressed in pDCs and recognizes unmethylated 2'-deoxyribo (cytidine-phosphate-guanosine) (CpG) DNA motifs that are characteristic of viral DNA. TLRs contain N-terminal extracellular leucine-rich repeats that recognize PAMPs. TLR3, TLR7, and TLR9 signal from within an endosomal compartment, and all TLRs, except TLR3, initiate intracellular signaling by recruiting the adaptor protein myeloid differentiation factor 88 (Myd88), which activates transcriptional factors, such as NF-κB, cJun, and interferon regulatory factors (IRFs). Conversely, TLR3 activates NF-κB, cJun, and IRFs via the adaptor molecule called Toll/IL-1 receptor domain-containing adaptor inducing IFN-β (TRIF). In each case, activity along these TLR pathways leads to increased production of type I interferons (IFN-α/β) by infected as well as by bystander cells (Seth et al. 2006).

In addition to TLRs, two cytoplasmic DExD/H box RNA helicases, retinoic acid-inducible gene-I (RIG-I) and melanoma-differentiation-associated gene 5 (Mda5), recognize dsRNA and trigger activation of NF-κB, IRF3, and IRF7 (Kawai and Akira 2006). In common with TLR3, these RNA helicases are expressed in cells other than pDCs and are critical for production of type I IFNs (Kawai and Akira 2006). RIG-I and Mda5 activate IFN signaling pathways through protein–protein interactions between the caspase recruitment-like domains (CARDs) with the virus-induced signaling adaptor (VISA; also called IPS-1, MAVS, and CARDIF). Signaling through VISA contributes to the innate immune responses against several RNA viruses, including influenza, Newcastle disease virus, vesicular stromatitis virus (VSV), and Sendai virus (Chang et al. 2006; Kumar et al. 2006; Sun et al. 2006).

Whether the induction of PRRs by viruses differs between males and females has only recently been considered. For example, the *Tlr7* gene is encoded on the X chromosome and may escape X inactivation, resulting in more copies of *Tlr7* in females than in males (Pisitkun et al. 2006). Exposure to TLR7 ligands in vitro causes higher production of IFN-α in cells from women than in cells from men, even though *Tlr7* mRNA levels are similar between the sexes suggesting that escape from X inactivation is not involved (Berghofer et al. 2006). Conversely, stimulation with CpG, a TLR9 ligand, shows no sex bias in IFN-α production (Berghofer et al. 2006). Current data also indicate that numbers of pDCs do not differ between males and females (Berghofer et al. 2006). Regardless of mechanism, the ability of females to produce more IFN-α in response to TLR ligation has a substantial impact on both virus clearance and on immune responses during infection. Animal models further illustrate that elevated detection of self-antigens by TLR7 and subsequent production of type I IFNs leads to development of autoimmune diseases, including systemic lupus erythematosus and insulin-dependent diabetes mellitus (Theofilopoulos et al. 2005; Pisitkun et al. 2006). Thus, the same mechanisms mediating development of sex differences in susceptibility to autoimmunity may contribute to dimorphic responses to viruses.

4.2.2 Type 1 Interferons

RNA and DNA viruses induce type 1 IFNs through both TLR-dependent and -independent mechanisms (Hornung et al. 2004; Varani et al. 2007). Many host cells can produce type I IFNs; pDCs, in particular, produce high concentrations of IFN-α/β following viral infection (Haeryfar, 2005). Unlike other cells, DCs do not have to be virally-infected to produce IFN-α/β (Haeryfar, 2005). The activity of DCs can be directly influenced by sex steroids, including estrogens and progestins as outlined in Chap. 3 of this book. Type I IFNs are not directly antiviral, but rather their antiviral activity is carried out by inducing the expression of several IFN-stimulated genes (ISGs), including IFN-γ, Mx, and protein kinase R (Weber et al. 2004). Type I IFNs also upregulate major histocompatibility complex (MHC) class I and class II antigens, cause cell cycle arrest, stimulate T cell and natural killer cell activity, activate proapoptotic factors, and inhibit antiapoptotic factors (Fitzgerald-Bocarsly et al. 2008). As documented in Chap. 3, the activity of innate immune cells differs between the sexes and is affected by sex hormones. The goal of this chapter will be to illustrate that innate recognition and adaptive responses to viruses differ between the sexes and are altered by sex hormones.

4.2.3 Adaptive Immune Responses to Viruses

Humoral immune responses (i.e., antibody production by B cells) are typically elevated in females when compared with males (Falter et al. 1991; Gomez et al. 1993). Cell-mediated immune responses also differ between males and females. T cells, in particular, CD4 helper T cells (Th cells), are functionally and phenotypically heterogeneous and can be differentiated based on the cytokines they release. Reliance on subsets of Th cells (i.e., Th1, Th2, or Th17 cells) to overcome infection differs between males and females with females reportedly exhibiting higher Th2 responses (i.e., higher IL-4, IL-5, IL-6, and IL-10 production) than males (Bijlsma et al. 1999; Roberts et al. 2001). There also are reports of females having higher Th1 responses (i.e., higher concentrations of IFN-γ) than males (Araneo et al. 1991; Barrat et al. 1997). Sex differences in Th cell responses may mediate sex differences in response to viruses. It will be demonstrate in this chapter that sex differences in cytokine responses to infection play a critical role in determining susceptibility to viruses. The ultimate goal of this chapter is to illustrate that the innate recognition and response to viruses as well as downstream adaptive immune responses differ between males and females during viral infections. Secondarily, evidence illustrating that sex steroids modulate immune responses to viruses and contribute to dimorphic responses during infection will be provided.

4.3 Sex Differences in Response to Viral Infection

Males and females differ in their susceptibility to a variety of viral pathogens (Table 4.1). Susceptibility to viral infections often is reduced among females because females typically mount higher immune responses than males. Immune responses to viruses can vary with changes in hormone concentrations caused by

Table 4.1 Viruses for which sex and/or sex steroids affect the intensity (I), prevalence (P), or mortality (M) following infection

Virus	Host	Measure	Sex difference	Effect of hormone manipulation	Reference
Arboviruses (Group A)	*Mus musculus*	I	M > F	T↑	(Giron et al. 1973)
Coxsackievirus	*Mus musculus*	I	M > F	T↑	(Lyden et al. 1987)
Encephalomyocarditis virus	*Mus musculus*	I	M > F	T↑, C↓, E2↓	(Friedman et al. 1972)
Equine arteritis virus	*Equus caballus*	I	M > F	T↑, C↓	(Little et al. 1992; McCollum et al. 1994)
Friend virus	*Mus musculus*	I	M > F	E2↓	(Mirand et al. 1967; Bruland et al. 2003)
Junin virus	*Calomys musculinus*	P	M > F		(Mills et al. 1992; Mills et al. 1994)
Hantaviruses (multiple species)	*Homo sapiens*	P	M > F		(White et al. 1996; Williams et al. 1997; Ferrer et al. 1998; Armien et al. 2004)
Puumala virus	*Myodes glareolus*	P	M > F		(Bernshtein et al. 1999; Olsson et al. 2002)
Bayou virus	*Oryzomys palustris*	P	M > F		(McIntyre et al. 2005)
Sin Nombre virus	*Peromyscus baylii*	P	M > F		(Mills et al. 1997)
Sin Nombre virus	*P. maniculatus*	P	M > F		(Childs et al. 1994; Weigler et al. 1996; Mills et al. 1997; Pearce–Duvet et al. 2006; Douglass et al. 2007)
Seoul virus	*Rattus norvegicus*	I	M > F	C↓,O↑	(Klein et al. 2000, 2001; Klein et al. 2002; Hannah et al. 2008)
Sin Nombre virus	*Reithrodontomys megalotis*	P	M > F		(Mills et al. 1997)
Black Creek Canal virus	*Sigmodon hispidus*	P	M > F		(Glass et al. 1998)
Herpes simplex virus type 1	*Mus musculus*	I, M	M > F		(Han et al. 2001)
Herpes simplex virus type 2	*Homo sapiens*	I, P	M < F	P4↑	(Fleming 1997; Obasi 1999; Langenberg, 1999; Mertz 1992; Gillgrass 2003)
	Mus musculus	I	M < F	P4↑	(Teepe 1990)
Human immunodeficiency virus	*Homo sapiens*	I	M > F	E2↑	(Farzadegan et al. 1998; Sterling et al. 2001; Napravnik et al. 2002; Katagiri et al. 2006)

(continued)

Table 4.1 (continued)

Virus	Host	Measure	Sex difference	Effect of hormone manipulation	Reference
Influenza A virus	Mus musculus	I	M < F		(Nohara et al. 2002)
Measles	Homo sapiens	M	M < F		(Garenne, 1994)
Mousepox	Mus musculus	I	M > F	O↑	(Brownstein et al. 1991; Brownstein and Gras, 1995)
Polyoma virus	Mus musculus	I	M > F	T↑, C↓	(Lamey et al. 1985)
Theiler's virus	Mus musculus	I	M > F		(Bihl et al. 1999; Butterfield et al. 2003)
Vesicular stomatitis virus	Mus musculus	I	M > F		(Barna et al. 1996)
West Nile virus	Homo sapiens	I	M > F		(O'Leary et al. 2004; Jean et al. 2007)
	Equus caballus	I,P	M > F	T↑, C↓	(Porter et al. 2003; Epp et al. 2007)

I intensity, *P* prevalence, *M* mortality, *E2* estradiol, *P4* progesterone, *T* testosterone, *C* castration of males, *O* ovariectomy of females natural fluctuations over the menstrual or estrous cycle, from contraception use, and during pregnancy (Brabin 2002). Although behavioral factors can influence exposure to viruses, several studies illustrate that physiological differences between males and females cause dimorphic responses to infection.

4.3.1 Sexually Transmitted Viruses

Human immunodeficiency virus (HIV) and herpes simplex virus-2 (HSV-2) are important sexually transmitted pathogens. The amount of circulating HIV RNA in plasma is lower in women than in men, despite the fact that CD4 T cell counts and progression to acquired immunodeficiency syndrome (AIDS) are similar between the sexes (Farzadegan et al. 1998; Sterling et al. 2001; Napravnik et al. 2002). HIV loads in women, often, are below the cutoff value for initiation of antiretroviral therapy (Sterling et al. 2001). Due to viral loads being a factor used in the current guidelines for the initiation of antiretroviral therapy, questions have been raised as to whether sex differences in HIV RNA levels may result in delayed treatment of women with HIV (Sterling et al. 2001). This observation is, especially, disconcerting because the number of people living with HIV/AIDS is expanding faster for women than for men worldwide, with the most noticeable gap occurring in regions experiencing an AIDS epidemic, such as subSaharan Africa (Quinn and Overbaugh 2005).

Postmenopasal women and women taking reproductive hormones orally as contraceptives may be more susceptible to HIV-1 infection and may shed more virus from genital tissues than premenopasal women and women not on contraceptives, suggesting a positive association with progesterone and a negative association with estrogen (Clemetson et al. 1993; Aaby et al. 1996; Mostad et al. 1997; Baeten et al. 2007; Leclerc et al. 2008). HIV-positive women using oral contraceptives have a more rapid progression of disease correlating to the enhanced virus shedding associated with this form of contraceptive. The more aggressive disease course was not observed in women using the intrauterine contraceptive device (Stringer et al. 2007).

Experimental studies using Simian immunodeficiency virus (SIV) indicate that 100% of nonhormone-treated ovariectomized female macaques develop disease compared with 0% of estrogen-treated animals (Smith et al. 2000). Further, progesterone treatment of menstruating female macaques results in a 7.7-fold increase in SIV infection (Marx et al. 1996). Normal endogenous hormonal fluctuations occurring during the reproductive cycle also are sufficient to affect viral transmission. Macaques infected during the luteal phase of the cycle, when progesterone levels are elevated, have a higher infection rate than females infected during the follicular phase (i.e., when estrogen levels are elevated and progesterone is low) (Sodora et al. 1998). Many studies of humans illustrate variations in HIV shedding from cervical and vaginal tissues throughout the menstrual cycle with a significant increase in HIV RNA in genital secretions after the periovulatory period (Hanna

1999; Benki et al. 2004, 2008). Sex steroids may directly affect genital tract physiology, as progesterone thins the vaginal epithelium and decreased vaginal epithelium thickness correlates with increased SIV load (Marx et al. 1996; Poonia et al. 2006).

HIV-1 uses CD4 as its major receptor and one of a variety of chemokine receptors as coreceptors. HIV strains using the CXCR4 coreceptor are able to undergo reverse transcription, integration, and progeny virus production in the female reproductive tract whereas HIV-1 variants using CCR5 as a coreceptor fail to undergo early virus replication events (Howell et al. 2005). CD4, CCR5, and CXCR4 are all expressed in the female reproductive tract and expression levels vary through the different phases of the menstrual cycle (Yeaman et al. 2003) with CD4 and CCR5 being expressed when estrogen levels are high and progesterone levels are low, whereas the expression of CD4 and CCR5 is low and expression of CXCR4 is high in the secretory phase of the cycle when both progesterone and estrogen levels are high. Thus, the ability to become infected with HIV may depend upon the relative virus receptor/coreceptor levels expressed in the genital tract. Other immunological parameters also vary in the genital tract during the menstrual cycle. For example, concentrations of IL-1β, IL-4, IL-6, IL-8, IL-10, CCL4 (MIP-1β), CCL5 (RANTES), TGF-β, and TNFR-II are all elevated during menses, but not at other times in the vagina (Al-Harthi et al. 2001). Plasma levels of these cytokines/chemokines do not vary with the menstrual cycle. Most importantly, vaginal levels of IL-1 β, IL-6, and IL-8 correlate with virus load in the reproductive tract suggesting that immunological variations occurring during the menstrual cycle can affect local virus shedding (Al-Harthi et al. 2001).

Genital herpes is one of the best examples for a sex bias in sexually transmitted diseases. There is a higher prevalence of herpes infection in women than men in all age groups and seronegative women have a faster virus acquisition rate and a higher incidence of symptomatic infections than men (Mertz et al. 1992; Fleming et al. 1997; Langenberg et al. 1999; Obasi et al. 1999). Although HSV-2 shedding is increased in women using oral hormone-based contraceptives, virus shedding does not vary across the menstrual cycle (Mostad et al. 2000a; Mostad et al. 2000b).

In mice, intravaginal inoculation of female mice with HSV-2 in diestrus results in 75% mortality compared with 33% in proestrus, 16% in estrus, and 9% in metestrus, and exogenous progesterone treatment also increases HSV susceptibility (Teepe et al. 1990; Parr et al. 1994). Vaginal concentrations of IgA to HSV glycoprotein B are substantially higher in estrus than in diestrus while IgG antibodies are higher in diestrus compared with estrus (Gallichan and Rosenthal, 1996). Progesterone suppresses both cellular and humoral immunity to HSV which primarily explains the increased viral susceptibility observed in women (Gillgrass et al. 2003).

4.3.2 Picornaviruses

Many picornavirus-induced diseases, at least in mice, show a sex bias with worse disease in males or pregnant females and little disease in nonpregnant females. This is

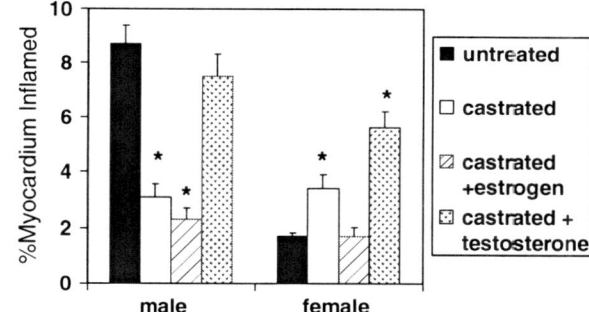

Fig. 4.1 Sex hormones modulate Coxsackievirus B3-induced myocarditis. Male and female BALB/c mice were untreated or castrated/ovariectomized at 3 weeks cf age and implanted with 7 mg testosterone or estrogen. Mice were infected with 10^4 PFU virus 2 weeks after hormone implantation. Animals were euthanized 7 days after infection and myocarditis was evaluated by image analysis for the percent of the ventricular myocardium inflamed. Results are mean $\pm$SEM of 5–6 animals per group. *Significantly different from untreated mice of the same sex at $p \leq 0.05$ using the Wilcoxon Ranked Score test. Adapted from Huber et al. (1982)

true for Thieler's-virus-induced encephalitis, encephalomyocarditis-virus-induced diabetes, and coxsackievirus-induced myocarditis (Friedman et al. 1972; Morrow et al. 1980; Giron and Patterson 1982; Huber et al. 1982; Fuller et al. 2007). It is also true of infectious avian encephalomyelitis virus infection of chickens, where testosterone increases susceptibility of the birds to infection (Westbury and Sinkovic 1978). In each of the mouse models, castration of males or treatment with estrogen reduces disease susceptibility whereas treatment of females with androgens (testosterone) and progesterone enhances the susceptibility to disease (Fig. 4.1) (Lyden and Huber 1984). There is variability in susceptibility to coxsackievirus B3 infection within nonpregnant female mice although females are far more resistant to viral disease than males (Schwartz et al. 2004). When females are infected during specific phases of the estrus cycle, animals infected during diestrus, when progesterone levels are high and estrogen levels are low, develop significantly more severe disease than females infected during estrus, when progesterone levels are low and estrogen levels are high (Schwartz et al. 2004). Thus, even the more moderate hormonal fluctuations of normal cycling females can substantially influence the pathogenicity of this picornavirus infection. Two important questions are how do sex-associated hormones impact virus susceptibility and is sex-dependent variation in viral pathogenesis as observed in mice relevant to clinical disease? Although many viruses can cause clinical myocarditis, enteroviruses of the picornavirus family, in particular coxsackieviruses, are important etiologic agents (Bowles et al. 2003). The male: female ratio of clinical myocarditis is 2:1 in humans, indicating a male dominance to the disease and, as with the mouse model, susceptibility increases in women during the third trimester of pregnancy and in the postpartum period (Woodruff 1980). This provides circumstantial evidence that the mouse model of coxsackievirus B3-induced myocarditis may accurately reflect the human disease and that sex-associated hormones influence disease susceptibility in humans as they do in mice.

One of the major factors which might determine sex bias in coxsackievirus infections is sex-biased expression of the virus receptors on host cells. Coxsackieviruses use two known receptors. Coxsackievirus–adenovirus receptor (CAR) is a member of the immunoglobulin superfamily and is located in tight junctions (junction-associated molecule) (Bazzoni 2006). These molecules are not only necessary for the structural formation and stabilization of tight junctions, primarily in epithelial and endothelial cells, but also function in promoting leukocyte transmigration. CAR binds immunoglobulin and B lymphocytes and therefore might be involved in transmigration of these lymphocytes (Carson 2001; Carson and Chapman 2001). DAF is more widely distributed in tissues and cells than CAR and its main function is to protect cells against complement-mediated lysis (Medof et al. 1987). Unlike CAR, DAF is expressed in nonattached cells, such as leukocytes, and may be responsible for the infection and replication of coxsackieviruses in B cells, DCs, and activated T cells (Anderson et al. 1996; Liu et al. 2000a). Phytoestrogens, such as the isoflavone genistein, are able to induce CAR expression in human bladder cancer cells in vitro (Pong et al. 2006). Although this does not directly prove that sex-associated hormones modulate expression of this receptor, it is highly suggestive that this might happen. Better evidence for hormonal regulation of viral receptor expression exists for DAF. The human C3 promoter has estrogen response elements (EREs) suggesting hormonal influence of expression of complement (Vik et al. 1991) and estrogen treatment significantly increases C3 mRNA expression in vivo in mice (Li et al. 2002). As C3 expression is augmented with estrogen, so is the expression of DAF, a complement regulatory molecule meant to inhibit activated C3 (Song et al. 1996). To determine if modulation of DAF expression correlates to susceptibility of cells to coxsackievirus B3 infection, young women undergoing their natural menstrual cycle were followed through multiple cycles and evaluated in luteal and follicular phases for both DAF expression on peripheral blood lymphocytes and for susceptibility of the lymphocytes to coxsackievirus B3 infection. DAF expression in the follicular phase and the ability of virus to replicate in lymphocytes correlate strongly with estrogen levels (Sartini et al. 2004).

Sex-associated hormones also impact the immune response to coxsackievirus infections. Male mice and female mice infected during diestrus (i.e., when progesterone is high and estrogen is low) develop predominantly Th1 cell responses, whereas females infected during estrus (i.e., when estrogen is high and progesterone is low) develop primarily Th2 responses to infection (Huber et al. 1999; Schwartz et al. 2004; Frisancho-Kiss et al. 2007). Females infected during estrus generate FoxP3$^+$ regulatory T cells whereas these cells are absent when mice are infected during the other phases of the estrus cycle (Huber 2008). Sex steroids modulate regulatory T cell responses (Polanczyk et al. 2005; Tai et al. 2008) by promoting conversion of CD4$^+$CD25$^-$ cells to the regulatory CD4$^+$CD25$^+$FoxP3$^+$ phenotype via estrogen receptors on the regulatory cell precursor. Regulatory T cells inhibit CD4$^+$ Th1 cell responses (Nie et al. 2007), which are essential to coxsackievirus B3-induced myocarditis susceptibility (Huber et al. 2002a; Huber et al. 2002b). Exogenous administration of 17β-estradiol to females infected during diestrus upregulates cells and effectively abrogates both the Th1 cell response and myocarditis susceptibility. The effects of sex steroids on regulatory T cells in the coxsackievirus model may be mediated directly through hormone interaction with T regulatory cell precursors or indirectly through effects on other cell populations. $CD4^+CD25^+FoxP3^+$ regulatory T cells express certain TLR, most notably TLR4, TLR5, and TLR8 (Caramalho et al. 2003; Peng et al. 2005). Additionally, TLR3 signaling upregulates PD-L1, a negative regulatory molecule in DCs that suppresses $CD4^+$ cell responses (Groschel et al. 2008). The effects of sex-associated hormones on virus receptor expression and virus load may be mediated by effects on TLR signaling that may facilitate increased virus replication. Future studies should consider the effects of sex steroids on the expression of TLR8, TLR3, and RNA helicases as these receptors respond to ssRNA and dsRNA during coxsackievirus infections. In addition to receptors for RNA, male mice infected with coxsackievirus express elevated levels of TLR4 mRNA, IFN-γ, and reduced regulatory T cells whereas infected females have increased T cell Ig mucin-3 (TIM-3), IL-4, and regulatory T cells (Frisancho-Kiss et al. 2007). Inhibition of TIM-3 in males enhances TLR4 expression and reduces regulatory T cells, whereas inhibition of TLR4 signaling increases TIM-3 expression. Thus, TLR4 participates in coxsackieviral pathology by affecting regulatory T cell responses and this effect is sex dependent. Although the primary ligand for TLR4 is lipopolysaccharide, it is now clear that multiple endogenous ligands exist for TLRs besides microbial molecules (Tsan and Gao 2004). Infection with lysogenic viruses, such as picornaviruses, which can cause wide-spread cell lysis and release of endogenous molecules could promote more diverse TLR signaling than would be expected from the pathogen itself.

4.3.3 Hantaviruses

Globally, hantaviruses are one of the most widely distributed zoonotic pathogens that are maintained in the environment by rodents. Spillover of hantaviruses from rodents to humans causes hantavirus cardiopulmonary syndrome or hemorrhagic fever with renal syndrome (HFRS), depending on the species of virus (Klein and Calisher 2007). Reported human hantavirus infections in the Americas and Europe, as well as field observations of several rodent-virus systems indicate that more males than females are infected with hantaviruses (Childs et al. 1994; Weigler et al. 1996; White et al. 1996; Mills et al. 1997; Williams et al. 1997; Glass et al. 1998; Mills et al. 1998; Bernshtein et al. 1999). Sex differences in hantavirus infection only become apparent after puberty, suggesting that sex steroid hormones may underlie the dimorphism in infection (Childs et al. 1988; Mills et al. 1997). Sex steroids can modulate sex differences in infection through effects on the immune system or on the expression of behaviors (e.g., aggression) that increase the likelihood of being exposed to hantaviruses (Zuk and McKean 1996; Root et al. 1999; Klein 2000, 2004). Recent data from HFRS patients reveal sex-specific patterns in the acute immune response to Puumala virus (PUUV) infection, in which women produce higher levels of IL-9, FGF-2, and GM-CSF and lower levels of IL-8 and IP-10 than men (Klingstrom et al. 2008). Although a similar proportion of men and women have antibodies against PUUV (Ahlm et al. 1994), men are more likely to develop symptoms of disease (i.e., be hospitalized) during PUUV infection than are women (Vapalahti et al. 2003; Klingstrom et al. 2008). Whether sexually dimorphic immune responses during hantavirus infection cause differences in the severity of disease between men and women requires further investigation.

Laboratory studies of Norway rats inoculated with Seoul virus (SEOV; i.e., the hantavirus that naturally infects Norway rats) reveal that when given the same challenge, male and female rats are equally likely to become infected (Klein et al. 2000). After inoculation, however, male rats have more copies of SEOV RNA in the lungs for a longer duration of time and shed significantly more virus in saliva than females (Klein et al. 2000, 2001; Klein et al. 2002; Hannah et al. 2008). Additionally, large-scale genomic analyses reveal that the expression of genes that encode for immunological proteins associated with innate antiviral defenses, proin-flammatory responses, T cell responsiveness, and antibody production is higher in females than in males (Klein et al. 2004). The induction of PRRs (*Tlr7* and *Rig-I*), expression of antiviral genes (*Myd88*, *Visa*, *Jun*, *Irf7*, *Ifnβ*, *Ifnar1*, *Jak2*, *Stat3*, and *Mx2*), and production of Mx protein also is elevated in the lungs of intact females compared with intact males (Fig. 4.2) (Easterbrook and Klein, 2008; Hannah et al. 2008). Similarly, immunocompetence, as measured by swelling in response to PHA, is higher in female than in male deer mice during Sin Nombre virus infection (Lehmer et al. 2007). Conversely, the production of regulatory factors, including *Fox3* and TGF-β, is elevated in the lungs of SEOV-infected males as compared with females (Fig. 4.2) (Easterbrook and Klein 2008). These sexually dimorphic immune responses may be dependent on the levels of estradiol in females and testosterone in males, as gonadectomy reverses these differences (Hannah et al. 2008). Elevated antiviral immune responses in females may contribute to less efficient SEOV replication and shedding in females than in males.

Steroid hormones can bind to their respective receptors, that then translocate to the nucleus and bind to hormone response elements (HREs) in the promoter region of hormone-responsive genes, thereby influencing gene transcription. The extent to which genes associated with antiviral defenses are transcriptionally regulated by sex steroids is not well characterized (but see Fox et al. 1991; Brahmachary et al. 2006). To test the hypothesis that genes associated with antiviral defenses against hantaviruses are transcriptionally regulated by steroids, computational analyses were employed to identify HREs (i.e., EREs, androgen response elements (AREs), progesterone response elements, and glucocorticoid response elements (GREs)) in promoters of antiviral genes in rats. Putative AREs and EREs have been identified in the promoters of several of antiviral genes, including *Tlr3*, *Tlr7*, *Myd88*, *Irf7*, *Jun*, *Hsp70*, *Ifnar1*, and *Mx2*, suggesting that sex steroids may directly affect dimorphic antiviral responses against SEOV infection (Hannah et al. 2008).

Hantaviruses are transmitted through passage of virus in saliva during aggressive encounters (Glass et al. 1988; Hinson et al. 2004) and male rodents shed more virus

Fig. 4.2 The expression of *Ifnβ* and *Tnfα* is elevated and the expression of *Tgfβ* is reduced in the lungs of female, but not male, rats during SEOV infection. Expression of *Ifnβ*, *Tnfα*, and *Tgfβ* in the lungs was measured days 0, 3, 15, 30, and 40 p.i. by real-time RT-PCR. Gene expression is displayed as relative to expression in same sex uninfected rats (*gray line*) and the expression of each cytokine was normalized to *Gapdh*. Significantly different between male and female rats during SEOV infection (*), $p < 0.05$ using two-way ANOVAs (Easterbrook and Klein 2008)

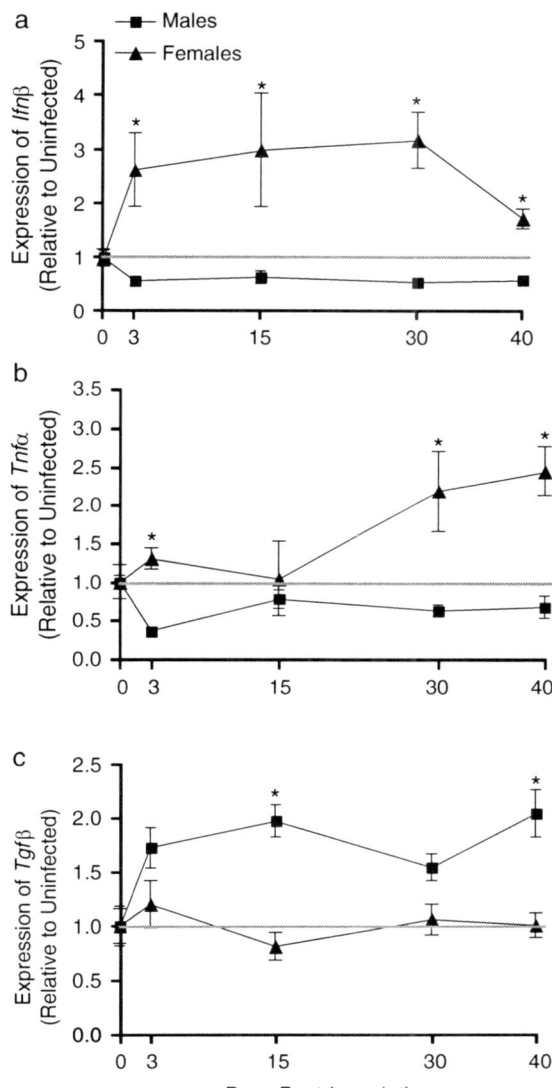

in saliva and are more likely to be infected with hantaviruses than females (Klein, 2003; Hinson et al. 2004). Wild-caught male rats that engage in elevated levels of aggression are more likely to be infected with SEOV and have higher circulating testosterone concentrations than males that engage in less aggression (Easterbrook et al. 2007). Suppressed antiviral defenses combined with the propensity to engage in aggression, both of which are mediated by testosterone, may make males better transmitters of hantaviruses than females.

4.3.4   Encephalitis-Causing Viruses

Several viruses can cross the blood–brain barrier to infect the central nervous system and cause inflammation. Pathology following infection with encephalitis-causing viruses may be mediated by replication of the viruses or by the host immune response mounted to clear infection. Thus, there is a balance between protection and development of pathology that might differ between the sexes and be mediated by sex hormones. In response to infection with lymphocytic choriomeningitis virus, estrogen treatment increases CD4$^+$ and CD8$^+$ T cell responses and increases susceptibility to fatal immune-related meningitis (Muller et al. 1995). In contrast, heightened immune cell activity, including production of nitric oxide synthase 2 and expression of MHC II, correlates with reduced spread of VSV in the central nervous system of female mice (Barna et al. 1996).

West Nile virus (WNV) is carried by mosquitoes and reportedly infects humans, horses, and birds. In humans and horses, infection can cause severe neuroinvasive disease and males are more likely to develop severe CNS symptoms of disease and die from infection than females (Porter et al. 2003; O'Leary et al. 2004; Epp et al. 2007; Jean et al. 2007). Interestingly, among horses, fatality rates are higher among stallions than either geldings (castrated males) or mares, suggesting that testosterone may be involved in the cause of the death (Epp et al. 2007). Whether androgens affect behaviors that increase the likelihood of infection seems improbably as rates of infection do not consistently differ between the sexes; thus, androgen-dependent effects on immunity remains a viable hypothesis for the dimorphic development of encephalomyelitis and fatality rates following exposure to WNV.

4.3.5   Paramyxoviruses

In humans, although males are more susceptible than females to many viral agents, measles is one virus that causes significantly higher mortality in females than in males. Although reviewed more extensively in Chap. 11, we will briefly note that a deaths caused by exposure to measles from 1950–1989 revealed that female-biased mortality is apparent among infants (<1–4 years), children (5–14 years), and adults (15–44 years) worldwide (Garenne 1994). The observed sex difference in response to measles virus implies that males and females might respond differently to attenuated viral vaccines. The standard measles vaccine is a low titer viral vaccine that is offered to infants at 9 months of age. One problem with the standard measles vaccine is that it does not protect infants against infection during the period between when maternal antibody begins to decline and immunization occurs (i.e., from 4–9 months of age). Thus, administration of a high titer measles vaccine to infants <9 months of age was initiated by the World Health Organization in the late 1980s in regions of West Africa. In response to a high titer measles vaccine, mortality rates were consistently higher for girls than for boys, which lead to termination of the vaccine trials (Knudsen et al. 1996). Whether girls and boys differed in their immunological responses to the vaccine has not been reported, but is discussed in Chap. 11 of this book.

4.4 Genetic Factors

4.4.1 Sex Chromosomes

Sex determination in mammals is mediated by the *Sry* gene on the Y chromosome which causes the formation of testes that produce and release testosterone. In the presence of both the *Sry* gene and testosterone, male-typic development ensues (Canning and Lovell-Badge 2002). In the absence of the Y chromosome and, hence, the *Sry* gene, ovaries develop (Canning and Lovell-Badge 2002). Early hypotheses about the role of host genes in resistance to disease initially speculated that female resistance and male susceptibility to infectious diseases were related to genes on sex chromosomes (Lenz 1931; Purtilo and Sullivan 1979). As male mammals are heterogametic and there are genes on the X chromosome that regulate immune function (e.g., *Tlr7* and *Foxp3*), deleterious recessive alleles are more likely to be expressed in males than in females. Moreover, even small differences in the effects of alleles are more likely to be evident in males than in females because the phenotype of females results from the average effect of two alleles (Burgoyne et al. 2001). Although sex differences in physiology may be caused by direct effects of sex steroids, an alternative hypothesis is that genes on the X chromosome, the Y chromosome, or both alter the expression of sexually dimorphic phenotypes (via direct, nonhormonal mechanisms).

Whether sex chromosomal genes modulate sex differences in the development of the immune system and susceptibility to viruses has not been reported. Recent studies, however, have addressed whether sex differences in susceptibility to autoimmune disease are mediated by sex chromosomes, by examining responses to experimentally induced autoimmunity in mice with the *Sry* gene either deleted (XY-*Sry*) or translocated to an autosomal region (XX*Sry*). Utilization of these mice enables investigators to separate gonadal sex (i.e., the presence of ovaries or testes) from sex chromosome complement (i.e., XX or XY). Susceptibility to experimental autoimmune encephalitis (EAE) and the severity of lupus in this mouse model is contingent on the presence of the XX sex chromosome complement (Palaszynski et al. 2005; Smith-Bouvier et al. 2008). The effect of sex chromosome complement on susceptibility to EAE is associated with reduced production of Th2 cytokines, including IL-4, IL-5, and IL-13, and increased expression of IL-13Rα2 on macrophages and DCs of XX mice (Smith-Bouvier et al. 2008). Consequently, the *Il13rα2* gene is expressed on the X chromosome (Donaldson et al. 1998); whether additional genes on sex chromosomes, including *Foxp3* and *Tlr7*, modulate sex differences in the recognition and response to viruses requires consideration.

4.4.2   Disease Susceptibility Genes

Several disease-resistance genes have been identified in vertebrate animals. Disease-resistance genes confer the advantage of more efficient immune responses, reduced replication of viruses, and less severe pathology associated with infection. The most well-studied disease-resistance genes encode proteins in the immunoglobulin (Ig) superfamily, including: (1) class I MHC glycoproteins; (2) class II MHC glycoproteins; (3) proteins in the T cell receptor complex; and (4) Ig molecules on B-cells. Genes of the Ig superfamily are important mediators of both cell-mediated and humoral immunity and are affected by sex steroid hormones as noted in Chap. 3. Presumably, if males are more susceptible to infection than females, then estrogens may enhance and androgens may suppress the expression of disease-resistance genes. Polymorphisms in genes that encode for immunological proteins can affect susceptibility to viruses. For example, sexually dimorphic polymorphisms in the IL-1β gene predict severity and persistence of Hepatitis C virus in humans (Wang et al. 2003).

Sex steroid hormones can alter the expression of disease-specific genes. In many cases, the expression of these autosomal genes determines the lethality of infection. Ectromelia virus (i.e., mousepox) is a naturally occurring lethal pathogen in mice. Susceptibility to mousepox has been mapped to disease-related loci on autosomal chromosomes. Using recombinant inbred strains of mice, four loci have been identified, *Rmp1-4* (resistance to mousepox loci), that confer resistance to mousepox. The effects of loci *Rmp2* (on chromosome 2) and *Rmp4* (on chromosome 1) on resistance to infection differ between the sexes, in which these loci confer greater resistance in female than in male congenic mice. If congenic mice are neonatally gonadectomized at 4–7 days of age and infected with mousepox as adults, then the sex difference in resistance is abolished; gonadectomized males and females are equally susceptible to mousepox (Brownstein and Gras 1995). Neonatal ovariectomy increases female susceptibility to mousepox; whereas, castration of neonatal males has little effect on susceptibility to infection (i.e., castrated males are as susceptible as intact males) (Brownstein and Gras 1995). Thus, estrogens may enhance as opposed to androgens suppressing genetic resistance to mousepox.

Two susceptibility loci, termed *Tmevp2* and *Tmevp3* (short for Theiler's muine encephalomyelitis virus persistence), have been mapped to Chromosome 10, control persistence of Theiler's virus in the CNS of mice, and contribute to the greater viral loads observed in males than in females (Fig. 4.3) (Bihl et al. 1999). Additional analyses using composite interval mapping to identify sex-dependent quantitative trait loci that control the severity of Theiler's virus disease in mice reveal sexually dimorphic loci on chromosomes 1, 5, 15, and 16 in males and on chromosome 1 in females (Butterfield et al. 2003). Whether sex differences in susceptibility to other viruses can be traced to disease-specific genes and/or loci requires additional evaluation.

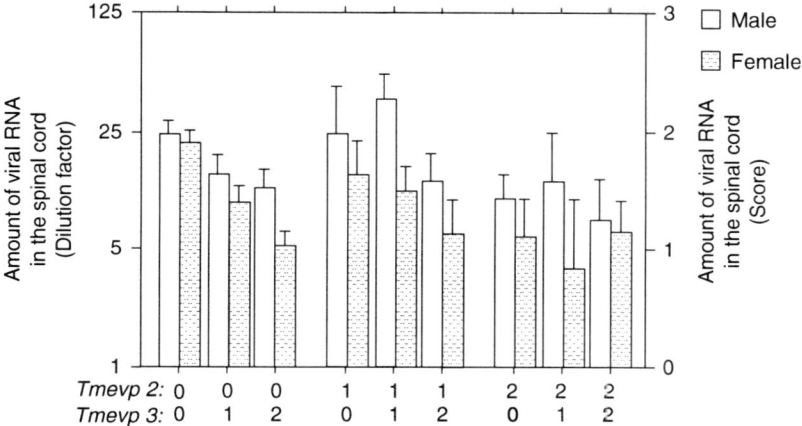

Fig. 4.3 Allelic segregation of the *Tmevp2* and *Tmevp3* loci has differential effects on Theiler's virus persistence in the central nervous system depending on the sex of the mice. These data illustrate the effect of sex on the number of resistant (B10.S) alleles for the *Tmevp2* and *Tmevp3* loci in mice. The amount of viral RNA in the spinal cord was measured 45 days postinoculation and is expressed as the highest RNA dilution that gave a hybridization signal or the score of viral persistence. Data adapted from Bihl et al. (1999)

4.4.3 Virus Genes

As noted above, steroid hormones can bind to their respective receptors, that then bind to HREs in the promoter region of hormone-responsive genes, thereby influencing gene transcirption. Several retroviruses contain the core motif for HREs in their sequence; thus, steroid hormones, including glucocorticoids, progesterone, and androgens, can directly regulate transcription of viruses, including HIV, bovine leukemia virus, Friend murine leukemia virus, and mouse mammary tumor virus (Cato et al. 1986; Darbre et al. 1986; Otten et al. 1988; Kolesnitchenko and Snart 1992; Niermann and Buehring 1997). At least in response to Friend murine leukemia virus, the presence of a GRE in the virus is one determinant of sex differences in susceptibility to infection (Bruland et al. 2003).

4.5 Conclusions

Males and females differ in their susceptibilities to viral infections. Sex-biased susceptibility to infection is evolutionarily well-conserved as differences are observed between males and females of several mammalian species and in response to a wide array of both DNA and RNA viruses (Table 4.1). The nature of the sexual dimorphism during viral infection depends on both the host and virus. Although males are more susceptible than females to many viruses, there are exceptions such as HSV-2 and measles virus for which males are more resistant than females and endocrine–immune interactions might underlie this sex reversal.

The immune systems of males and females differ. Most studies characterizing sex differences in immune responses to viruses have focused on type I IFNs and adaptive immune responses, with particular attention paid to the Th1/Th2 dichotomy. Although this approach has yielded valuable information about the causes of sex differences in viral infections, future studies should move beyond this paradigm and begin to consider other host responses to infection. In particular, the role of Th17 cells in mediating inflammatory responses that are common during virus infection should be examined. As females typically mount higher inflammatory responses during infection, we would predict the activity of Th17 cells, including expression of *RorγT* and production of IL-17, might be elevated in females compared with males. The balance between host immune responses resulting in protection against viruses and development of immunopathology might be differentially regulated between the sexes. Dimorphic responses of regulatory T cells, as have been shown in responses to hantaviruses and picornaviruses, should be considered in future studies.

Endocrine–immune interactions play a fundamental role mediating sexually dimorphic responses to viruses. Sex steroid concentrations differ dramatically between the sexes, and consequently to date, most studies have focused on characterizing the role of sex steroids as mediators of sex differences in viral infection. Future studies must continue to examine whether other steroid and peptide hormones, including glucocorticoids, contribute to sex differences in viral infection (Easterbrook and Klein 2008). Future studies must also consider whether natural hormonal fluctuations associated with puberty, pregnancy, or menstruation affect responses to viruses, as has been demonstrated during HIV, HSV, and coxsackievirus infection. Further, whether changes in host endocrine and immune responses following infection are mediated by the host or by the virus must be considered to fully elucidate the nature of these dynamic relationships. Finally, in addition to hormonal mechanisms, genetic differences between the sexes may directly modify responses to infection, as has been shown for autoimmune diseases.

The burden of viruses as a major cause of global infectious diseases is well recognized. Missing from current initiatives is an understanding of the host-related causes of sex differences in response to viral infection. To date, many studies of human, nonhuman primate, and rodent viral infections either do not report the sex of their subjects, report using sex-matched subjects, or use only one sex. Increased awareness about sex differences in response to viruses will impact both our understanding of host–virus interactions and the development of antiviral treatments that are successful in both sexes. By illustrating that males and females respond differently to viruses, this implies to us that the sexes may respond differently to antiviral treatments, including vaccines.

References

Aaby P, Ariyoshi K, Buckner M, Jensen H, Berry N, Wilkins A, Richard D, Larsen O, Dias F, Melbye M, Whittle H (1996) Age of wife as a major determinant of male-to-female transmission of HIV-2 infection: a community study from rural West Africa. Aids 10:1585–1590

Ahlm C, Linderholm M, Juto P, Stegmayr B, Settergren B (1994) Prevalence of serum IgG antibodies to Puumala virus (haemorrhagic fever with renal syndrome) in northern Sweden. Epidemiol Infect 113:129–136

Al-Harthi L, Kovacs A, Coombs RW, Reichelderfer PS, Wright DJ, Cohen MH, Cohn J, Cu-Uvin S, Watts H, Lewis S, Beckner S, Landay A (2001) A menstrual cycle pattern for cytokine levels exists in HIV-positive women: implication for HIV vaginal and plasma shedding. Aids 15:1535–1543

Anderson D, Wilson J, Carthy C, Yang D, Kandolf R, McManus B (1996) Direct interactions of coxsackievirus B3 with immune cells in the splenic compartment of mice susceptible or resistant to myocarditis. J Virol 70:4632–4645

Araneo BA, Dowell T, Diegel M, Daynes RA (1991) Dihydrotestosterone exerts a depressive influence on the production of interleukin-4 (IL-4), IL-5, and gamma-interferon, but not IL-2 by activated murine T cells. Blood 78:688–699

Armien B, Pascale JM, Bayard V, Munoz C, Mosca I, Guerrero G, Armien A, Quiroz E, Castillo Z, Zaldivar Y, Gracia F, Hjelle B, Koster F (2004) High seroprevalence of hantavirus infection on the Azuero peninsula of Panama. Am J Trop Med Hyg 70:682–687

Baeten JM, Benki S, Chohan V, Lavreys L, McClelland RS, Mandaliya K, Ndinya-Achola JO, Jaoko W, Overbaugh J (2007) Hormonal contraceptive use, herpes simplex virus infection, and risk of HIV-1 acquisition among Kenyan women. Aids 21:1771–1777

Barna M, Komatsu T, Bi Z, Reiss CS (1996) Sex differences in susceptibility to viral infection of the central nervous system. J Neuroimmunol 67:31–39

Barrat F, Lesourd B, Boulouis HJ, Thibault D, Vincent-Naulleau S, Gjata B, Louise A, Neway T, Pilet C (1997) Sex and parity modulate cytokine production during murine ageing. Clin Exp Immunol 109:562–568

Bazzoni G (2006) Endothelial tight junctions: permeable barriers of the vessel wall. Thromb Haemost 95:36–42

Benki S, Mostad SB, Richardson BA, Mandaliya K, Kreiss JK, Overbaugh J (2004) Cyclic shedding of HIV-1 RNA in cervical secretions during the menstrual cycle. J Infect Dis 189:2192–2201

Benki S, Mostad SB, Richardson BA, Mandaliya K, Kreiss JK, Overbaugh J (2008) Increased Levels of HIV-1-Infected Cells in Endocervical Secretions After the Luteinizing Hormone Surge. J Acquir Immune Defic Syndr 47:529–534

Bergelson JM, Mohanty JG, Crowell RL, St John NF, Lublin DM, Finberg RW (1995) Coxsackievirus B3 adapted to growth in RD cells binds to decay-accelerating factor (CD55). J Virol 69:1903–1906

Berghofer B, Frommer T, Haley G, Fink L, Bein G, Hackstein H (2006) TLR7 ligands induce higher IFN-alpha production in females. J Immunol 177:2088–2096

Bernshtein AD, Apekina NS, Mikhailova TV, Myasnikov YA, Khlyap LA, Korotkov YS, Gavrilovskaya IN (1999) Dynamics of Puumala hantavirus infection in naturally infected bank voles (Clethrinomys glareolus). Arch Virol 144:2415–2428

Bihl F, Brahic M, Bureau JF (1999) Two loci, Tmevp2 and Tmevp3, located on the telomeric region of chromosome 10, control the persistence of Theiler's virus in the central nervous system of mice. Genetics 152:385–392

Bijlsma JW, Cutolo M, Masi AT, Chikanza IC (1999) The neuroendocrine immune basis of rheumatic diseases. Immunol Today 20:298–301

Bowles NE, Ni J, Kearney DL, Pauschinger M, Schultheiss HP, McCarthy R, Hare J, Bricker JT, Bowles KR, Towbin JA (2003) Detection of viruses in myocardial tissues by polymerase chain reaction. evidence of adenovirus as a common cause of myocarditis in children and adults. J Am Coll Cardiol 42:466–472

Brabin L (2002) Interactions of the female hormonal environment, susceptibility to viral infections, and disease progression. AIDS Patient Care STDS 16:211–221

Brahmachary M, Schonbach C, Yang L, Huang E, Tan SL, Chowdhary R, Krishnan SP, Lin CY, Hume DA, Kai C, Kawai J, Carninci P, Hayashizaki Y, Bajic VB (2006) Computational promoter analysis of mouse, rat and human antimicrobial peptide-coding genes. BMC Bioinformatics 7(Suppl 5):S8

Brownstein DG, Gras L (1995) Chromosome mapping of Rmp-4, a gonad-dependent gene encoding host resistance to mousepox. J Virol 69:6958–6964

Brownstein DG, Bhatt PN, Gras L, Jacoby RO (1991) Chromosomal locations and gonadal dependence of genes that mediate resistance to ectromelia (mousepox) virus-induced mortality. J Virol 65:1946–1951

Bruland T, Lavik LA, Dai HY, Dalen A (2003) A glucocorticoid response element in the LTR U3 region of Friend murine leukaemia virus variant FIS-2 enhances virus production in vitro and is a major determinant for sex differences in susceptibility to FIS-2 infection in vivo. J Gen Virol 84:907–916

Burgoyne PS, Lovell-Badge R, Rattigan A (2001) Evidence that the testis determination pathway interacts with a non-dosage compensated, X-linked gene. Int J Dev Biol 45:509–512

Butterfield RJ, Roper RJ, Rhein DM, Melvold RW, Haynes L, Ma RZ, Doerge RW, Teuscher C (2003) Sex-specific quantitative trait loci govern susceptibility to Theiler's murine encephalomyelitis virus-induced demyelination. Genetics 163:1041–1046

Canning CA, Lovell-Badge R (2002) Sry and sex determination: how lazy can it be? Trends Genet 18:111–113

Caramalho I, Lopes-Carvalho T, Ostler D, Zelenay S, Haury M, Demengeot J (2003) Regulatory T cells selectively express toll-like receptors and are activated by lipopolysaccharide. J Exp Med 197:403–411

Carson SD (2001) Receptor for the group B coxsackieviruses and adenoviruses: CAR. Rev Med Virol 11:219–226

Carson SD, Chapman NM (2001) Coxsackievirus and adenovirus receptor (CAR) binds immunoglobulins. Biochemistry 40:14324–14329

Cato AC, Miksicek R, Schutz G, Arnemann J, Beato M (1986) The hormone regulatory element of mouse mammary tumour virus mediates progesterone induction. Embo J 5:2237–2240

Chang TH, Liao CL, Lin YL (2006) Flavivirus induces interferon-beta gene expression through a pathway involving RIG-I-dependent IRF-3 and PI3K-dependent NF-kappaB activation. Microbes Infect 8:157–171

Chellappan S, Kraus VB, Kroger B, Munger K, Howley PM, Phelps WC, Nevins JR (1992) Adenovirus E1A, simian virus 40 tumor antigen, and human papillomavirus E7 protein share the capacity to disrupt the interaction between transcription factor E2F and the retinoblastoma gene product. Proc Natl Acad Sci USA 89:4549–4553

Cheung TC, Humphreys IR, Potter KG, Norris PS, Shumway HM, Tran BR, Patterson G, Jean-Jacques R, Yoon M, Spear PG, Murphy KM, Lurain NS, Benedict CA, Ware CF (2005) Evolutionarily divergent herpesviruses modulate T cell activation by targeting the herpesvirus entry mediator cosignaling pathway. Proc Natl Acad Sci USA 102:13218–13223

Childs JE, Glass GE, Korch GW, LeDuc JW (1988) The ecology and epizootiology of hantaviral infections in small mammal communities of Baltimore: A review and synthesis. Bulletin of the Society of Vector Biology 13:113–122

Childs JE, Ksiazek TG, Spiropoulou CF, Krebs JW, Morzunov S, Maupin GO, Gage KL, Rollin PE, Sarisky J, Enscore RE et al (1994) Serologic and genetic identification of Peromyscus maniculatus as the primary rodent reservoir for a new hantavirus in the southwestern United States. J Infect Dis 169:1271–1280

Choe H, Farzan M, Sun Y, Sullivan N, Rollins B, Ponath PD, Wu L, Mackay CR, LaRosa G, Newman W, Gerard N, Gerard C, Sodroski J (1996) The beta-chemokine receptors CCR3 and CCR5 facilitate infection by primary HIV-1 isolates. Cell 85:1135–1148

Clemetson DB, Moss GB, Willerford DM, Hensel M, Emonyi W, Holmes KK, Plummer F, Ndinya-Achola J, Roberts PL, Hillier S et al (1993) Detection of HIV DNA in cervical and vaginal secretions. Prevalence and correlates among women in Nairobi, Kenya. Jama 269:2860–2864

Collman RG, Yi Y, Liu QH, Freedman BD (2000) Chemokine signaling and HIV-1 fusion mediated by macrophage CXCR4: implications for target cell tropism. J Leukoc Biol 68:318–323

D'Addario M, Ahmad A, Xu JW, Menezes J (1999) Epstein-Barr virus envelope glycoprotein gp350 induces NF-kappaB activation and IL-1beta synthesis in human monocytes-macrophages involving PKC and PI3-K. Faseb J 13:2203–2213

D'Addario M, Ahmad A, Morgan A, Menezes J (2000) Binding of the Epstein-Barr virus major envelope glycoprotein gp350 results in the upregulation of the TNF-alpha gene expression in monocytic cells via NF-kappaB involving PKC, PI3-K and tyrosine kinases. J Mol Biol 298:765–778

D'Addario M, Libermann TA, Xu J, Ahmad A, Menezes J (2001) Epstein-Barr Virus and its glycoprotein-350 upregulate IL-6 in human B-lymphocytes via CD21, involving activation of NF-kappaB and different signaling pathways. J Mol Biol 308:501–514

Darbre P, Page M, King RJ (1986) Androgen regulation by the long terminal repeat of mouse mammary tumor virus. Mol Cell Biol 6:2847–2854

Daughaday CC, Brandt WE, McCown JM, Russell PK (1981) Evidence for two mechanisms of dengue virus infection of adherent human monocytes: trypsin-sensitive virus receptors and trypsin-resistant immune complex receptors. Infect Immun 32:469–473

Del Corno M, Liu QH, Schols D, de Clercq E, Gessani S, Freedman BD, Collman RG (2001) HIV-1 gp120 and chemokine activation of Pyk2 and mitogen-activated protein kinases in primary macrophages mediated by calcium-dependent, pertussis toxin-insensitive chemokine receptor signaling. Blood 98:2909–2916

DiMaio D, Lai CC, Klein O (1998) Virocrine transformation: the intersection between viral transforming proteins and cellular signal transduction pathways. Annu Rev Microbiol 52:397–421

Dobner T, Horikoshi N, Rubenwolf S, Shenk T (1996) Blockage by adenovirus E4orf6 of transcriptional activation by the p53 tumor suppressor. Science 272:1470–1473

Domingo E (2000) Viruses at the edge of adaptation. Virology 270:251–253

Domingo E, Holland J (1997) RNA virus mutations and fitness for survival. Am Rev Microbiol 51:151–178

Donaldson DD, Whitters MJ, Fitz LJ, Neben TY, Finnerty H, Henderson SL, O'Hara RM Jr, Beier DR, Turner KJ, Wood CR, Collins M (1998) The murine IL-13 receptor alpha 2: molecular cloning, characterization, and comparison with murine IL-13 receptor alpha 1. J Immunol 161:2317–2324

Douglass RJ, Calisher CH, Wagoner KD, Mills JN (2007) Sin nombre virus infection of deer mice in montana: characteristics of newly infected mice, incidence, and temporal pattern of infection. J Wildl Dis 43:12–22

Easterbrook JD, Klein SL (2008) Corticosteroids modulate Seoul virus infection, regulatory T-cell responses and matrix metalloprotease 9 expression in male, but not female, Norway rats. J Gen Virol 89:2723–2730

Easterbrook JD, Kaplan JB, Glass GE, Pletnikov MV, Klein SL (2007) Elevated testosterone and reduced 5-HIAA concentrations are associated with wounding and hantavirus infection in male Norway rats. Horm Behav 52(4):474–481

Eigen M (1993) Viral quasispecies. Sci Am 269:42–49

Epp T, Waldner C, West K, Townsend H (2007) Factors associated with West Nile virus disease fatalities in horses. Can Vet J 48:1137–1145

Falter H, Persinger MA, Reid K (1991) Sex differences in primary humoral responses of albino rats to human serum albumin. Immunol Lett 28:143–145

Farzadegan H, Hoover DR, Astemborski J, Lyles CM, Margolick JB, Markham RB, Quinn TC, Vlahov D (1998) Sex differences in HIV-1 viral load and progression to AIDS. Lancet 352:1510–1514

Feng Y, Broder CC, Kennedy PE, Berger EA (1996) HIV-1 entry cofactor: functional cDNA cloning of a seven-transmembrane, G protein-coupled receptor. Science 272:872–877

Ferrer JF, Jonsson CB, Esteban E, Galligan D, Basombrio MA, Peralta-Ramos M, Bharadwaj M, Torrez-Martinez N, Callahan J, Segovia A, Hjelle B (1998) High prevalence of hantavirus infection in Indian communities of the Paraguayan and Argentinean Gran Chaco. Am J Trop Med Hyg 59:438–444

Fitzgerald-Bocarsly P, Dai J, Singh S (2008) Plasmacytoid dendritic cells and type I IFN: 50 years of convergent history. Cytokine Growth Factor Rev 19:3–19

Fleming DT, McQuillan GM, Johnson RE, Nahmias AJ, Aral SO, Lee FK, St Louis ME (1997) Herpes simplex virus type 2 in the United States, 1976 to 1994. N Engl J Med 337:1105–1111

Fox HS, Bond BL, Parslow TG (1991) Estrogen regulates the IFN-gamma promoter. J Immunol 146:4362–4367

Friedman SB, Grota LJ, Glasgow LA (1972) Differential susceptibility of male and female mice to encephalomyocarditis virus: effects of castration, adrenalectomy, and the administration of sex hormones. Infect Immun 5:637–644

Frisancho-Kiss S, Davis SE, Nyland JF, Frisancho JA, Cihakova D, Barrett MA, Rose NR, Fairweather D (2007) Cutting edge: cross-regulation by TLR4 and T cell Ig mucin-3 determines sex differences in inflammatory heart disease. J Immunol 178:6710–6714

Fuller A, Yahikozawa H, So EY, Dal Canto M, Koh CS, Welsh CJ, Kim BS (2007) Castration of male C57L/J mice increases susceptibility and estrogen treatment restores resistance to Theiler's virus-induced demyelinating disease. J Neurosci Res 85:871–881

Gallichan WS, Rosenthal KL (1996) Effects of the estrous cycle on local humoral immune responses and protection of intranasally immunized female mice against herpes simplex virus type 2 infection in the genital tract. Virology 224:487–497

Garenne M (1994) Sex differences in measles mortality: a world review. Int J Epidemiol 23:632–642

Gavrilovskaya IN, Shepley M, Shaw R, Ginsberg MH, Mackow ER (1998) beta3 Integrins mediate the cellular entry of hantaviruses that cause respiratory failure. Proc Natl Acad Sci USA 95:7074–7079

Gillgrass AE, Ashkar AA, Rosenthal KL, Kaushic C (2003) Prolonged exposure to progesterone prevents induction of protective mucosal responses following intravaginal immunization with attenuated herpes simplex virus type 2. J Virol 77:9845–9851

Giron DJ, Patterson RR (1982) Effect of steroid hormones on virus-induced diabetes mellitus. Infect Immun 37:820–822

Giron DJ, Allen PT, Pindak FF, Schmidt JP (1973) Further studies on the influence of steroids on viral infection in mice. Infect Immun 8:151–155

Gitlin L, Barchet W, Gilfillan S, Cella M, Beutler B, Flavell RA, Diamond MS, Colonna M (2006) Essential role of mda-5 in type I IFN responses to polyriboinosinic:polyribocytidylic acid and encephalomyocarditis picornavirus. Proc Natl Acad Sci USA 103:8459–8464

Glass GE, Childs JE, Korch GW, LeDuc JW (1988) Association of intraspecific wounding with hantaviral infection in wild rats (Rattus norvegicus). Epidemiol Infect 101:459–472

Glass GE, Livingstone W, Mills JN, Hlady WG, Fine JB, Biggler W, Coke T, Frazier D, Atherley S, Rollin PE, Ksiazek TG, Peters CJ, Childs JE (1998) Black Creek Canal Virus infection in Sigmodon hispidus in southern Florida. Am J Trop Med Hyg 59:699–703

Gomez E, Ortiz V, Saint-Martin B, Boeck L, Diaz-Sanchez V, Bourges H (1993) Hormonal regulation of the secretory IgA (sIgA) system: estradiol- and progesterone-induced changes in sIgA in parotid saliva along the menstrual cycle. Am J Reprod Immunol 29:219–223

Groschel S, Piggott KD, Vaglio A, Ma-Krupa W, Singh K, Goronzy JJ, Weyand CM (2008) TLR-mediated induction of negative regulatory ligands on dendritic cells. J Mol Med 86:443–455

Haeryfar SM (2005) The importance of being a pDC in antiviral immunity: the IFN mission versus Ag presentation? Trends Immunol 26:311–317

Han X, Lundberg P, Tanamachi B, Openshaw H, Longmate J, Cantin E (2001) Gender influences herpes simplex virus type 1 infection in normal and gamma interferon-mutant mice. J Virol 75:3048–3052

Hanna L (1999) The menstrual cycle and viral load. Beta 12:18

Hannah MF, Bajic VB, Klein SL (2008) Sex differences in the recognition of and innate antiviral responses to Seoul virus in Norway rats. Brain Behav Immun 22(4):503–516

Hardarson HS, Baker JS, Yang Z, Purevjav E, Huang CH, Alexopoulou L, Li N, Flavell RA, Bowles NE, Vallejo JG (2007) Toll-like receptor 3 is an essential component of the innate stress response in virus-induced cardiac injury. Am J Physiol Heart Circ Physiol 292: H251–H258

Hinson ER, Shone SM, Zink MC, Glass GE, Klein SL (2004) Wounding: the primary mode of Seoul virus transmission among male Norway rats. Am J Trop Med Hyg 70:310–317

Hornung V, Schlender J, Guenthner-Biller M, Rothenfusser S, Endres S, Conzelmann KK, Hartmann G (2004) Replication-dependent potent IFN-alpha induction in human plasmacytoid dendritic cells by a single-stranded RNA virus. J Immunol 173:5935–5943

Howell AL, Asin SN, Yeaman GR, Wira CR (2005) HIV-1 infection of the female reproductive tract. Curr HIV/AIDS Rep 2:35–38

Huber S, Job L, Auld K (1982) Influence of sex hormones on coxsackie B3 virus infection in Balb/c mice. Cell Immunol 67:173–179

Huber S, Kupperman J, Newell M (1999) Hormonal regulation of CD4+ T-cell responses in coxsackievirus B3-induced myocarditis in mice. J Virol 73:4689–4695

Huber S, Shi C, Budd RC (2002a) Gammadelta T cells promote a Th1 response during coxsackievirus B3 infection in vivo: role of Fas and Fas ligand. J Virol 76:5487–6494

Huber SA, Sartini D, Exley M (2002b) Vgamma4(+) T cells promote autoimmune CD8(+) cytolytic T-lymphocyte activation in coxsackievirus B3-induced myocarditis in mice: role for CD4(+) Th1 cells. J Virol 76:10785–10790

Huber SA (2008) Coxsackievirus B3-induced myocarditis: infection of females during the estrus phase of the ovarian cycle leads to activation of T regulatory cells. Virology 378:292–298

Jean CM, Honarmand S, Louie JK, Glaser CA (2007) Risk factors for West Nile virus neuroinvasive disease, California, 2005. Emerg Infect Dis 13:1918–1920

Jung JU, Stager M, Desrosiers RC (1994) Virus-encoded cyclin. Mol Cell Biol 14:7235–7244

Katagiri D, Hayashi H, Victoriano AF, Okamoto T, Onozaki K (2006) Estrogen stimulates transcription of human immunodeficiency virus type 1 (HIV-1). Int Immunopharmacol 6:170–181

Kawai T, Akira S (2006) Innate immune recognition of viral infection. Nat Immunol 7:131–137

Klein SL (2000) The effects of hormones on sex differences in infection: from genes to behavior. Neurosci Biobehav Rev 24:627–638

Klein SL (2003) Parasite manipulation of the proximate mechanisms that mediate social behavior in vertebrates. Physiol Behav 79:441–449

Klein SL (2004) Hormonal and immunological mechanisms mediating sex differences in parasite infection. Parasite Immunol 26:247–264

Klein SL, Calisher CH (2007) Emergence and persistence of hantaviruses. Curr Top Microbiol Immunol 315:217–252

Klein SL, Bird BH, Glass GE (2000) Sex differences in Seoul virus infection are not related to adult sex steroid concentrations in Norway rats. J Virol 74:8213–8217

Klein SL, Bird BH, Glass GE (2001) Sex differences in immune responses and viral shedding following Seoul virus infection in Norway rats. Am J Trop Med Hyg 65:57–63

Klein SL, Marson AL, Scott AL, Ketner G, Glass GE (2002) Neonatal sex steroids affect responses to Seoul virus infection in male but not female Norway rats. Brain Behav Immun 16:736–746

Klein SL, Cernetich A, Hilmer S, Hoffman EP, Scott AL, Glass GE (2004) Differential expression of immunoregulatory genes in male and female Norway rats following infection with Seoul virus. J Med Virol 74:180–190

Klingstrom J, Lindgren T, Ahlm C (2008) Hantavirus infection induces higher plasma levels of IL-9, FGF-2 and GM-CSF, and lower levels of IL-8 and IP-10, in females compared to males. Clin Vaccine Immunol doi:10.1128/CVI.00035-08

Knudsen KM, Aaby P, Whittle H, Rowe M, Samb B, Simondon F, Sterne J, Fine P (1996) Child mortality following standard, medium or high titre measles immunization in West Africa. Int J Epidemiol 25:665–673

Kolesnitchenko V, Snart RS (1992) Regulatory elements in the human immunodeficiency virus type 1 long terminal repeat LTR (HIV-1) responsive to steroid hormone stimulation. AIDS Res Hum Retroviruses 8:1977–1980

Kubarek L, Jagodzinski PP (2007) Epigenetic up-regulation of CXCR4 and CXCL12 expression by 17 beta-estradiol and tamoxifen is associated with formation of DNA methyltransferase 3B4 splice variant in Ishikawa endometrial adenocarcinoma cells. FEBS Lett 581:1441–1448

Kumar H, Kawai T, Kato H, Sato S, Takahashi K, Coban C, Yamamoto M, Uematsu S, Ishii KJ, Takeuchi O, Akira S (2006) Essential role of IPS-1 in innate immune responses against RNA viruses. J Exp Med 203:1795–1803

Lamey PJ, Ferguson MM, Marshall W (1985) Sex hormone involvement in the development of experimental virally induced murine salivary gland tumors. J Oral Pathol 14:414–421

Langenberg AG, Corey L, Ashley RL, Leong WP, Straus SE (1999) A prospective study of new infections with herpes simplex virus type 1 and type 2. Chiron HSV Vaccine Study Group. N Engl J Med 341:1432–1438

Leclerc PM, Dubois-Colas N, Garenne M (2008) Hormonal contraception and HIV prevalence in four African countries. Contraception 77:371–376

Lehmer EM, Clay CA, Wilson E, St Jeor S, Dearing MD (2007) Differential resource allocation in deer mice exposed to sin nombre virus. Physiol Biochem Zool 80:514–521

Lenz F (1931) Morbidic hereditary factors. Macmillan Press, New York

Leopold PL, Crystal RG (2007) Intracellular trafficking of adenovirus: many means to many ends. Adv Drug Deliv Rev 59:810–821

Li CF, Ross FP, Cao X, Teitelbaum SL (1995) Estrogen enhances alpha v beta 3 integrin expression by avian osteoclast precursors via stabilization of beta 3 integrin mRNA. Mol Endocrinol 9:805–813

Li SH, Huang HL, Chen YH (2002) Ovarian steroid-regulated synthesis and secretion of complement C3 and factor B in mouse endometrium during the natural estrous cycle and pregnancy period. Biol Reprod 66:322–332

Little TV, Holyoak GR, McCollum WH, Timoney PJ (1992) Output of equine arteritis virus from persistently infected stallions is testosterone-dependent. In:Proceedings of the sixth international conference on equine infectious diseases pp 225–229

Liu P, Aitken K, Kong YY, Opavsky MA, Martino T, Dawood F, Wen WH, Kozieradzki I, Bachmaier K, Straus D, Mak TW, Penninger JM (2000a) The tyrosine kinase p56lck is essential in coxsackievirus B3-mediated heart disease. Nat Med 6:429–434

Liu QH, Williams DA, McManus C, Baribaud F, Doms RW, Schols D, De Clercq E, Kotlikoff MI, Collman RG, Freedman BD (2000b) HIV-1 gp120 and chemokines activate ion channels in primary macrophages through CCR5 and CXCR4 stimulation. Proc Natl Acad Sci USA 97:4832–4837

Liu J, Miwa T, Hilliard B, Chen Y, Lambris JD, Wells AD, Song WC (2005) The complement inhibitory protein DAF (CD55) suppresses T cell immunity in vivo. J Exp Med 201:567–577

Lyden D, Huber S (1984) Aggravation of coxsackievirus group B type 3-induced myocarditis and increase in cellular immunity to myocyte antigens in pregnant BALB/c mice and animals treated with progesterone. Cell Immunol 87:462–472

Lyden DC, Olszewski J, Feran M, Job LP, Huber SA (1987) Coxsackievirus B-3-induced
    myocarditis. Effect of sex steroids on viremia and infectivity of cardiocytes. Am J Pathol
    126:432–438
Maddon PJ, Dalgleish AG, McDougal JS, Clapham PR, Weiss RA, Axel R (1986) The T4 gene
    encodes the AIDS virus receptor and is expressed in the immune system and the brain. Cell
    47:333–348
Marx PA, Spira AI, Gettie A, Dailey PJ, Veazey RS, Lackner AA, Mahoney CJ, Miller CJ,
    Claypool LE, Ho DD, Alexander NJ (1996) Progesterone implants enhance SIV vaginal
    transmission and early virus load. Nat Med 2:1084–1089
McCollum WH, Little TV, Timoney PJ, Swerczek TW (1994) Resistance of castrated male horses
    to attempted establishment of the carrier state with equine arteritis virus. J Comp Pathol
    111:383–388
McIntyre NE, Chu YK, Owen RD, Abuzeineh A, De la Sancha N, Dick CW, Holsomback T,
    Nisbett RA, Jonsson C (2005) A longitudinal study of Bayou virus, hosts, and habitat. Am
    J Trop Med Hyg 73:1043–1049
Medof ME, Walter EI, Rutgers JL, Knowles DM, Nussenzweig V (1987) Identification of the
    complement decay-accelerating factor (DAF) on epithelium and glandular cells and in body
    fluids. J Exp Med 165:848–864
Medzhitov R, Janeway CA Jr (2002) Decoding the patterns of self and nonself by the innate
    immune system. Science 296:298–300
Melchjorsen J, Jensen SB, Malmgaard L, Rasmussen SB, Weber F, Bowie AG, Matikainen S,
    Paludan SR (2005) Activation of innate defense against a paramyxovirus is mediated by RIG-I
    and TLR7 and TLR8 in a cell-type-specific manner. J Virol 79:12944–12951
Mertz GJ, Benedetti J, Ashley R, Selke SA, Corey L (1992) Risk factors for the sexual transmis-
    sion of genital herpes. Ann Intern Med 116:197–202
Mills JN, Ellis BA, McKee KT Jr, Calderon GE, Maiztegui JI, Nelson GO, Ksiazek TG, Peters CJ,
    Childs JE (1992) A longitudinal study of Junin virus activity in the rodent reservoir of
    Argentine hemorrhagic fever. Am J Trop Med Hyg 47:749–763
Mills JN, Ellis BA, Childs JE, McKee KT Jr, Maiztegui JI, Peters CJ, Ksiazek TG, Jahrling PB
    (1994) Prevalence of infection with Junin virus in rodent populations in the epidemic area of
    Argentine hemorrhagic fever. Am J Trop Med Hyg 51:554–562
Mills JN, Ksiazek TG, Ellis BA, Rollin PE, Nichol ST, Yates TL, Gannon WL, Levy CE,
    Engelthaler DM, Davis T, Tanda DT, Frampton JW, Nichols CR, Peters CJ, Childs JE
    (1997) Patterns of association with host and habitat: antibody reactive with Sin Nombre
    virus in small mammals in the major biotic communities of the southwestern United States.
    Am J Trop Med Hyg 56:273–284
Mills JN, Johnson JM, Ksiazek TG, Ellis BA, Rollin PE, Yates TL, Mann MO, Johnson MR,
    Campbell ML, Miyashiro J, Patrick M, Zyzak M, Lavender D, Novak MG, Schmidt K, Peters
    CJ, Childs JE (1998) A survey of hantavirus antibody in small-mammal populations in selected
    United States National Parks. Am J Trop Med Hyg 58:525–532
Mirand EA, Back N, Grace JT Jr, Buffet R (1967) Effect of pituitary and gonadal hormones on
    Friend Virus Disease in mice. Proc Soc Exp Biol Med 124:1055–1059
Misse D, Esteve PO, Renneboog B, Vidal M, Cerutti M, St Pierre Y, Yssel H, Parmentier M, Veas
    F (2001) HIV-1 glycoprotein 120 induces the MMP-9 cytopathogenic factor production that is
    abolished by inhibition of the p38 mitogen-activated protein kinase signaling pathway. Blood
    98:541–547
Mo R, Chen J, Grolleau-Julius A, Murphy HS, Richardson BC, Yung RL (2005) Estrogen
    regulates CCR gene expression and function in T lymphocytes. J Immunol 174:6023–6029
Morrow PL, Freedman A, Craighead JE (1980) Testosterone effect on experimental diabetes
    mellitus in encephalomyocarditis (EMC) virus infected mice. Diabetologia 18:247–249
Mostad SB, Overbaugh J, DeVange DM, Welch MJ, Chohan B, Mandaliya K, Nyange P, Martin
    HL Jr, Ndinya-Achola J, Bwayo JJ, Kreiss JK (1997) Hormonal contraception, vitamin A
    deficiency, and other risk factors for shedding of HIV-1 infected cells from the cervix and
    vagina. Lancet 350:922–927

Mostad SB, Kreiss JK, Ryncarz A, Chohan B, Mandaliya K, Ndinya-Achola J, Bwayo JJ, Corey L (2000a) Cervical shedding of herpes simplex virus and cytomegalovirus throughout the menstrual cycle in women infected with human immunodeficiency virus type 1. Am J Obstet Gynecol 183:948–955

Mostad SB, Kreiss JK, Ryncarz AJ, Mandaliya K, Chohan B, Ndinya-Achola J, Bwayo JJ, Corey L (2000b) Cervical shedding of herpes simplex virus in human immunodeficiency virus-infected women: effects of hormonal contraception, pregnancy, and vitamin A deficiency. J Infect Dis 181:58–63

Muller D, Chen M, Vikingsson A, Hildeman D, Pederson K (1995) Oestrogen influences CD4+ T-lymphocyte activity in vivo and in vitro in beta 2-microglobulin-deficient mice. Immunology 86:162–167

Munk C, Zielonka J, Constabel H, Kloke BP, Rengstl B, Battenberg M, Bonci F, Pistello M, Lochelt M, Cichutek K (2007) Multiple restrictions of human immunodeficiency virus type 1 in feline cells. J Virol 81:7048–7060

Napravnik S, Poole C, Thomas JC, Eron JJ Jr (2002) Gender difference in HIV RNA levels: a meta-analysis of published studies. J Acquir Immune Defic Syndr 31:11–19

Nassal M (2008) Hepatitis B viruses: Reverse transcription a different way. Virus Res 134(1–2):235–249

Nie CQ, Bernard NJ, Schofield L, Hansen DS (2007) CD4+ CD25+ regulatory T cells suppress CD4+ T-cell function and inhibit the development of Plasmodium berghei-specific TH1 responses involved in cerebral malaria pathogenesis. Infect Immun 75:2275–2282

Niermann GL, Buehring GC (1997) Hormone regulation of bovine leukemia virus via the long terminal repeat. Virology 239:249–258

Nieva JL, Agirre A (2003) Are fusion peptides a good model to study viral cell fusion? Biochim Biophys Acta 1614:104–115

Nohara K, Izumi H, Tamura S, Nagata R, Tohyama C (2002) Effect of low-dose 2, 3, 7, 8-tetrachlorodibenzo-p-dioxin (TCDD) on influenza A virus-induced mortality in mice. Toxicology 170:131–138

Obasi A, Mosha F, Quigley M, Sekirassa Z, Gibbs T, Munguti K, Todd J, Grosskurth H, Mayaud P, Changalucha J, Brown D, Mabey D, Hayes R (1999) Antibody to herpes simplex virus type 2 as a marker of sexual risk behavior in rural Tanzania. J Infect Dis 179:16–24

O'Leary DR, Marfin AA, Montgomery SP, Kipp AM, Lehman JA, Biggerstaff BJ, Elko VL, Collins PD, Jones JE, Campbell GL (2004) The epidemic of West Nile virus in the United States, 2002. Vector Borne Zoonotic Dis 4:61–70

Olsson GE, White N, Ahlm C, Elgh F, Verlemyr AC, Juto P, Palo RT (2002) Demographic factors associated with hantavirus infection in bank voles (Clethrionomys glareolus). Emerg Infect Dis 8:924–929

Otten AD, Sanders MM, McKnight GS (1988) The MMTV LTR promoter is induced by progesterone and dihydrotestosterone but not by estrogen. Mol Endocrinol 2:143–147

Palaszynski KM, Smith DL, Kamrava S, Burgoyne PS, Arnold AP, Voskuhl RR (2005) A yin-yang effect between sex chromosome complement and sex hormones on the immune response. Endocrinology 146:3280–3285

Parr MB, Kepple L, McDermott MR, Drew MD, Bozzola JJ, Parr EL (1994) A mouse model for studies of mucosal immunity to vaginal infection by herpes simplex virus type 2. Lab Invest 70:369–380

Pearce-Duvet JM, St Jeor SC, Boone JD, Dearing MD (2006) Changes in sin nombre virus antibody prevalence in deer mice across seasons: the interaction between habitat, sex, and infection in deer mice. J Wildl Dis 42:819–824

Peng G, Guo Z, Kiniwa Y, Voo KS, Peng W, Fu T, Wang DY, Li Y, Wang HY, Wang RF (2005) Toll-like receptor 8-mediated reversal of CD4+ regulatory T cell function. Science 309: 1380–1384

Pisitkun P, Deane JA, Difilippantonio MJ, Tarasenko T, Satterthwaite AB, Bolland S (2006) Autoreactive B cell responses to RNA-related antigens due to TLR7 gene duplication. Science 312:1669–1672

Polanczyk MJ, Hopke C, Huan J, Vandenbark AA, Offner H (2005) Enhanced FoxP3 expression and Treg cell function in pregnant and estrogen-treated mice. J Neuroimmunol 170:85–92

Pong RC, Roark R, Ou JY, Fan J, Stanfield J, Frenkel E, Sagalowsky A, Hsieh JT (2006) Mechanism of increased coxsackie and adenovirus receptor gene expression and adenovirus uptake by phytoestrogen and histone deacetylase inhibitor in human bladder cancer cells and the potential clinical application. Cancer Res 66:8822–8828

Poonia B, Walter L, Dufour J, Harrison R, Marx PA, Veazey RS (2006) Cyclic changes in the vaginal epithelium of normal rhesus macaques. J Endocrinol 190:829–835

Porter MB, Long MT, Getman LM, Giguere S, MacKay RJ, Lester GD, Alleman AR, Wamsley HL, Franklin RP, Jacks S, Buergelt CD, Detrisac CJ (2003) West Nile virus encephalomyelitis in horses: 46 cases (2001). J Am Vet Med Assoc 222:1241–1247

Purtilo DT, Sullivan JL (1979) Immunological bases for superior survival of females. Am J Dis Child 133:1251–1253

Quinn TC, Overbaugh J (2005) HIV/AIDS in women: an expanding epidemic. Science 308:1582–1583

Roberts CW, Walker W, Alexander J (2001) Sex-associated hormones and immunity to protozoan parasites. Clin Microbiol Rev 14:476–488

Roivainen M, Piirainen L, Hovi T, Virtanen I, Riikonen T, Heino J, Hyypia T (1994) Entry of coxsackievirus A9 into host cells: specific interactions with alpha v beta 3 integrin, the vitronectin receptor. Virology 203:357–365

Root JJ, Calisher CH, Beaty BJ (1999) Relationships of deer mouse movement, vegetative structure, and prevalence of infection with Sin Nombre virus. J Wildl Dis 35:311–318

Sartini D, Moussawi M, Sallam R, Bernstein I, Huber S (2004) Correlation between serum estradiol in the follicular phase of the ovarian cycle and decay acceleration factor (DAF) expression on red blood cells and coxsackiervirus B-3 induced hemagglutination in young cycling women. Am J Reprod Immunol 51:180–187

Schwartz J, Sartini D, Huber S (2004) Myocarditis susceptibility in female mice depends upon ovarian cycle phase at infection. Virology 330:16–23

Seth RB, Sun L, Chen ZJ (2006) Antiviral innate immunity pathways. Cell Res 16:141–147

Smith SM, Baskin GB, Marx PA (2000) Estrogen protects against vaginal transmission of simian immunodeficiency virus. J Infect Dis 182:708–715

Smith-Bouvier DL, Divekar AA, Sasidhar M, Du S, Tiwari-Woodruff SK, King JK, Arnold AP, Singh RR, Voskuhl RR (2008) A role for sex chromosome complement in the female bias in autoimmune disease. J Exp Med 205(5):1099–1108

Sodora DL, Gettie A, Miller CJ, Marx PA (1998) Vaginal transmission of SIV: assessing infectivity and hormonal influences in macaques inoculated with cell-free and cell-associated viral stocks. AIDS Res Hum Retroviruses 14(Suppl 1):S119–S123

Song WC, Deng C, Raszmann K, Moore R, Newbold R, McLachlan JA, Negishi M (1996) Mouse decay-accelerating factor: selective and tissue-specific induction by estrogen of the gene encoding the glycosylphosphatidylinositol-anchored form. J Immunol 157:4166–4172

Sterling TR, Vlahov D, Astemborski J, Hoover DR, Margolick JB, Quinn TC (2001) Initial plasma HIV-1 RNA levels and progression to AIDS in women and men. N Engl J Med 344:720–725

Stringer EM, Kaseba C, Levy J, Sinkala M, Goldenberg RL, Chi BH, Matongo I, Vermund SH, Mwanahamuntu M, Stringer JS (2007) A randomized trial of the intrauterine contraceptive device vs hormonal contraception in women who are infected with the human immunodeficiency virus. Am J Obstet Gynecol 197(144):e141–e148

Sun X, Funk CD, Deng C, Sahu A, Lambris JD, Song WC (1999) Role of decay-accelerating factor in regulating complement activation on the erythrocyte surface as revealed by gene targeting. Proc Natl Acad Sci USA 96:628–633

Sun Q, Sun L, Liu HH, Chen X, Seth RB, Forman J, Chen ZJ (2006) The specific and essential role of MAVS in antiviral innate immune responses. Immunity 24:633–642

Tai P, Wang J, Jin H, Song X, Yan J, Kang Y, Zhao L, An X, Du X, Chen X, Wang S, Xia G, Wang B (2008) Induction of regulatory T cells by physiological level estrogen. J Cell Physiol 214:456–464

Teepe AG, Allen LB, Wordinger RJ, Harris EF (1990) Effect of the estrous cycle on susceptibility of female mice to intravaginal inoculation of herpes simplex virus type 2 (HSV-2). Antiviral Res 14:227–235

Theofilopoulos AN, Baccala R, Beutler B, Kono DH (2005) Type I interferons (alpha/beta) in immunity and autoimmunity. Annu Rev Immunol 23:307–336

Tolou H, Nicoli J, Chastel C (2002) Viral evolution and emerging viral infections: what future for the viruses? A theoretical evaluation based on informational spaces and quasispecies. Virus Genes 24:267–274

Tsan MF, Gao B (2004) Endogenous ligands of Toll-like receptors. J Leukoc Biol 76:514–519

Vapalahti O, Mustonen J, Lundkvist A, Henttonen H, Plyusnin A, Vaheri A (2003) Hantavirus infections in Europe. Lancet Infect Dis 3:653–661

Varani S, Cederarv M, Feld S, Tammik C, Frascaroli G, Landini MP, Soderberg-Naucler C (2007) Human cytomegalovirus differentially controls B cell and T cell responses through effects on plasmacytoid dendritic cells. J Immunol 179:7767–7776

Vik DP, Amiguet P, Moffat GJ, Fey M, Amiguet-Barras F, Wetsel RA, Tack BF (1991) Structural features of the human C3 gene: intron/exon organization, transcriptional start site, and promoter region sequence. Biochemistry 30:1080–1085

Wang Y, Kato N, Hoshida Y, Yoshida H, Taniguchi H, Goto T, Moriyama M, Otsuka M, Shiina S, Shiratori Y, Ito Y, Omata M (2003) Interleukin-1beta gene polymorphisms associated with hepatocellular carcinoma in hepatitis C virus infection. Hepatology 37:65–71

Weber F, Kochs G, Haller O (2004) Inverse interference: how viruses fight the interferon system. Viral Immunol 17:498–515

Weigler BJ, Ksiazek TG, Vandenbergh JG, Levin M, Sullivan WT (1996) Serological evidence for zoonotic hantaviruses in North Carolina rodents. J Wildl Dis 32:354–357

Westbury HA, Sinkovic B (1978) The pathogenesis of infectious avian encephalomyelitis. 3. The relationship between viraemia, invasion of the brain by the virus, and the development of specific serum neutralising antibody. Aust Vet J 54:76–80

White DJ, Means RG, Birkhead GS, Bosler EM, Grady LJ, Chatterjee N, Woodall J, Hjelle B, Rollin PE, Ksiazek TG, Morse DL (1996) Human and rodent hantavirus infection in New York State: public health significance of an emerging infectious disease. Arch Intern Med 156: 722–726

Wickham TJ, Mathias P, Cheresh DA, Nemerow GR (1993) Integrins alpha v beta 3 and alpha v beta 5 promote adenovirus internalization but not virus attachment. Cell 73:309–319

Wickham TJ, Filardo EJ, Cheresh DA, Nemerow GR (1994) Integrin alpha v beta 5 selectively promotes adenovirus mediated cell membrane permeabilization. J Cell Biol 127:257–264

Williams RJ, Bryan RT, Mills JN, Palma RE, Vera I, De Velasquez F, Baez E, Schmidt WE, Figueroa RE, Peters CJ, Zaki SR, Khan AS, Ksiazek TG (1997) An outbreak of hantavirus pulmonary syndrome in western Paraguay. Am J Trop Med Hyg 57:274–282

Woodruff J (1980) Viral myocarditis. Am J Pathol 101:425–483

Woodward TL, Mienaltowski AS, Modi RR, Bennett JM, Haslam SZ (2001) Fibronectin and the alpha(5)beta(1) integrin are under developmental and ovarian steroid regulation in the normal mouse mammary gland. Endocrinology 142:3214–3222

Yeaman GR, Howell AL, Weldon S, Demian DJ, Collins JE, O'Connell DM, Asin SN, Wira CR, Fanger MW (2003) Human immunodeficiency virus receptor and coreceptor expression on human uterine epithelial cells: regulation of expression during the menstrual cycle and implications for human immunodeficiency virus infection. Immunology 109:137–146

Zuk M, McKean KA (1996) Sex differences in parasite infections: patterns and processes. Int J Parasitol 26:1009–1023

Chapter 5
Sex Differences in Innate Immune Responses to Bacterial Pathogens

Jennifer A. Rettew, Ian Marriott, and Yvette M. Huet

Abstract Sex-based differences in innate immune responses to bacterial infection are evident in human patients and animal models of disease. Females are less susceptible to the development of bacterial infections and subsequent bacteremia and/or sepsis, while males exhibit a greater incidence of such infections and are more likely to develop fatal sequellae. While the precise effects and mechanisms of action remain to be determined, it is apparent that sex steroid hormones can have direct effects on the expression and function of key bacterial pattern recognition receptors on innate immune cells. Changes in the expression of these receptors are likely to have profound effects on the production of the inflammatory mediators responsible for the lethal nature of septic shock and may underlie the observed sexual dimorphism demonstrated in immune responses to bacterial endotoxins.

5.1 Introduction

It has long been known that sex is a contributing factor in the incidence and progression of disorders associated with immune system dysregulation (e.g., development of autoimmune diseases). There also is an increased appreciation and interest in the fundamental role that sex plays in susceptibility to infectious disease. In this chapter, we illustrate that the responses to bacterial infections and endotoxin differ based on sex as well as reproductive status. In discussing sex biases in response to bacterial infection, it is important to note that disease severity and outcome following bacterial infection are often dependent on the host inflammatory responses elicited by endotoxins produced by many bacterial species. Systemic inflammatory response syndrome (SIRS) describes the physiological

Y.M. Huet (✉)
Department of Biology, 9201 University City Boulevard, University of North Carolina at Charlotte, Charlotte, NC 28223
e-mail: ymhuet@uncc.edu

S.L. Klein and C.W. Roberts (eds.), *Sex Hormones and Immunity to Infection*,
DOI 10.1007/978-3-642-02155-8_5, © Springer-Verlag Berlin Heidelberg 2010

changes associated with an overactive and systemic host response that can be due to either an infectious stimulus, such as endotoxin, or the physiological response to challenges, such as hemorrhage. Patients with SIRS exhibit symptoms, such as fever or hypothermia, tachycardia, tachypnea, and white blood cell count abnormalities (Martel 2002). The term SIRS encompasses sepsis, bacteremia, and endotoxemia. Sepsis occurs when organisms, at a local site of infection, proliferate and gain access to the blood stream via tissue damage and/or invasion mechanisms. Bacteremia and endotoxemia occur when bacteria and endotoxins (e.g., lipopolysaccharide (LPS)), respectively, are present in the blood stream. LPS is a structural component of the cell wall of Gram-negative bacteria that, while not actively secreted by these organisms, is often released into the extracellular milieu of the host following bacterial lysis. The systemic circulation of microbes and/or endotoxin often leads to a systemic inflammatory response and frequently sepsis. Severe sepsis can lead to septic shock, which is categorized by a catastrophic drop in blood pressure that results in diminished perfusion of tissues, hypoxia, and dysfunction of organs, including the kidneys, liver, lungs, and central nervous system (CNS). This loss of function may lead to multiple organ failure and death (Martel 2002). In fact, sepsis and the multiple organ failure associated with septic shock are the most common cause of late postinjury death in surgical intensive care units (Sauaia et al. 1995).

Sex-based differences in the immune response to bacterial infection are evident at multiple levels. Both innate and adaptive immunity are sexually dimorphic in human patients and mouse models of bacterial disease. While disparities between men and women in B cell activity and antibody production in response to bacterial infection and vaccination are well known, marked differences also exist between males and females in the frequency, severity, and outcome of severe sepsis and septic shock. In this chapter, the sex-based differences in susceptibility to bacterial infection and the mechanisms that may account for such a sexual dimorphism are reviewed.

5.2 Sex-Based Differences in Susceptibility to Bacterial Infection and Sepsis

5.2.1 Males Exhibit Greater Incidence and Severity of Bacteremia, Endotoxemia, and Higher Tissue Bacterial Burdens Following Infection Than Females

Animal studies provide evidence that males exhibit greater susceptibility to bacterial challenge than their female counterparts. Sex-based differences have been observed in susceptibility of mice to *Mycobacterium marinum* infection, with males showing higher disease severity, bacterial burden, and mortality than infected female mice (Yamamoto et al. 1991). Similarly, females infected with *Helicobacter pylori* show a delayed onset of intestinal dysplasia and develop less intestinal inflammation and histopathology relative to infected males (Ohtani et al. 2007). This sex difference is not limited to bacterial burden, as animal models of endotoxemia indicate that female mice demonstrate higher survival rates than males when subjected to severe sepsis. For example, administration of *Vibrio vulnificus*-derived LPS leads to endotoxic shock in male rats, with a mortality rate of 82%. In contrast, females treated in the same manner exhibit only a 21% mortality rate following LPS challenge (Merkel et al. 2001). Studies of *V. vulnificus* in mice have yielded similar results. In one study, all female mice survived LPS-induced sepsis, but only 70% of their male counterparts survived a similar treatment (Laubach et al. 1998). In the cecal ligation and puncture-induced model of sepsis, survival is greater in female mice (44%) relative to males (5%) (Kahlke et al. 2002).

Findings in human patients appear to correspond with those of animal studies, with men exhibiting greater susceptibility to bacterial infection than women. A study of patients with bacteremia at Boston City Hospital in 1972 found that the incidence of infection was significantly higher in male patients than in females (McGowan et al. 1975). In addition, men exhibit increased mortality associated with nosocomial infection compared with females (Dinkel and Lebok 1994). Being male is a major risk factor for bacterial infection following severe injury, with one study showing that male patients exhibit a 58% greater risk of developing major bacterial infections following trauma than females (Offner et al. 1999).

Sex differences in the incidence and/or severity of bacterial infection are paralleled by similar differences in the development of severe sepsis and septic shock. *V. vulnificus* infection, which occurs following the ingestion of raw or undercooked seafood, elicits endotoxic shock in humans with a fatality rate of almost 60%; 80% of *V. vulnificus*-associated mortality, however, is observed in males (Oliver 1989). In addition, reviews of hospital cases reveal that significantly fewer female patients are referred to the intensive care unit, and of all patients referred, men develop severe septic shock more frequently than women (Dosset et al. 2008; Wichmann et al. 2000). Furthermore, the outcome following the development of sepsis also differs between the sexes, with men exhibiting greater mortality than women (Schroder et al. 1998).

Females are less susceptible to the development of bacterial infections and subsequent bacteremia and/or sepsis, while males exhibit greater incidence and severity of bacterial infections and are far more likely to develop lethal sequellae. Although these sex-based differences have been appreciated for many years, the mechanisms accounting for such sex differences are only now becoming apparent. The divergence in sepsis severity may stem from the differences in circulating endotoxin levels. Female rats exhibit significantly lower endotoxin levels following cecal ligation and puncture than similarly treated males (Erikoglu et al. 2005). Contradictory results, however, have been reported (Kono et al. 2000). Sex-based differences in sepsis severity might also be caused by the relative production of key proinflammatory cytokines that precipitate the lethal consequences of bacterial septic shock.

5.2.2 Sex-Based Differences in the Production of Inflammatory Cytokines and Development of Bacterial Septic Shock

The development of sepsis is driven by the overproduction of cytokines, including tumor necrosis factor (TNF)-α, interleukin (IL)-1β, and IL-6 (Blackwell and Christman 1996). The importance of these molecules in the development of septic shock is underscored by the observation that sera levels of IL-6, TNF-α, and IL-8 are significantly lower in sepsis patients who survive than in those who succumb to this condition (Dosset et al. 2008; Majetschak et al. 2000). The rapid onset and progression of septic shock are testaments to the role played by innate immunity in the development of this often self-destructive host response. Macrophages can detect bacteria and endotoxins via pattern recognition receptors (PRRs) to produce large amounts of proinflammatory cytokines. The lethal nature of septic shock is mediated, in large part, by the widespread activation of macrophages and the subsequent overproduction of proinflammatory mediators.

A sexual dimorphism exists in the circulating levels of proinflammatory cytokines following bacterial infection and/or septic shock. For example, male sepsis patients have higher circulating levels of TNF-α than do female patients and this observation correlates with a worse prognosis (Schroder et al. 1998). In mice, *Escherichia coli*-derived LPS elicits higher circulating levels of IL-6 in males than in similarly treated female mice (Marriott et al. 2006). Interestingly, female sepsis patients exhibit higher levels of the antiinflammatory cytokine IL-10 than do age and disease severity matched male patients (Schroder et al. 1998). While these studies support the contention that the overproduction of proinflammatory cytokines, diminished production of antiinflammatory cytokines, or both correlate with susceptibility to septic shock in males, some studies have failed to detect such sex-based differences (May et al. 2008; Schroder et al. 1998) and the reason for this discrepancy is not clear.

Sex-based differences exist at the level of proinflammatory cytokine production by antigen presenting cells. Sex differences in the production of proinflammatory cytokines and chemokines are reported following in vitro LPS challenge of isolated human peripheral monocytes and macrophages isolated from rodents. For example, peripheral monocytes from male patients produce twofold higher levels of TNF-α than female-derived cells following LPS challenge (Asai et al. 2001). Peripheral blood cells derived from young adult men produce significantly more TNF-α following LPS challenge than do similarly treated cells derived from young adult women (Moxley et al. 2002). Peritoneal macrophages isolated from male mice subjected to cecal ligation-induced sepsis secrete greater amounts of TNF-α than do female-derived macrophages from similarly treated mice (Kahlke et al. 2002).

These sex differences are not limited to TNF-α. Male-derived macrophages produce significantly larger amounts of the proinflammatory chemokine CXCL10 (IP-10) than do macrophages from female mice (Marriott et al. 2006). This effect seems to be chemokine-specific, as no sex-based difference is observed in the production of CCL2 (MCP-1) (Marriott et al. 2006). In addition, in vitro studies using peritoneal macrophages show that cells derived from young male mice produce higher levels of IL-1β and IL-6 following LPS challenge than do similarly treated cells derived from females (Kahlke et al. 2000). Results for IL-6 are contradictory, as LPS-stimulated peripheral monocytes from females produce more IL-6 than do monocytes from males, even though these same cells from females produce less TNF-α than do cells from males (Asai et al. 2001). Furthermore, macrophages isolated from female mice after thermal injury produce higher levels of IL-6 upon LPS stimulation than do those isolated from males (Kovacs et al. 2002). Similarly, hypoxia-stimulated IL-6 secretion is greater in Kupffer cells derived from females (i.e., the resident macrophages of the liver) than in those derived from males (Zheng et al. 2006). Although these results contradict the sex differences in TNF-α production, interpreting changes in IL-6 levels is complicated by its role in the development of helper T cell type 2 (Th2) responses.

Sex-based differences are reported in the production of antiinflammatory cytokines. For example, peritoneal macrophages isolated from male mice produce significantly lower amounts of the potentially immunosuppressive prostanoid PGE_2 than do cells derived from females (Marriott et al. 2006). Cells from female mice also can be induced to secrete more prostanoids than males following adjuvant administration (Du et al. 1984), burn injury (Gregory et al. 2000), or collagen-induced arthritis (Leslie et al. 1987). Furthermore, female-derived splenic macrophages also secrete higher levels of the antiinflammatory cytokine IL-10 than do cells derived from males (Kahlke et al. 2000). There are, however, studies that have failed to detect such sex differences in IL-10 secretion (Asai et al. 2001). One study of cecal ligation and puncture-induced sepsis found that exogenous IL-10 treatment increases survival in males, but has no effect on survival in females (Kahlke et al. 2002).

Sex differences in LPS-induced inflammatory mediator production observed in monocytes and macrophages are apparent in other sentinel cells, such as neutrophils (Spitzer and Zhang 1996), and may also include nonleukocytes cells. LPS-induced lung inflammation in male mice is associated with higher levels of TNF-α than that produced in the airway fluid of female mice (Card et al. 2006). In addition, gastric tissue from male mice infected with *H. pylori* has a higher expression of TNF-α than does tissue from infected females (Ohtani et al. 2007).

The majority of empirical data supports the hypothesis that females are protected from potentially lethal endotoxic shock in two ways: (1) by producing lower levels of proinflammatory mediators that precipitate the systemic effects associated with sepsis; and (2) by elevating the production of antiinflammatory and regulatory molecules that serve to attenuate endotoxin-mediated systemic inflammation. Despite the consensus that male- and female-derived immune cells differentially produce proinflammatory mediators following bacterial or endotoxin exposure, the mechanisms that underlie these differences remain poorly defined.

5.2.2.1 Sex-Based Differences in the Expression and Functionality of Receptors for Conserved Bacterial Motifs

The recent discovery of a family of PRRs with a high degree of homology to the Toll family of proteins in *Drosophila* has shed light on the means by which innate

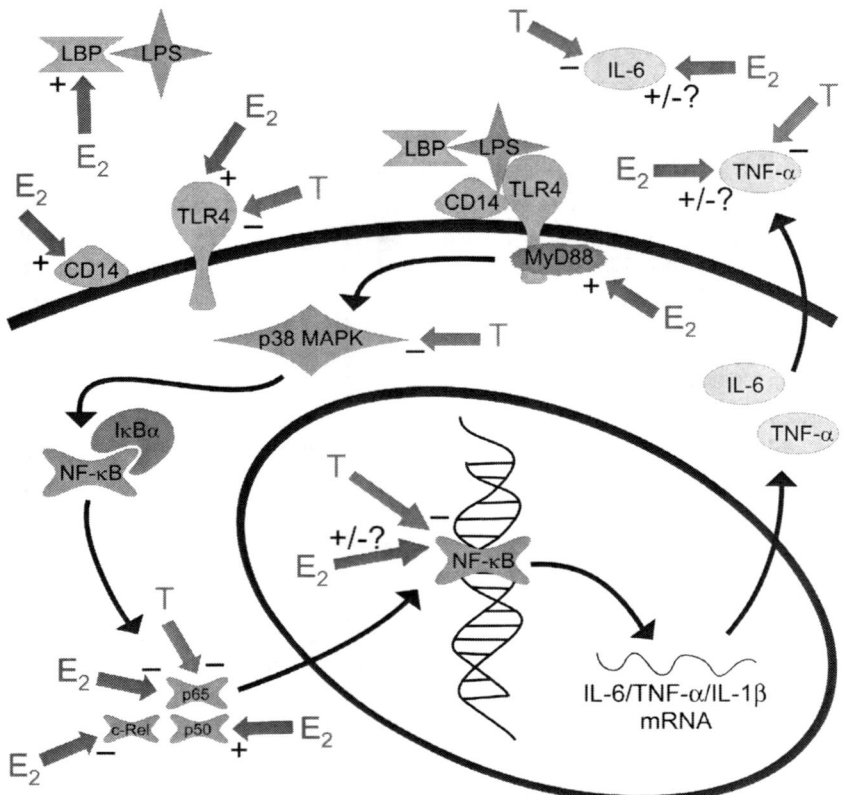

Fig. 5.1 Mechanisms underlying LPS-induced proinflammatory cytokine production by host cells and hypothesized points of regulation by sex steroid hormones. LBP: lipopolysaccharide binding protein, LPS: lipopolysaccharide, TLR4: Toll-like receptor 4, p38 MAPK: mitogen-activated protein kinase p38, NF-κB: nuclear factor kappa B, E_2: estrogen, T: testosterone. Plus sign indicates a stimulatory effect, whereas negative sign indicates an inhibitory effect

immune cells recognize a wide range of pathogens without the need for prior exposure (Wright 1999; Medzhitov and Janeway 2000). To date, at least eleven members of the Toll-like family of receptors (TLR) have been discovered in mice and humans. These receptors detect the presence of conserved microbial motifs and initiate the production of cytokines, chemokines, and costimulatory molecule expression (Barton and Medzhitov. 2003).

As shown in Fig. 5.1, the detection of LPS involves Toll-like receptor 4 (TLR4) and several coreceptors and binding proteins. LPS binding protein (LBP) is a type I acute phase protein that catalyzes the monomerization of LPS and mediates its transfer to CD14 (Schumann et al. 1990). CD14 is a coreceptor for LPS and is either expressed within the plasma membrane of innate immune cells, such as macrophages, or is present as a soluble molecule in the surrounding milieu. Cellular responses are initiated through binding with TLR4, as neither CD14 nor LBP contains a cytoplasmic component (Poltorak et al. 1998; Chow et al. 1999; Lien et al. 2000). TLR4 is expressed by immune cells, including monocytes, macrophages, and dendritic cells (DCs) (Medzhitov et al. 1997). TLR4 also is expressed on nonleukocytic cells capable of proinflammatory cytokine production, including epithelial cells in a variety of tissues (Diamond et al. 2000; Kumar et al. 2004; Quintar et al. 2006; Hornef et al. 2002) smooth muscle cells (Quintar et al. 2006) endothelial cells (Rock et al. 1998) and resident CNS cells, such as astrocytes (Lewis et al. 2008). All these cells can therefore respond to LPS and have the potential to contribute to a systemic inflammatory response. Ligation of TLR4, facilitated by CD14 and LBP, initiates an intracellular signaling cascade that results in the activation of NF-κB, a pivotal transcription factor in the regulation of inflammatory cytokine expression. As such, widespread activation of cells via TLR4 can precipitate the overproduction of soluble immune mediators that can lead to the development of a systemic inflammatory response that underlies the lethal nature of septic shock (Palsson-McDermott and O'Neill 2004).

Sex differences exist at the level of expression of these LPS receptors and could therefore underlie the observed differences in LPS-mediated responses in male- and female-derived cells with regard to inflammatory mediator production. Male mice have significantly higher circulating levels of LBP after LPS challenge than do female mice and macrophages isolated from male mice express higher levels of cell surface CD14 than do female-derived cells (Marriott et al. 2006). The sex difference in CD14 exists at the level of protein expression rather than at the level of gene transcription, as levels of mRNA encoding CD14 are not significantly different between male- and female-derived cells (Marriott et al. 2006). Other studies of CD14 expression reveal that peritoneal macrophages from males show a tendency to express enhanced levels of cell surface CD14 compared with female-derived cells; whereas the expression of CD14 on splenic macrophages is similar between male- and female-derived cells (Eisenmenger et al. 2004).

Cells from males not only express higher levels of receptors for LPS than their female counterparts, but macrophages isolated from male mice express more cell surface TLR4 than female-derived macrophages. Sex difference in TLR4 occurs at the level of protein expression, as expression of mRNA for this PRR is not significantly different between sexes (Marriott et al. 2006). In contrast, hypoxia induces higher expression of TLR4 mRNA in Kupffer cells from male mice compared with female mice (Zheng et al. 2006). Peritoneal macrophages from males also exhibit a tendency to express more cell surface TLR4 following trauma hemorrhage than do cells from similarly treated females (Eisenmenger et al. 2004). Conversely, the expression of TLR4 in naive macrophages or splenic macrophages following trauma hemorrhage is similar between the sexes.

Sex-based differences in LPS responses also may stem from the differences in the intracellular signaling pathways initiated following TLR4 ligation. Hypoxia initiates a MyD88-dependent signaling cascade resulting in IL-6 release from female-derived Kupffer cells. In contrast, male-derived Kupffer cells rely on a MyD88-independent signaling pathway for production of this cytokine (Zheng et al. 2006). Consistent with this finding, female-derived Kupffer cells express higher levels of MyD88 than do cells from males (Zheng et al. 2006). Although Src protein levels are higher in female-derived liver macrophages, hypoxia decreases such expression in female-derived cells and elicits an increased Src production in male-derived Kupfer cells (Zheng et al. 2006). Ligation of TLR4 initiates MAP kinase signaling and the activation of NF-κB. Female-derived peritoneal macrophages show greater p38 MAP kinase phosphorylation, and hence, activation following LPS challenge than do macrophages from males (Angele et al. 2003). Furthermore, NF-κB activity in Kupffer cells is three times higher in females following alcohol-induced liver injury than in similarly treated males (Kono et al. 2000).

Taken together, evidence is accumulating that sex differences exist at the level of expression of the PRRs that detect bacterial components and influence the signaling pathways that result in the production of proinflammatory cytokines. Sex differences in the activation of TLRs and their coreceptors potentially underlie the sexual dimorphism observed in immune responses during sepsis and/or septic shock. The higher expression of TLR4 and its coreceptors on sentinel cells from males and the more efficient LPS-induced signaling pathways in female cells appear to be contradictory. However, TLR stimulation induces the production of both proinflammatory and antiinflammatory cytokines. Thus, sex differences in the activity of signaling pathways downstream of TLR4 could result in higher antiinflammatory cytokine production rather than elevated release of proinflammatory mediators. More research is required to validate such a hypothesis.

5.3 Sex Steroid Hormones Affect Susceptibility to Bacterial Infection Through Effects on Innate Immune Cells

5.3.1 Susceptibility to Bacterial Infection and Sepsis Differ With Age and Reproductive Status

The differences in immune responses between males and females are often assumed to be a consequence of sex steroid hormones. Circumstantial evidence for such a hypothesis comes from the documented differences in susceptibility to bacterial infection and sepsis/septic shock observed with age and changes in reproductive status. Sex differences in susceptibility and mortality due to sepsis are not present in prepubescent children over the age of one, which is correlated with low levels of sex steroid hormones (Bindl et al. 2003). In infants under the age of one, boys are more susceptible to developing severe sepsis than girls, at levels similar to that seen in adult males (Bindl et al. 2003). Whether sex differences in severity of sepsis in infants are due to the presence of residual sex hormones that are present at high levels during fetal development and that rapidly decrease within the first year of life requires evaluation.

Susceptibility to sepsis also changes with old age, which may be a result of decreased concentrations of circulating sex hormones that may change susceptibility to bacterial challenge. In a mouse model of aging, Kahlke and coworkers (2000) found that peritoneal macrophages from aged male and female mice release lower levels of the proinflammatory cytokines IL-1β and IL-6 and greater levels of the antiinflammatory cytokine IL-10 upon LPS stimulation compared with macrophages from young mice. Differences in cytokine production between males and females that are apparent in young mice are not present in aged mice (Kahlke et al. 2000). In humans, elderly sepsis patients do not display sex differences in either the incidence of septic shock or shock-associated mortality as seen in younger patients (Angstwurm et al. 2005).

Major changes in sex steroid production occur during menopause in women and a growing body of evidence suggests that changes in immune cell populations and functions occur at this time. Ovaries are functional in humans from the teenage years through to the 5th decade of life. During perimenopause, circulating estrogen concentrations fluctuate greatly from low (<120 pM) to high (2 μM), but do not become significantly reduced until close to the final menstrual period (Sherman and Korenman 1975; Metcalf 1988; Shideler et al. 1989).

Plasma estrogen concentrations are undetectable in postmenopausal women and progesterone levels are consistently less than 2 nM (Rannevik et al. 1995). Ovariectomy of mice, which models surgically-induced menopause, results in a decreased immune response and increased mortality following bacterial infection. Ovariectomized female mice show an increased mortality following cecal ligation and puncture-induced sepsis in comparison with age-matched intact females (Knoferl et al. 2002). Similarly, ovariectomized female mice exhibit twice the lung bacterial burden following *M. avium* infection as compared with age-matched intact females (Tsuyuguchi et al. 2001). Mortality associated with LPS challenge also is significantly elevated in ovariectomized female rats (Merkel et al. 2001). Importantly, these effects are reversed following exogenous estrogen replacement, confirming that the elevated bacterial burden and susceptibility to endotoxin administration in aged females are directly attributable to the loss of endogenous estrogen (Tsuyuguchi et al. 2001; Merkel et al. 2001).

In humans, LPS sensitivity and endotoxin-associated mortality increase with age, such that postmenopausal women display a higher incidence of sepsis than do premenopausal women (Meyers et al. 1989; Beery 2003). Peripheral monocytes isolated from postmenopausal women produce significantly higher levels of inflammatory cytokines following LPS challenge than do similarly stimulated cells from premenopausal women (Majetschak et al. 2000; Moxley et al. 2004). Majetschak and colleagues (2000) show that monocytes derived from postmenopausal women secrete proinflammatory cytokines at the same level as do monocytes from males. In contrast, LPS-stimulated monocytes from postmenopausal women can produce more TNF-α than similarly treated male-derived cells, which may reflect that half of the postmenopausal women monitored were receiving hormone replacement therapy (Moxley et al. 2004).

Normal mucosal flora vary following menopause, with the subsequent fall in estrogen levels. Reductions in circulating estrogen levels correlate with the loss of *Lactobacilli* species in the vagina, which causes an increase in pH and coliform microorganisms, promoting the growth of pathogenic bacteria and predisposing patients to infection (Gupta et al. 2006). Hormone replacement therapy reverses this effect and women not receiving hormone replacement therapy have greater incidences of *E. coli* and bacteroids (Gupta et al. 2006). Such a hormone-dependent change in mucosal flora has important implications for the development of chronic inflammatory diseases. Consequently, postmenopausal women receiving hormone replacement therapy demonstrate a reduced incidence of inflammatory bowel disease compared with women not receiving exogenous hormones (Kane and Reddy 2008).

Age and reproductive status can significantly influence the severity of bacterial infections and the incidence of sepsis. The lack of sex differences in sepsis susceptibility in prepubescent children and the increased sepsis susceptibility of postmenopausal females that is reversed following hormone replacement, strongly indicate that hormones underlie the observed sex differences in bacteremia and sepsis/septic shock.

5.3.2   Effects of Sex Steroid Hormones on Susceptibility to Bacterial Infection

Variations in sepsis susceptibility with changes in age and reproductive status imply a role for sex steroid hormones in the incidence and outcome of bacterial infections. There are, however, numerous studies indicating that sex hormones influence acute immune responses to bacteria challenge. In general, estrogen is considered to be "immunoprotective" during bacterial infections. This generalization fails to delineate whether estrogen acts as an immunoenhancer to combat bacterial infection or as an immunosuppressor that protects against the overactive and damaging immune response associated with sepsis. Testosterone, on the other hand, is widely accepted to be immunosuppressive and causes an increased susceptibility of both males and females to bacterial infection. For example, exogenous testosterone administration increases susceptibility of female mice to *M. marinum* infection, whereas castration of males (i.e., removal of endogenous testosterone) attenuates this infection (Yamamoto et al. 1991).

Gonadectomized male C57BL/6 mice exhibit elevated sensitivity to endotoxin challenge indicating that removal of testosterone may result in elevated immune responses (Rettew et al. 2008). While this observation would seem to contradict the finding that males are more susceptible to sepsis than females, several studies show that both male and female sepsis patients exhibit abnormally low circulating levels of testosterone (Christeff et al. 1992; Fourrier et al. 1994). Further, death of males due to septic shock is associated with extremely low levels of this androgen

(Christeff et al. 1988). To date, the mechanisms underlying these observations are not clear; LPS, however, inhibits testosterone synthesis (Reddy et al. 2006), which in turn could result in a more robust inflammatory immune response and hence septic shock.

Some animal studies, however, contradict the suppressive effects of testosterone on sepsis-induced mortality. In A/J mice, gonadectomized males are less susceptible to sepsis than intact males and this effect is reversed following the administration of dihydrotestosterone (DHT) (Torres et al. 2005). Furthermore, castrated neonatal mice are less susceptible to sepsis as adults than their sham-operated counterparts, suggesting that this outcome is due to prenatal androgen exposure (Bernhardt et al. 2007).

A small number of studies report no effect of estrogen receptor (ER) agonists on monocyte responses to intracellular Gram-positive bacterial pathogens, such as *Listeria monocytogenes* (Cristofaro et al. 2006; Opal et al. 2005). There is, however, an overwhelming body of evidence indicating that estrogens, in particular 17β-estradiol, are protective against bacterial infection and sepsis susceptibility. For example, estrogens increase resistance to streptococcal infections (Nicol et al. 1964). 17β-estradiol administration increases survival and decreases the oxidative stress along the rat gastrointestinal tract following intraperitoneal LPS challenge (Sener et al. 2005). Similarly, rats treated with intramuscular injections of 17β-estradiol and progesterone have lower endotoxemia and exhibit less liver and lung congestion following cecal ligation and puncture-induced sepsis, regardless of sex. Conversely, females receiving 17β-estradiol and progesterone have less liver inflammation and less lung edema than similarly treated males (Erikoglu et al. 2005). Consistent with these observations, female rats treated with the ERα ligand WAY204,688 or the ERβ ligand WAY202,196 have lower circulating endotoxin levels and higher survival rates following *Pseudomonas aeruginosa* infection than do untreated animals (Cristofaro et al. 2006; Opal et al. 2005). Furthermore, treatment of male and female mice with either ERα or ERβ agonists increases survival following cecal ligation and puncture. Conversely, treatment with an ERβ agonist, but not an ERα agonist, decreases bacteremia in these animals (Cristofaro et al. 2006; Opal et al. 2005).

In addition to experiments employing the administration of exogenous estrogen or ER agonists, numerous studies reveal that endogenous estrogen affects bacterial infection outcome and sepsis. Ovariectomy markedly increases the severity of *M. avium* infection and this effect is reversed following estrogen replacement (Tsuyuguchi et al. 2001). Similarly, ovariectomy increases mortality following cecal ligation and puncture-induced sepsis in female mice (Knoferl et al. 2002). Ovariectomy also exacerbates *H. pylori* infection, such that the severity of infection becomes comparable to that seen in males and this effect is reversed following exogenous estrogen treatment of ovariectomized females (Ohtani et al. 2007).

The removal of endogenous estrogens following ovariectomy also increases mortality associated with LPS challenge in rats and this effect is absent in ovariectomized females that receive exogenous estrogen (Merkel et al. 2001). Androgenized females have significantly higher mortality rate following LPS challenge than sham-treated females (Merkel et al. 2001). Finally, *E. coli* infections in dogs are most prevalent during diestrus, when estrogens are at their lowest levels (Sugiura et al. 2004).

Generally, estrogens are protective against bacterial infection and septic shock. High estrogen levels, however, such as those that might be seen in late proestrus and early estrus, can exacerbate such infections. For example, 17β-estradiol-treated female mice exhibit greater bacteremia and death following gonococcal infection than untreated animals (Kita et al. 1985). In addition, ovariectomized mice receiving high dose exogenous estrogen replacement demonstrate a higher incidence of *E. coli* infection in the kidney due to urinary tract infection than do untreated mice (Curran et al. 2007). Furthermore, acute injection of the estrogen estriol increases LPS-associated mortality in rats (Ikejima et al. 1998).

Finally, and perhaps more importantly, the increased incidence and severity of sepsis in humans correlate with circulating estrone and 17β-estradiol levels. Female patients admitted to an intensive care unit for sepsis have 10–20 times higher levels of estrogens than do nonsepsis patients; whereas male sepsis patients exhibit 3–5 times higher estrogen levels (Fourrier et al. 1994). These results are in agreement with another recent study demonstrating that 17β-estradiol levels are significantly higher in nonsurviving than surviving sepsis patients, regardless of sex (Dosset et al. 2008). Furthermore, the probability of death due to septic shock is lowest when circulating 17β-estradiol levels are within the physiological range (0.01–0.37 ng/ml in females and 0.02–0.06 ng/ml in males), but increases sharply when levels are beyond this range (May et al. 2008). Although all male sepsis patients have elevated estrone and 17β-estradiol levels in the first 2 days following admission to an intensive care unit, men who subsequently succumb to septic shock have maintained and even elevated estrogen levels when compared with sepsis survivors (Christeff et al. 1992). The correlation between circulating estrone levels and sepsis severity suggests that the levels of this hormone could predict the outcome for sepsis patients. Increased estrone and 17β-estradiol levels associated with sepsis do not coincide with increased LH or FSH levels, indicating that bacterial infection increases synthesis of estrogens in the gonads (Fourrier et al. 1994).

Taken together, these findings indicate that while testosterone tends to increase susceptibility to bacterial infection, estrogen generally provides protection. However, super-physiological levels of estrogen and progesterone may exert a contrary effect and negatively influence the outcome following sepsis, which is summarized in Fig. 5.2.

5.3.3 Effects of Sex Steroid Hormones on Immune Cell Function

Sex differences and the effects of sex steroids on innate immune responses to bacterial infections are widely reported. As discussed in Chap. 3, receptors for sex steroid hormones are present in a variety of leukocytes. Given that testosterone

Fig. 5.2 Summary of the effects of gonadectomy and subsequent hormone replacement on sepsis susceptibility. Black bars indicate level of susceptibility to endotoxin exposure in intact males and females (Basal), whereas dark bars indicate susceptibility following gonadectomy of males (GDX) or ovariectomy of females (OVX). Arrows indicate changes in susceptibility following exposure to testosterone (T) or low or high doses of estradiol (E2^{lo} and E2^{hi}, respectively)

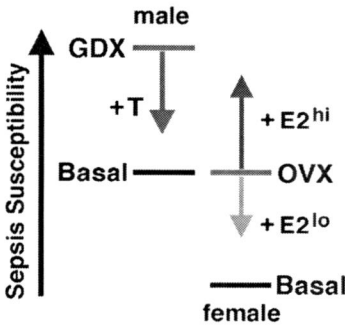

suppresses immune responses to bacterial challenge and the overproduction of cytokines underlies the lethal nature of sepsis, androgens might directly alter the secretion and production of these inflammatory mediators through androgen receptor (AR)-mediated mechanisms. Consistent with this hypothesis, androgens suppress the responses of immune cells to endotoxin challenge. The removal of endogenous testosterone following castration increases LPS-induced IL-1β and IL-6 production by mouse peritoneal and splenic macrophages (Wichmann et al. 1997). The in vitro treatment of murine peritoneal macrophages with testosterone decreases the secretion of IL-1β following LPS challenge (Savita and Rai 1998) and causes a modest reduction in TNF-α secretion by such cells isolated from male but not female mice (Chao et al. 1995). Furthermore, we have recently demonstrated that acute administration of testosterone in vitro reduces the production of TNF-α by RAW 264.7 macrophage-like cells and primary macrophages (Rettew et al. 2008). The effect of testosterone is apparent in cells isolated from gonadectomized males, but not in cells derived from sham-operated animals or gonadectomized males that receive in vivo testosterone replacement (Rettew et al. 2008). Androgen-mediated suppression of endotoxin-induced inflammatory cytokine production is not limited to macrophages as in vitro DHT treatment reduces IL-6 expression in LPS-stimulated umbilical cord endothelial cells via the AR (Norata et al. 2006).

In addition to inflammatory cytokines, testosterone attenuates the production of other immune molecules associated with bacteria clearance. Notably, macrophages isolated from castrated male mice produce more nitric oxide (NO) upon LPS stimulation indicating that endogenous testosterone may suppress NO release by these cells. Consistent with this conclusion, the in vitro treatment of either primary macrophages or RAW 264.7 macrophage-like cells with testosterone decreases the expression of inducible nitric oxide synthase and NO release (Friedl et al. 2000; Savita and Rai 1998). The immunosuppressive effects of testosterone are not due to the peripheral aromatization and conversion of testosterone to 17β-estradiol, because DHT, a nonaromatizable androgen, also has suppressive effects on immune responses to LPS; further, the actions of testosterone are not affected by the exposure of cells to ER antagonists (Norata et al. 2006).

The immunosuppressive actions of testosterone are specific to certain bacterial challenges and may be sex-restricted. For example, testosterone reportedly does not depress immune function in healthy animals but can suppress immune responses following insults, such as trauma-hemorrhage (Angele et al. 1999). Further, gonadally intact female mice treated with DHT have higher TNF-α responses to LPS than do untreated females or even gonadally intact male mice (Card et al. 2006). While the reasons for these discrepancies are presently unclear, the suppressive effects of testosterone on some antiinflammatory responses may be involved. Gonadectomy increases sera IL-10 levels after LPS challenge compared with levels in intact males (Torres et al. 2005) and DHT attenuates IL-10 production by LPS-stimulated macrophages (Angele et al. 1999).

The consensus of current studies is that testosterone is immunosuppressive and generally decreases proinflammatory cytokine production by immune cells; conversely, estrogen may increase the production of these mediators by cells, such as macrophages, thereby decreasing susceptibility to bacterial infection. The expression of proinflammatory cytokines, including *Tnfα*, within the brain following intracerebral injection of LPS is attenuated in ovariectomized animals: an effect that is reversed following exogenous estrogen administration (Soucy et al. 2005). In addition, an intermediate dose of 17β-estradiol (0.1 ng/mL) increases TNF-α secretion by rat peritoneal macrophages stimulated with LPS (Chao et al. 1995) and acute, in vivo estriol treatment increases sera TNF-α levels and Kupffer cell TNF-α mRNA expression in rats challenged with LPS (Ikejima et al. 1998). Exposure to 17β-estradiol or estriol increases sera TNF-α levels in female mice following in vivo LPS challenge: an effect that is antagonized by the ER antagonist tamoxifen (Zuckerman et al. 1995, 1996). Furthermore, treatment of isolated human peripheral monocytes with 17β-estradiol increases TNF-α production from male-derived cells (Asai et al. 2001). Peritoneal macrophages isolated after acute, in vivo LPS treatment demonstrate a modest increase in the expression of TNF-α mRNA after in vivo estriol exposure (Zuckerman et al. 1996).

In addition to TNF-α, estrogen can augment the production of other inflammatory cytokines, including IL-1β and IL-6. Treatment of human peripheral monocytes with 17β-estradiol increases IL-6 production from male-derived cells both before and after LPS stimulation (Asai et al. 2001). Although there is no difference in maximal IL-6 production in female mice following estrogen treatment, those treated with estriol produce IL-6 with faster kinetics following LPS challenge than that seen in untreated females, and this effect is inhibited by tamoxifen, which can antagonize ERs (Zuckerman et al. 1996). Furthermore, the in vitro treatment of mouse peritoneal macrophages with 17β-estradiol increases LPS-stimulated IL-1β production (Savita and Rai 1998). The in vivo treatment of rodents with 17β-estradiol increases sera NO levels (Ikejima et al. 1998) and splenocyte production of IL-1α and IL-1β following in vitro restimulation (Dai et al. 2007). Lastly, estrogen may promote immune responses by limiting the production of the antiinflam-matory cytokine IL-10, as human peripheral monocytes isolated from males produce lower amounts of IL-10 following in vitro exposure to 17β-estradiol (Asai et al. 2001).

Having made the case for estrogens enhancing the production of inflammatory mediators, it might be surprising to learn that a preponderance of evidence contra-dicts this hypothesis. For example, lower levels of IL-6 and TNF-α have been observed in the peritoneal fluid following cecal ligation and puncture in mice treated with the ER agonist WAY202,196 (Cristofaro et al. 2006). Acute intraperi-toneal treatment with 17β-estradiol attenuates LPS-induced elevations in sera TNF-α levels in rats (Sener et al. 2005). Both low ($<10^{-5}$ ng/mL) and high (>1 ng/mL) doses of 17β-estradiol decrease the secretion of TNF-α by LPS-treated rat peritoneal macrophages (Chao et al. 1995). 17β-estradiol decreases LPS-induced *Tnfα* expression by murine macrophages (Salem et al. 2000) and human periph-eral monocytes (Vlotides et al. 2007; Asai et al. 2001). This effect is not limited to monocytes and macrophages, as cells in gastric tissue infected by *H. pylori* produce more TNF-α following ovariectomy: an effect that is reversed by 17β-estradiol replacement (Ohtani et al. 2007). Furthermore, in vitro 17β-estradiol treatment of astrocytes decreases LPS-induced *Tnfα* expression (Kipp et al. 2007; Lewis et al. 2008), which is inhibited by exposure to the ER antagonist ICI 182,780 (Kipp et al. 2007).

Estrogen decreases the production of IL-6 following bacterial challenge or exposure to endotoxin. Acute in vivo exposure to 17β-estradiol decreases IL-6 levels following endotoxin challenge (Zuckerman et al. 1995) and macrophages from 17β-estradiol-treated mice express lower levels of *Il6* mRNA than do macro-phages from untreated mice, following in vitro restimulation with LPS (Zuckerman et al. 1995). In addition to immune sentinel cells, other cell types capable of responding to LPS have been shown to produce lower levels of IL-6 following estrogen treatment. For example, human retinal pigment epithelial cells have lower LPS-induced IL-6 responses following exposure to 17β-estradiol, which is reversed following treatment with the ER antagonist ICI 182,780 (Paimela et al. 2007).

In addition to TNF-α and IL-6, estrogens decrease the production of bacterial- or endotoxin-induced production of other inflammatory mediators, including IL-1β, inflammatory chemokines, and NO. Resident CNS cells that exert sentinel immune functions, including microglia and astrocytes, exhibit diminished LPS-induced IL-1β production following exposure to either 17β-estradiol or ER agonists (Lewis et al. 2008). Similarly, gastric tissue from ovariectomized females infected with *H. pylori* contains higher levels of IL-1β than do tissues from infected intact females, which is reversed following 17β-estradiol replacement (Ohtani et al. 2007). CXCL8 (IL-8) secretion attracts immune cells to the site of bacterial infection and excess infiltration of leukocytes, including neutrophils, can contribute to the pathogenesis of sepsis. Treatment of peripheral monocytes isolated from women with 17β-estradiol results in a dose-dependent decrease in LPS-induced expression and secretion of CXCL8 (Pioli et al. 2007). 17β-estradiol also decreases the expression of *Cxcl2* (MIP-2) in LPS-treated macrophages (Ghisletti et al. 2005). This effect is driven by the effects of estrogen on NF-κB function, as 17β-estradiol decreases p65 binding to the *Cxcl2* promoter (Ghisletti et al. 2005). Finally, ovariectomy elevates the production of NO by macrophages upon stimulation with LPS, indicating that endogenous estrogen may decrease NO secretion. Consequently, the in vitro treatment of macrophages (Savita and Rai 1998) and microglia (Vegato et al. 2000) with 17β-estradiol diminishes LPS-induced NO release: an effect reversed following inhibition of ER activity (Savita and Rai 1998).

The hypothesis that estrogen limits inflammatory responses is further supported by the enhanced production of the antiinflammatory cytokine IL-10 following exposure to 17β-estradiol. For example, 17β-estradiol treatment of gonadectomized male mice increases IL-10 secretion from LPS-stimulated Kupffer cells (Angele et al. 1999). Furthermore, exogenous 17β-estradiol treatment causes a significant increase in IL-10 production in ovariectomized female mice infected with *H. pylori* (Ohtani et al. 2007). Further, long-term in vivo treatment of male mice with 17β-estradiol increases splenocyte IL-10 production following in vitro concanavalin A stimulation (Dai et al. 2007).

Taken together, testosterone limits the production of key inflammatory mediators consistent with an immunosuppressive role for this hormone. In contrast, the effects of estrogen appear to be far more controversial. While data are available to support that estrogens enhance immune responses, a large body of data also illustrate a suppressive action of estrogens on inflammatory cytokine and chemokine release by a number of host immune cell types. Although the causes for these disparate results are unclear, we hypothesize that the bipotential effects of estrogen are dependent on the particular challenges faced by the host, the kinetics of the immune response, and the effective dose of estrogen. For example, while intermediate doses of estrogen enhance cytokine release, very low or very high doses can have an opposite effect. We must consider also the fine balance between immunosuppressive versus immunoprotective actions of estrogen. Thus, in order to protect an individual during sepsis, decreased production of potent proinflammatory cytokines, such as TNF-α, IL-6, and IL-1, may be desirable. This is an important area for further research.

5.3.4 Effects of Sex Hormones on the Induction of Pattern Recognition Receptors and Signal Transduction Pathways

As discussed in the preceding sections of this chapter, acute immune responses to bacterial pathogens are initiated via the recognition of conserved microbial motifs, including LPS and bacterial lipoproteins, by TLR4 and TLR2, respectively. In addition, recognition of LPS is facilitated by coreceptors such as CD14 and molecules, including LBP. Recognition of bacterial components by these receptors initiates MAP kinase cascades and NF-κB activation via adaptor molecules, including MyD88. The activation of the transcriptional factor NF-κB affects transcription of pro- and antiinflammatory cytokines precipitating the "cytokine storm" that underlies bacterial septic shock (Miyake et al. 2004; Palsson-McDermott and O'Neill 2004).

In accordance with the inhibitory effects of androgens on cytokine production, testosterone attenuates LPS-induced activation of p38 MAP kinase, but not ERK1/2 or JNK/SAPK activation in macrophages (Benten et al. 2004). Alternatively, DHT interferes with the ability of NF-κB to bind to DNA promoters in umbilical cord endothelial cells (Norata et al. 2006). Estrogen also can inhibit NF-κB activity. For example, treatment with the ER agonist WAY204,688 decreases cellular NF-κB reporter activity (Opal et al. 2005) and acute 17β-estradiol treatment decreases DNA binding of NF-κB (Paimela et al. 2007; Ghisletti et al. 2005). 17β-estradiol appears to regulate NF-κB activity at the level of nuclear transloca-tion. 17β-estradiol retains p65/Rel-A and c-Rel in the cytoplasm of immune cells following stimulation (Ghisletti et al. 2005; Dai et al. 2007), regardless of IκBα synthesis, phosphorylation, or degradation (Ghisletti et al. 2005), and appears to target the microtubule-associated transport system used by NF-κB subunits (Ghisletti et al. 2005). 17β-estradiol does not inhibit the nuclear translocation of the p50 subunit, but increases the expression of Bcl-3, a protein that binds to p50 homodimers to permit transcription of NF-κB responsive genes (Dai et al. 2007). As such, this may be a mechanism underlying the seemingly paradoxical ability of estrogen to both inhibit and promote inflammatory gene expression.

Although sex steroid hormones may have specific effects on the function of signal transduction components, it is clear that endocrine-induced changes in the level of expression of PRRs, their coreceptors, or adaptor molecules may have a profound influence on downstream cellular events, including NF-κB activation and cytokine production. Sex steroid hormones can directly alter the expression of innate receptors for bacterial components, as summarized in Fig. 5.1. Exposure to testosterone in vitro decreases cell surface levels of TLR4 on RAW 264.7 macro-phage-like cells and primary murine peritoneal macrophages in a time and dose dependant manner (Rettew et al. 2008). Furthermore, gonadectomy increases cell surface TLR4 expression on isolated monocytes: an effect that is reversed follow-ing in vivo testosterone replacement (Rettew et al. 2008). Similarly, gonadectomy increases the induction of TLR4 following *E. coli* infection of rat prostate cells compared with cells from intact males or gonadectomized males with testosterone replacement (Quintar et al. 2006).

Although the effects of testosterone on PRR activity are consistent, the effects of estrogens on PRR expression are more complicated as acute administration of estrogen fails to alter such expression in vitro. For example, acute estrogen treat-ment of macrophage-like cell lines (Vegato et al. 2004; Vlotides et al. 2007), primary murine macrophages (Rettew et al. 2009); Kupffer cells (Ikejima et al. 1998), or LPS-challenged human monocytes (Pioli et al. 2007) fails to significantly alter TLR4 or CD14 expression. In contrast, estrogen significantly affects PRR expression in vivo. The in vivo administration of estriol elevates the expression of mRNA encoding CD14 and LBP in Kupffer cells (Ikejima et al. 1998) and ovariectomy results in a markedly lower *Tlr2* transcription in the CNS in response to LPS compared with that seen in intact females or ovariectomized females treated with exogenous estrogen (Soucy et al. 2005). Furthermore, our recent studies indicate that removal of endogenous estrogens following ovariectomy results in a modest but significant decrease in TLR4 cell surface expression and circulating LBP levels. Interestingly, 17β-estradiol replacement in ovariectomized animals increases the levels of TLR4 and LBP to levels above those seen in intact females (Rettew et al. 2009). Finally, expression of MyD88 is increased following an in vivo treatment of male mice with 17β-estradiol (Zheng et al. 2006). While hormone replacement therapy does not alter the levels of TLR4 expression in circulating monocytes isolated from elderly post-menopausal women undergoing exercise training (Flynn et al. 2003), the results of this study are difficult to interpret due to the independent effects of aging and exercise.

5.4   Summary and Conclusions

Sex-based differences in innate immune responses to bacterial infection are evident in human patients and rodent models of disease. Females are less susceptible to the development of bacterial infections and subsequent bacteremia and sepsis, whereas males exhibit a greater incidence of such infections and are more likely to develop fatal sequellae. Females are protected from septic shock in two ways: (1) by producing lower levels of the proinflammatory cytokines responsible for the lethal nature of septic shock; (2) by elevating the production of antiinflammatory molecules. Although evidence is accumulating that sex differences exist at the level of expression of PRRs for bacterial motifs and downstream signaling pathways, the mechanism of action is controversial as cells from males express higher levels of TLR4 and its coreceptors, whereas LPS-induced signaling pathways in female cells appear to be more efficient at evoking cellular responses than are those in male cells.

The observed differences in susceptibility to bacteria and endotoxin between the sexes may be caused by the actions of sex steroids. Both age and reproductive status significantly influence the severity of bacterial infections and the incidence of sepsis. Testosterone also increases susceptibility to infection in rodent models, whereas estrogen generally reduces susceptibility. Androgens, however, limit the production of key inflammatory mediators by immune cells following bacterial endotoxin exposure. Conversely, the role played by estrogens is debated, with evidence for both enhancement and suppression of immune responses. A possible mechanism for the effects of sex hormones on immune responses to bacterial pathogens may be that testosterone and estrogen exert direct effects on the expression of PRRs, including TLR4 and its coreceptors, and can modulate the signal transduction pathways employed by these receptors to initiate cytokine expression. Consequently, sex hormones affect proinflammatory cytokine expression.

Although sex steroid hormones can elicit demonstrable changes in the expression and function of microbial PRRs and can significantly alter the production of soluble immune mediators responsible for lethal septic shock, linking these seemingly paradoxical effects to the sex differences in susceptibility to bacterial infection and sepsis severity remains challenging. While the precise mechanisms remain to be determined, we propose a scenario, summarized in Fig. 5.3, in which male

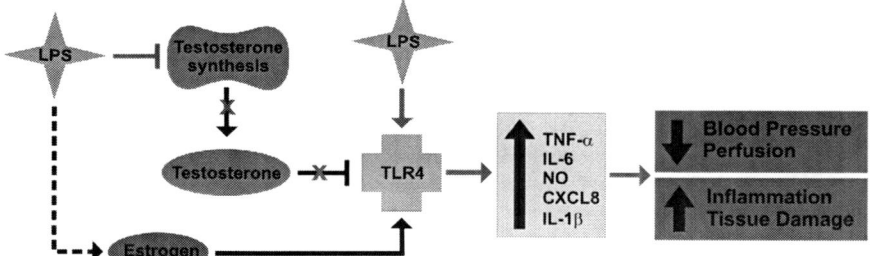

Fig. 5.3 Putative mechanism underlying elevated susceptibility to bacterial septic shock in males compared with females. LPS inhibits testosterone synthesis thereby removing the inhibitory effect of testosterone on TLR4 expression on sentinel cells. Elevation of estrogen levels (perhaps via a direct action of LPS) increases TLR4 expression. LPS is recognized via TLR4 on these sensitized cells resulting in a massive release of inflammatory mediators precipitating septic shock

sepsis patients exhibit reduced levels of immunosuppressive testosterone and highly elevated levels of estrogen that, at least in males, leads to an excessive and damaging systemic immune response. Together, reduced testosterone and elevated estrogen in males would facilitate the expression of TLR4, CD14, and LBP rendering sentinel immune cells more sensitive to bacterial LPS and leading to markedly elevated levels of soluble inflammatory mediators thereby precipitating septic shock. Further research is required to validate such a hypothesis and to resolve the paradoxical effects of physiological levels of estrogens in females.

References

Angele MK, Knoferl MW, Schwacha MG, Ayala A, Cioffi WG, Bland KI, Chaudry IH (1999) Sex steroid regulate pro- and anti-inflammatory cytokine release by macrophages after trauma-hemorrhage. Cell Physiol 46:C35–C42

Angele MK, Nitsch S, Knoferl MW, Ayala A, Angele P, Schildberg FW, Jauch KW, Chaudry IH (2003) Sex-specific p38 MAP kinase activation following trauma-hemorrhage: involvement of testosterone and estradiol. Am J Physiol Endocrinol Metab 285:E189–E196

Angstwurm MWA, Gaertner R, Schopohl J (2005) Outcome in elderly patients with severe infection is influenced by sex hormones but not gender. Crit Care Med 33:2786–2791

Asai K, Hiki N, Mimura Y, Ogawa T, Unou K, Kaminishi M (2001) Gender differences in cytokine secretion by human peripheral blood mononuclear cells: role of estrogen in modulating LPS-induced cytokine secretion in an ex vivo septic model. Shock 16:340–343

Barton GM, Medzhitov R (2003) Toll-like receptor signaling pathways. Science 300:1524–1525

Beery TA (2003) Sex differences in infection and sepsis. Crit Care Nurs Clin North Am 15:55–62

Benten WPM, Vuo Z, Krucken J, Wunderlich F (2004) Rapid effects of androgens in macrophages. Steriods 69:585–590

Bernhardt JA, d'Acampora AJ, Tramonte R, Serafilm JDM (2007) Effect of post-natal castration on sepsis mortality in rats. Acta Cir Bras 22:22–29

Bindl L, Burderus S, Dahlem P, Demirakca S, Goldner M, Huth R, Kohl M, Krause M, Kuhl P, LaschP LK, Merz U, Moeller J, Mohamad Y, Peters M, Porz W, Vierzig A, Ruchard J, Scharf J, Varnholt V (2003) Gender-based differences in children with sepsis and ARDS: the ESPNIC ARDS database group. Intens Care Med 29:1770–1773

Blackwell TS, Christman JW (1996) Sepsis and cytokines: current status. Br J Anaesth 77:
110–117

Card JW, Carey MA, Bradbury A, DeGraff LM, Morgan DL, Moorman MP, Flake GP, Zeldin DC
(2006) Gender differences in murine airway responsiveness and lipopolysaccharide-induced
inflammation. J Immunol 177:621–630

Chao TC, Van Alten PJ, Greager JA, Walter RJ (1995) Steroid sex hormones regulate the release
of tumor necrosis factor by macrophages. Cell Immunol 160:43–49

Chow JC, Young DW, Golenbock DT, Christ WJ, Gusovsky F (1999) Toll-like receptor 4
mediates lipopolysaccharide-induced signal transduction. J Biol Chem 274:10689–10692

Christeff N, Benassayag C, Carli-Vielle C, Carli A, Nunez EA (1988) Elevated Oestrogen and
reduced testosterone levels in the serum of male septic shock patients. J Steroid Biochem
29:435–440

Christeff N, Carli A, Benassayag C, Bleichner G, Vaxelaire J, Nunez EA (1992) Relationship
between changes in serum estrone levels and outcome in human males with septic shock. Circ
Shock 36:249–255

Cristofaro PA, Opal SM, Palardy JE, Parejo NA, Jhung J, Keith JC Jr, Harris HA (2006) WAY-
202196, a selective estrogen receptor-beta agonist, protects against death in experimental
septic shock. Crit Care Med 34:2188–2193

Curran EM, Hart-Van Tassell A, Judy BM, Nowicki B, Montgomery-Rice V, Estes DM, Nowicki
S (2007) Estrogen increases menopausal host susceptibility to experimental ascending urinary-
tract infection. J Infect Dis 195:680–683

Dai R, Phillips RA, Ahmed SA (2007) Despite inhibition of nuclear localization of NF-κB p65,
c-Rel, and RelB, 17-β estradiol up-regulates NF-κB signaling in mouse splenocytes: the
potential role of Bcl-3. J Immunol 179:1776–1783

Diamond G, Legarda D, Ryan LK (2000) The innate immune response of the respiratory epithelium. Immunol Rev 173:27–38

Dinkel RH, Lebok U (1994) A survey of nosocomial infections and their influence on hospital
mortality rates. J Hosp Infect 28:297–304

Dossett LA, Swenson BR, Evans HL, Bonatti H, Sawyer RG, May AK (2008) Serum estradiol
concentration as a predictor of death in critically ill and injured adults. Surgical Infections
9:41–48

Du JT, Vennos E, Ramey E, Ramwell PW (1984) Sex differences in arachidonate cyclo-oxygenase
products in elicited rat peritoneal macrophages. Biochim Biophys Acta 794:256–260

Eisenmenger SJ, Wichmann MW, Angele P, Faist E, Hatz R, Chaudry IH, Jauch KW, Angele MK
(2004) Differences in the expression of LPS-receptors are not responsible for the sex-specific
immune response after trauma and hemorrhagic shock. Cell Immunol 230:17–22

Erikoglu M, Sahin M, Ozer S, Avunduk MC (2005) Effect of gender on the severity of sepsis. Surg
Today 35:467–472

Flynn MG, McFarlin BK, Phillips MD, Steward LK, Timmerman KL (2003) Toll-like receptor 4
and CD14 mRNA expression are lower in resistive exercised-trained elderly women. J Appl
Physiol 95:1833–1842

Fourrier F, Jallot A, Leclerc L, Jourdain M, Racadot A, Chagnon JL, Rime A, Chopin C (1994)
Sex steroid hormones in circulatory shock, sepsis syndrome, and septic shock. Circ Shock
43:171–178

Friedl R, Brunner M, Moeslinger T, Spieckermann PG (2000) Testosterone inhibits expression of
inducible nitric oxide synthease in murine macrophages. Life Sci 68:417–429

Ghisletti S, Meda C, Maggi A, Vegato E (2005) 17β-estradiol inhibits inflammatory gene
expression by controlling NF-κB intracellular localization. Mol Cell Biol 25:2957–2968

Gregory MS, Duffner LA, Hahn EL, Tai HH, Faunce DE, Kovacs EJ (2000) Differential production of prostaglandin E(2) in male and female mice subjected to thermal injury contributes to
the gender difference in immune function: possible role for 15-hydroxyprostaglandic dehydrogenase. Cell Immunol 205:94–102

Gupta S, Kuman N, Singhal N, Kaur R, Manektala U (2006) Vaginal microflora in postmenopausal women on hormone replacement therapy. Indian J Pathol Microbiol 49:457–461

Hornef MW, Frisan T, Vandewalle A, Normark S, Richter-Dahlfors A (2002) Toll-like receptor 4 resides in the Golgi apparatus and colocalizes with internalized lipopolysaccharide in intestinal epithelial cells. J Exp Med 195:559–570

Ikejima K, Enomoto N, Iimuro Y, Ikejima A, Fang D, Xu J, Forman DT, Brenner DA, Thurman RG (1998) Estrogen increases sensitivity of hepatic Kupffer cells to endotoxin. Gastrointest Liver Physiol 37:G669–G676

Kahlke V, Angele MK, Ayala A, Schwacha MG, Cioffi WG, Bland KI, Chaudry IH (2000) Immune disfunction following trauma-haemorrhage: influence of gender and age. Cytokine 12:69–77

Kahlke V, Dohm C, Mees T, Brotzmann K, Schreiber S, Schroder J (2002) Early interleukin-10 treatment improves survival and enhances immune function only in males after hemorrhage and subsequent sepsis. Shock 18:24–28

Kane SV, Reddy D (2008) Hormonal replacement therapy after menopause is protective of disease activity in women with inflammatory bowel disease. Am J Gastroenterol 103:1193–1196

Kipp M, Karakaya S, Johann S, Kampmann E, Mey J, Beyer C (2007) Oestrogen and progesterone reduce lipopolysaccharide-induced expression of tumour necrosis factor-α and interleukin-18 in midbrain astrocytes. J Neuroendocrinol 19:819–822

Kita E, Takahashi S, Yasui K, Kashiba S (1985) Effect of estrogen (17β-estradiol) on the susceptibility of mice to disseminated gonococcal infection. Infect Immun 49:238–243

Knoferl MW, Angele MK, Diodato MD, Schwacha MG, Ayala A, Cioffi WG, Bland KI, Chaudry IH (2002) Female sex hormones regulate macrophage function after trauma-hemorrhage and prevent increased death rate from subsequent sepsis. Ann Surg 235:105–112

Kono H, Wheeler MD, Rusyn I, Lin M, Seabra V, Rivera CA, Bradford BU, Forman DT, Thurman RG (2000) Gender differences in early alcohol-induced liver injury: role of CD14, NF-kappaB, and TNF-alpha. Am J Physiol Gastrointest Liver Physiol 278:G652–G661

Kovacs EJ, Messinham KAN, Gregory MS (2002) Estrogen regulation of immune responses after injury. Mol Cell Endocrinol 193:129–135

Kumar MV, Nagineni CN, Chin MS, Hooks JJ, Detrick B (2004) Innate immunity in the retina: Toll-like receptor (TLR) signaling in human retinal pigment epithelial cells. J Neuroimmunology 153:7–15

Laubach VE, Foley PL, Shockey KS, Tribble CG, Kron IL (1998) Protective roles of nitric oxide and testosterone in endotoxemia: evidence from NOS-2-deficient mice. Am J Physiol 275:H2211–H2218

Leslie CA, Gonnerman WA, Cathcard TE (1987) Gender differences in eicosanoid production from macrophages of arthritis-susceptible mice. J Immunol 138:413–416

Lewis DK, Johnson AB, Stohlgren S, Harms A, Sohrabji F (2008) Effects of estrogen receptor agonists on regulation of the inflammatory response in astrocytes from young adult and middle-aged female rats. J Neuroimmunol 195:47–59

Lien E, Means TK, Heine H, Yoshimura A, Kusumoto S, Fukase K, Fenton MJ, Oikawa M, Qureshi N, Monks B, Finberg RW, Inagalls RR, Golenbock DT (2000) Toll-like receptor 4 imparts ligand-specific recognition of bacterial lipopolysaccharide. J Clin Invest 105:497–504

Majetschak M, Christensen B, Obertacke U, Waydhas C, Schindler AE, Nast-Kolb D, Schade FU (2000) Sex differences in posttraumatic cytokine release of endotoxin-stimulated whole blood: relationship to the development of severe sepsis. J Trauma 48:832–840

Marriott I, Bost KL, Huet-Hudson YM (2006) Sexual dimorphism in expression of receptors for bacterial lipopolysaccharides in murine macrophages: a possible mechanism for gender-based differences in endotoxic shock susceptibility. J Reprod Immunol 71:12–27

Martel MJ (2002) Hemorrhagic shock. J Obstet Gynaecol Can 24:504–511

May AK, Dossett LA, Norris PR, Hansen EN, Dorsett RC, Popovsky KA, Sawyer RG (2008) Estradiol is associated with mortality in critically ill trauma and surgical patients. Crit Care Med 36:62–68

McGowan JE Jr, Barnes MW, Finland M (1975) Bacteremia at Boston City Hospital: Occurance and mortality during 12 selected years (1935–1972), with special reference to hospital-acquired cases. J Infect Dis 132:316–335

Medzhitov R, Janeway CA Jr (2000) The Toll receptor family and microbial recognition. Trends Microbiol 8:452–456

Medzhitov R, Preston-Hurlburt P, Janeway CA Jr (1997) A human homologue of the Drosophila Toll protein signals activation of adaptive immunity. Nature 388:394–397

Merkel SM, Alexander S, Zufall E, Oliver JD, Huet-Hudson YM (2001) Essential role for estrogen in protection against *Vibrio vulnificus*-induced endotoxic shock. Infect Immun 69:6119–6122

Metcalf MG (1988) The approach of menopause: a New Zeland study. N Z Med J 101:103–106

Meyers BR, Sherman E, Mendelson MH, Velasquez G, Srulevitch-Chin E, Hubbard M, Hirschman SZ (1989) Bloodstream infections in the elderly. Am J Med 86:379–384

Miyake K (2004) Innate recognition of lipopolysaccharide by Toll-like receptor 4-MD-2. Trends Microbiol 12:186–192

Moxley G, Posthuma D, Carlson P, Estrada E, Han J, Benson LL, Neale MC (2002) Sexual dimorphism in innate immunity. Arthritis Rheum 46:250–258

Moxley G, Stern AG, Carlson P, Estrada E, Han J, Benson LL (2004) Premenopausal sexual dimorphism in lipopolysaccharide-stimulated production and secretion of tumor necrosis factor. J Rheumatol 31:686–694

Nicol T, Bilbey DL, Charles LM, Cordingley JL, Vernon-Roberts B (1964) Oestrogen: the natural stimulant of body defense. J Endocrinol 30:277–291

Norata GD, Tibolla G, Seccomandi PM, Poletti A, Catapano AL (2006) Dihydrotestosterone decreases tumor necrosis factor-α and lipopolysaccharide-induced inflammatory response in human endothelial cells. J Clin Endocr Metab 91:546–554

Offner PJ, Moore EE, Biffl WL (1999) Male gender is a risk factor for major infections after surgery. Arch Surg 134:935–940

Ohtani M, Garcia A, Rogers AB, Ge Z, Taylor NS, Xu S, Watanabe K, Marini RP, Whary MT, Wang TC, Fox JC (2007) Protective role of 17β-estradiol against the development of *Helicobacter pylori*-induced gastrice cancer in INS-GAS mice. Carcinogenesis 28:2597–2604

Oliver JD (1989) Vibrio vulnificus. In: Doyle MP (ed) Foodborne Bacterial Pathogens. Marcel-Dekker, New York, pp 569–600

Opal SM, Palardy JE, Cristofaro P, Parejo N, Jhung JW, Keith JC Jr, Chippari S, Caggiano TJ, Steffan RJ, Chadwick CC, Harnish DC (2005) The activity of pathway-selective estrogen receptor ligands in experimental septic shock. Shock 24:535–540

Paimela T, Ryhanen T, Mannermaa E, Ojala J, Kalesnykas G, Salminen A, Kaarniranta K (2007) The effect if 17β-estradiol on IL-6 secretion and NF-κB DNA-binding activity in human retinal pigment epithelial cells. Immunol Lett 110:139–144

Palsson-McDermott EM, O'Neill LA (2004) Signal transduction by the lipopolysaccharide receptor, Toll-like receptor-4. Immunology 113:153–162

Pioli PA, Jensen AL, Weaver LK, Amiel E, Shen Z, Shen L, Wira CR, Guyre PM (2007) Estradiol attenuates lipopolysaccharide-induced CXC chemokine ligand 8 production by human peripheral blood monocytes. J Immunol 179:6284–6290

Poltorak A, He X, Smirnova I, Liu MY, Van Huffel C, Du X, Birdwell D, Alejos E, Silva M, Galanos C, Freudenberg M, Ricciardi-Castagnoli P, Layton B, Beutler B (1998) Defective LPS signaling in C3H/HeJ and C57BL/10ScCr mice: mutations in the TLR4 gene. Science 282:2085–2088

Quintar AA, Roth FD, De Paul AL, Aoki A, Maldonado CA (2006) Toll-like receptor 4 in rat prostate: modulation by testosterone and acute bacterial infection in epithelial and stromal cells. Biol Reprod 75:664–672

Rannevik G, Jeppsson S, Johnell O, Bjerre B, Laurell-Borulf Y, Svanberg L (1995) A longitudinal study of the perimenopausal transition: altered profiles of steroid and pituitary hormones, SHBG and bone mineral density. Maturitas 21:103–113

Reddy MM, Mahipal SVK, Subhashini J, Reddy MC, Roy KR, Reddy GV, Reddy PRK, Reddanna P (2006) Bacterial lipopolysaccharide-induced oxidative stress in the impairment of steroidogenesis and spermatogenesis in rats. Reprod Toxicol 22:493–500

Rettew JA, Huet-Hudson YM, Marriott I (2008) Testosterone reduces macrophage expression in the mouse of toll-like receptor 4, a trigger for inflammation and innate immunity. Biol Reprod 78:432–437

Rettew JA, Huet-Hudson YM, Marriott I (2009) Estrogen augments cell surface TLR4 expression on murine macrophages and regulates sepsis susceptibility in vivo. Endocrinology 150:3877–3884

Rock FL, Hardiman G, Timans JC, Kastelein RA, Bazan JF (1998) A family of human receptors structurally related to Drosophila Toll. Proc. Natl, Acad. Sci. USA 95:588–593

Salem ML, Hossain MS, Nomoto K (2000) Mediation of the immunomodulatory effect of β-estradiol on inflammatory reponses by inhibition of recruitment and activation of inflammatory cells and their gene expression of TNF-α and IFN-γ. Int Arch Allergy Immunol 121:235–245

Sauaia A, Moore FA, Moore EE, Moser KS, Brennan R, Read RA, Pons PT (1995) Epidemiology of trauma deaths: a reassessment. J Trauma 38:185–193

Savita, Rai U (1998) Sex steroid hormones modulate the activation of murine peritoneal macrophages: receptor mediated modulation. Comp Biochem Physiol 119C:199–204

Schroder J, Kahlke V, Staubach KH, Zabel P, Stuber F (1998) Gender differences in human sepsis. Arch Surg 133:1200–1205

Schumann RR, Leong SR, Flaggs GW, Gray PW, Wright SD, Mathison JC, Tobias PS, Ulevitch RJ (1990) Structure and function of lipopolysaccharide binding protein. Science 249: 1429–1431

Sener G, Arbak S, Kurtaran P, Gedik N, Yegen BC (2005) Estrogen protects the liver and intestines against sepsis-induced injury in rats. J Surg Res 128:70–78

Sherman BM, Korenman SG (1975) Hormonal characteristics of the human menstrual cycle throughout reproductive life. J Clin Invest 55:699–706

Shideler SE, DeVane GW, Kalra PS, Benirschke K, Lasley BL (1989) Ovarian-pituitary hormone interactions during the perimenopause. Maturitas 11:311–339

Soucy G, Boivin G, Labrie F, Rivest S (2005) Estradiol is required for a proper immune response to bacterial and viral pathogens in the female brain. J Immunol 174:6391–6398

Spitzer JA, Zhang P (1996) Gender differences in neutrophil function and cytokine-induced neutrophil chemoattractant generation in endotoxic rats. Inflammation 20:485–498

Sugiura K, Nishikawa M, Ishiguro K, Tajima T, Inaba M, Torii R, Hatoya S, Wijewardana V, Kumagai D, Tamada H, Sawada T, Ikehara S, Inaba T (2004) Effect of ovarian hormones on periodical changes in immune resistance associated with estrous cycle in the beagle bitch. Immunobiology 209:619–627

Torres MB, Trentzsch H, Stewart D, Mooney ML, Fuentes JM, Saad DF, Reeves RH, De Maio A (2005) Protection from lethal endotoxin shock after testosterone depletion is linked to chromosome X. Shock 24:318–323

Tsuyuguchi K, Suzuki K, Matsumoto H, Tanaka E, Amitani R, Kuze F (2001) Effect of oestrogen on Mycobacterium avium complex pulmonary infection in mice. Clin Exp Immunol 123: 428–434

Vegato E, Pollio P, Ciana P, Maggi A (2000) Estrogen blocks inducible nitric oxide synthase accumulation in LPS-activated microglia cells. Exp Gerontol 35:1309–1316

Vegato E, Ghisletti S, Meda C, Etteri S, Belcredito S, Maggi A (2004) Regulation of the lipopolysaccharide signal transduction pathway by 17β-estradiol in macrophage cells. J Steroid Biochem 91:59–66

Vlotides G, Gaertner R, Angstwurm MWA (2007) Modulation of monocytic lipopolysaccharide-induced tissue factor expression and tumor necrosis factor alpha release by estrogen and calcitriol. Hormones 6:52–61

Wichmann MW, Ayala A, Chaudry IH (1997) Male sex steroids are responsible for depressing macrophage immune function after trauma-hemorrhage. Cell Physiol 42:C1335–C1340

Wichmann MW, Inthorn D, Andress HJ, Schildberg FW (2000) Incidence and mortality of severe sepsis in surgical intensive care patients: the influence of patient gender on disease process and outcome. Intens Care Med 26:167–172

Wright SD (1999) Toll, a new piece in the puzzle of innate immunity. J Exp Med 189:605–609

Yamamoto Y, Saito H, Setogawa T, Tomioka H (1991) Sex differences in host resistance to *Mycobacterium marinum* infection in mice. Infect Immun 59:4089–4096

Zheng R, Pan G, Thobe BM, Choudhry MA, Matsutani T, Samy TSA, Kang S, Bland KI, Chaudry IH (2006) MyD88 and Src are differentially regulated in Kupffer cells of males and proestrus females following hypoxia. Mol Med 12:65–73

Zuckerman SH, Bryan-Poole N, Evans GF, Short L, Glasebrook AL (1995) In vivo modulation of murine serum tumour nectosis factor and interleukin-6 levels during endotoxemia by oestrogen agonists and antagonists. Immunology 86:18–24

Zuckerman SH, Ahmari SE, Bryan-Poole N, Evans GF, Short L, Glasebrook AL (1996) Estriol: a potent regulator of TNF and IL-6 expression in a murine model of endotoxemia. Inflammation 20:581–597

Chapter 6
Sex Hormones and Regulation of Host Responses Against Parasites

James Alexander, Karen Irving, Heidi Snider, and Abhay Satoskar

Abstract Sex hormones play an influential role in the control of parasitic infection by their ability to modulate different components of both the innate and adaptive immune responses. The parasites themselves are phylogenetically diverse, target a range of different tissues, and have evolved numerous alternative strategies to evade or inhibit protective immune responses. Consequently, the influence of sex hormones on these infective agents can be complex. For example, while females exhibit greater resistance to infection by parasites including *Trypanosoma cruzi*, *Trypanosoma brucei*, *Giardia lamblia*, *Leishmania mexicana*, *Plasmodium chabaudi*, and *Trichinella spiralis*, their male counterparts are found to be more resistant to infection with *Trichomonas vaganalis*, *Toxoplasma gondii* and *Schistosoma mansoni*. This chapter will discuss: (1) the role of sex hormones in regulating the outcome of parasite infection, (2) mechanisms by which these hormones modulate host immune responses, and (3) the implications of these observations for the pathogenesis of human parasitic disease.

6.1 Parasites: Their Global Importance

The Dutch proverb "Big fleas have little fleas upon their backs to bite them, and little fleas have lesser fleas, and so *ad infinitum*" mimics a wonderful illustration of the universality and success of parasitism as a life strategy. Parasites consequently comprise a massive infectious disease problem for humans and domesticated animals. In particular, as major causes of mortality and morbidity, they are hugely detrimental to both the social and economic progress of the developing world. Malaria, for example, affects over 500,000,000 and kills at least 1,000,000 annually

J. Alexander (✉)
Strathcyde Institute of Pharmacy and Biomedical Sciences, University of Strathclyde, Glasgow, G4 0NR, UK
e-mail: j.alexander@strath.ac.uk

S.L. Klein and C.W. Roberts (eds.), *Sex Hormones and Immunity to Infection*, DOI 10.1007/978-3-642-02155-8_6, © Springer-Verlag Berlin Heidelberg 2010

in Africa, mainly pregnant woman and young children. Overall, the disease burden attributable to malaria is 42,000,000 disability adjusted life years (DALYS). The DALY is a health gap measure that extends the concept of potential years of life lost due to premature death to include equivalent years of "healthy" life lost by virtue of being in states of poor health or disability (Murray et al. 2002). Schistosomiasis and filariasis provide excellent examples of the chronic debilitating nature of many parasite infections, although mortality attributable to the diseases is negligible, the disease burdens equate to 1,930,000 and 5,600,000 DALYS, respectively. While generally considered a third-world problem, parasites can also impinge significantly on the health and well being as well as economies of developed countries. For example, in France where the incidence of adult-acquired toxoplasmosis approaches 80% of the population, the incidence of congenital toxoplasmosis is estimated to be 10 cases per 1,000 births in Paris (Desmonts and Courvreur, 1974). In the UK and USA, where approximately 30% of the population are infected, the incidence of congenital toxoplasmosis is 0.5 per 1,000 births (Williams et al. 1981) and 1–10 per 10,000 births (Lopez et al. 2000), respectively. In the USA, the estimated total medical costs and loss of productivity as a consequence of human toxoplasmosis, excluding AIDS patients was $2,628,000,000 per annum (US Department of Agriculture (1994) Charting the cost of food safety. Food Review). To compound the economic problem posed by this single parasite species, it infects all warm-blooded vertebrates including domesticated animals and is responsible for 25% of all abortions in sheep in the UK and Spain (Buxton, 1998; Pereira-Bueno et al. 2004). Despite high prevalence of parasite infections in the human population, obvert clinical disease is not always evident. Evolution has often resulted in a balance between parasite virulence and host resistance to infection and again the protozoan, *T. gondii* provides an excellent example whereby virtually all adult-acquired infections are mild to asymptomatic, despite infection being life long. However, perturbation of immune responses, such as those recorded in AIDS patients can upset the balanced host–parasite interplay, and under such circumstances, the infection can be fatal. The more subtle effects of sex hormones on the modulation of the immune response can also have significant outcomes. Indeed, susceptibility to disease mediated by *T. gondii* is increased during pregnancy while females are generally more susceptible to infection than males (reviewed by Roberts et al. 2001). A close examination of the literature reveals that the sexual dimorphism observed during *T. gondii* infection is not, in fact, unique to this parasite. To a greater or lesser extent, virtually all parasitic infections demonstrate differences in pathogenesis between males and females, which of course also contribute to the age-dependent differences often observed in disease susceptibility. Pregnancy and its associated progesterone production also significantly influence susceptibility to infection and this is discussed in detail in Chaps. 8 and 9. This chapter, therefore, provides an overview of those parasite infections where sex plays a significant role in disease severity and how hormones may function to modulate the immune response to determine disease outcomes.

6.2 Sex Differences in Susceptibility to Parasites

Unlike the example of toxoplasmosis above where males are more resistant to infection than females, sexual dimorphism in responses to parasite infection invariably reveals females to be more resistant than males, consequently giving rise to the so-called " female host supremacy paradigm" (reviewed in Chapter 7).

6.2.1 Male-biased Infections

A comprehensive list of those parasite infections where males are more susceptible than females is provided in Table 6.1. There are numerous examples from all phyla in which parasites are found that male hosts typically suffer from more severe infection than females. Traditionally, males have been associated with stronger cell-mediated and weaker humoral responses than females (Alexander and Stimson 1988) which would suggest that males have a more pronounced helper T cell type 1 (Th1) response than females that conversely would be anticipated to have a dominant helper T cell type 2 (Th2) response. However, what is clearly evident from the accumulated evidence is that there is no association between a dichotomy in male/female associated susceptibility and a bias in the Th1/Th2 responses. Thus, males are more susceptible to many *Leishmania* spp. than females where it is the general consensus that a Th1 response and IFN-γ production, in particular, is necessary for protection. Conversely, males are also more susceptible to *Trichinella spiralis*, but in this case worm expulsion is Th2 cytokine (i.e., IL-4/IL-13) dependent. In addition, males are also more susceptible to parasites where more complex immune responses are required, such as during malarial infection where protection is associated with the sequential generation of Th1 and Th2 responses.

6.2.2 Female-biased Infections

A list of those few parasite infections where females are more susceptible to infection than males is provided in Table 6.2 and discussed in more detail in Chapter 7 of this book. However, there is again no immediately obvious immunological indicator as to why this particular sex bias exists For example, while protective immunity against *L. major* and *T. gondii* is dependent on the generation of potent Th1 responses, protective immunity against *L. sigmondontis* is Th2 cytokine mediated (e.g., Le Goff et al. 2002). Furthermore, females are also more susceptible to infection with *Babesia microti* where immunity, as with *Plasmodium*

Table 6.1 Parasites to which males are more susceptible than females

	Parasite	Host species	Measure	References
Protozoa	Eimeria sciurorum	Sciurus vulgaris	↑I	Bertolino et al. (2003)
	Entamoeba histolytica	Homo sapiens	↑I ↑P	Acuna-Soto et al. (2000)
	Giardia muris	Mus musculus	↑PB	Faubert et al. (1985)
	Leishmania braziliensis	Homo sapiens	↑I	Jones et al. (1987)
	L. donovani	Homo sapiens	↑I	de Beer et al. (1991), Jahn (1986)
		Mus musculus	↑I	Zhang et al. (2001)
	L. infantum	Canis familiaris	↑I	Zaffaroni et al. 1999
	L. mexicana	Homo sapiens	↑P	Alexander (1988), Roberts et al. (1990)
		Mus musculus	↑I ↑P	Satoskar and Alexander (1995), Satoskar et al. (1998)
	L. major	Mus musculus	↑PB	Mock and Nacy (1988)
	L. vianna guyanensis	Mesocricetus auratus	↑I ↑P ↑PB	Travi et al. (2002)
	L. vianna panamensis	Mesocricetus auratus	↑I ↑P ↑PB	Travi et al. (2002)
	Plasmodium berghei	Mus musculus	↑I	Kamis and Ibrahim (1989)
	P. chabaudi	Mus musculus	↑I ↑P ↑M ↑PB	Benten et al. (1997), Mossmann et al. (1984)
	P. falciparum	Homo sapiens	↑I ↑PB	Weise (1979), Landgraf et al. (1994), Molineaux and Gramiccia (1979)
	P. mexicanum	Sceloporus occidentalis	↑I	Schall and Marghoob (1995)
	P. vivax	Homo sapiens	↑I	Moon and Cho (2001)
	Trypanosoma cruzi	Homo sapiens	↑I ↑P ↑M ↑PB	Brabin and Brabin (1992)
		Mus musculus	↑PB	Kierszenbaum et al. (1974), de Souza et al. (2001)
		Calomys callosus	↑PB	do Prado et al. (1998), do Prado et al. (1999)
	T. rhodesiense	Mus musculus	↑PB	Greenblatt and Rosenstreich (1984)
Nematoda	Angiostrongylus malaysiensis	Rattus norvegicus	↑I ↑PB	Kamis et al. (1992)
	Baylisascaris procyonis	Procyon lotor	↑I ↑PB	Evans (2001)
	Brugia malayi	Mus musculus	↑I ↑PB	Rajan et al. (1994)
		Meriones unguiculatus	↑PB	El Bihari and Ewert (1973)
	B. pahangi	Mus musculus	↑PB	Nakanishi et al. (1990), Nakanishi et al. (1989)
		Rattus norvegicus	↑I ↑PB	Bell et al. (1999)
	Capillaria muris	Clethrionomys glareolus	↑I ↑PB	Lewis and Twigg (1972)
	Dipetalonema vitae	Mesocricetus auratus	↑I↑PB	Reynouard et al. (1984)

Dirofilaria immitis	*Canis familiaris*	↑I	Selby et al. (1980)
	Felis catus	↑I	Kramer and Genchi (2002)
Elaphostrongylus rangiferi	*Rangifer tarandus tarandus*	↑PB	Halvorsen et al. (1984)
Heterakis spumosa	*Mus musculus*	↑PB	Harder et al. (1992)
Muspicea borreli	*Mus musculus*	↑I ↑PB	Spratt et al. (2002)
Necator americanus	*Homo sapiens*	↑I ↑PB	Behnke et al. (2000)
	Mesocricetus auratus	↑I ↑PB	Xue et al. (2005)
Nematospiroides dubius	*Apodemus sylvaticus*	↑I	Elton (1931)
Nippostrongylus brasiliensis	*Millardia meltada*	↑PB	Tiuria et al. (1994)
	Mesocricetus auratus	↑I↑PB	Solomon (1966)
Onchocerca volvulus	*Homo sapiens*	↑PB	Duerr et al. (2004)
Rhabdias bufonis	*Rana temporaria*	↑I	Lees and Bass (1960)
Strongyloides ratti	*Mus musculus*	↑I ↑PB	Kiyota et al. (1984)
	Rattus norvegicus	↑I ↑PB	Watanabe et al. (1999)
S. stercoralis	*Mus musculus*	↑I	Satoh et al. (2004)
S. venezuelensis	*Rattus norvegicus*	↑I ↑PB	Rivero et al. (2002)
Syphacia obvelata	*Apodemus sylvaticus*	↑I	Elton (1931)
	Mus musculus	↑I ↑PB	Okulewicz and Perec (2003), Taffs (1976)
S. stroma	*Apodemus sylvaticus*	↑PB	Behnke et al. (1999)
Trichinella spiralis	*Microtus pennsylvanicus*	↑PB	Klein et al. (1999)
	Mus musculus	↑PB	Reddington et al. (1981)
	Rattus norvegicus	↑PB	Reddington et al. (1981)
Toxocara spp.	*Homo sapiens*	↑I	Baboolal and Rawlins (2002)
Wuchereria bancrofti	*Homo sapiens*	↑I	Kazura et al. (1984)
Trematoda *Alaria taxideae*	*Martes americana*	↓PB	Poole et al. (1983)
Brachylaima cribbi	*Mus musculus*	↑PB	Butcher et al. (2002)
Crepidostomum farionis	*Salmo trutta*	↑I ↑PB	Thomas (1964)
Discocotyle sagittata	*Salmo trutta*	↑I ↑PB	Paling (1965)
Gorgoderina vitelliloba	*Rana temporaria*	↑I	Lees and Bass (1960)
Polystoma integerrimum	*Rana temporaria*	↑I	Lees and Bass (1960)
Pseudodiplorchis americanus	*Scaphiopus couchii*	↑PB	Tinsley (1989)

(continued)

Table 6.1 (continued)

Parasite		Host species	Measure	References
	Echinococcus multilocularis	*Mus musculus*	↑PB	Ohbayashi and Sakamoto (1966)
	Shistosoma haematobium	*Homo sapiens*	↑I ↑PB	Ansell et al. (2001)
	S. mansoni	*Mesocricetus auratus*	↑I ↑PB	Barrabes et al. (1980)
		Homo sapiens	↑I ↑PB ↑P	Degu et al. (2002)
	Amblyomma hebraeum	*Tragelaphus strepsiceros*	↑PB	Horak et al. (1987)
Arthropoda	*Hypoderma tarandi*	*Rangifer tarandus tarandus*	↑PB	Folstad et al. (1989)
	Polyplax spinulosa	*Rattus norvegicus*	↑I	Webster and MacDonald (1995)

Incidence (I); Parasite burden (PB); Pathology (P); Mortality (M)
(Modified from Klein, 2004)

Table 6.2 Parasites to which females are more susceptible than males

	Parasite	Host species	Measure	References
Protozoa	*Babesia microti*	*Mus musculus*	↑M ↑PB	Aguilar-Delfin et al. (2001)
	Toxoplasma gondii	*Mus musculus*	↑PB ↑M ↑P	Walker et al. (1997), Roberts et al. (1995)
	Leishmania major	*Mus musculus*	↑PB	Giannini (1986), Alexander (1988)
Nematoda	*Litomosoides sigmondontis*	*Mus musculus*	↑I ↑PB	Graham et al. (2005)
	Trichostrongylus retortaeformis	*Oryctolagus cuniculus*	↑I ↑PB	Dunsmore (1966)
Trematoda	*Schistosoma mansoni*	*Mus musculus*	↑PB ↑M ↑P	Eloi-Santos et al. (1992), Nakazawa et al. (1997)
	Hymenolepis nana	*Rattus norvegicus*	↑I	Webster and MacDonald (1995)
	Taenia crassiceps	*Mus musculus*	↑PB ↑P	Sciutto et al. (1991)
	T. solium	*Sus scrofa*	↑I	Morales et al. (2002)
Arthropoda	*Centruroides limpidus limpidus*	*Mus musculus*	↑M	Padilla et al. (2003)
	Notoedres muris	*Rattus norvegicus*	↑I	Webster and MacDonald (1995)

Incidence (I); Parasite burden (PB); Pathology (P); Mortality (M)
(Modified from Klein, 2004)

infections, is associated with sequential generation of Th1/Th2 responses (Chen et al. 2000). However, on closer inspection, similarities may exist as protective immunity against both *T. gondii* (Walker et al. 1997) and *B. microti* (Aguilar-Delfin et al. 2001 and 2003) is associated with potent Th1 innate responses as manifested by macrophage and NK cell IL-12 and IFN-γ production, respectively; male SCID mice are still more resistant than females to both infections. In addition, susceptibility to both *L. major* (Suffia et al. 2006) and *L. sigmondontis* (Taylor et al. 2005 and 2007) is associated with regulatory T cells producing IL-10 which is positively regulated via estrogen receptor (ER) α (Tai et al. 2008)

6.3   Sex Differences in Immune Responses to Parasites

There are surprisingly few studies where sex differences in immune responses to parasites have been quantified, and these are listed in Table 6.3. However, there is a wealth of documentation on how sex differences and sex hormones can influence the response of immune cells, which is detailed in Chapter 2. Thus, this section focuses on how sex differences and sex hormones affect immune responses during parasite infection.

Table 6.3 Immunological differences between the sexes in response to parasite infection

Parasite	Host Species	Males	Females	References
Leishmania	*Mus*	IgE↑	IFN-γ↑	Mock and Nacy
major	*musculus*	IgG1↑	DTH↑	(1988), Alexander
Leishmania		IL-5↑	IgG2a↑	(1988)
mexicana		TNF-α↑		
		IL-4 =		
		IL-10=		
		IL-12=		
Leishmania	*Mesocricetus*	IL-4 ↑		Travi et al. (2002)
guyanensis	*auratus*	IL-10		
Leishmania		TGF-β ↑		
panamensis		IFN-γ ↓		
Plasmodium	*Mus*	IgG ↓	IgG ↑	Benten et al. (1997),
chabaudi	*musculus*	IgG1↓	IgG1↑	Zhang et al. (2000)
		IgG2b↓	IFN-γ mRNA ↑	
		IL-2 mRNA↑	MHC Class II ↑	
Toxoplasma	*Mus*	IL-12↑	IL-10↑	Walker et al. (1997),
gondii	*musculus*	IFN-γ↑	Specific	Roberts et al. (1995)
		TNF-α↑	Proliferation ↓	
		IL-10 ↑		
Trypanosoma	*Mus*	Lytic		do Prado et al. (1998),
cruzi	*musculus*	Antibody↑		Schuster and Schaub
				(2001)
Schistosoma	*Homo*	IgG↑	IgA↑	Degu et al. (2002),
haematobium	*sapiens*	IgG1↑	TGF-β↑	Webster et al. (1997),
S. mansoni		IgG2a↑	IL-10↑	Abebe et al. (2001),
		IgG4↑	Specific	Remoue et al. (2001),
		IgE↑	Proliferation ↓	Naus et al. (2003)
		IFN-γ↑		
		TNF-α↑		
Taenia	*Mus*	IL-4↑	IL-10↑	Sciutto et al. (1991),
crassiceps	*musculus*	IFN-γ↑		Terrazas et al. (1998)

↑ Increase in production ↓ Decrease in production
(Modified from Klein, 2004)

6.3.1 Neutrophils

The best-characterized function of neutrophils is their preeminent role in the phagocytosis and killing of invading microorganisms via the generation of oxygen intermediates and the release of hydrolytic enzymes stored in their granules. Neutrophils, by promoting tissue injury, also contribute to the initiation of inflammation, an essential step in the launching of immunity. Neutrophils influence the course of several parasitic infections, including those mediated by *Leishmania donovani* (McFarlane et al. 2008) and *T. gondii* (Bliss et al. 1999) where they are key players in the development of a protective Th1 response, which increases the likelihood that disease outcome is subject to sex hormone modulation.

6.3.2 Mast Cells and Eosinophils

In contrast to neutrophil activation which is often associated with a protective Th1 response in a number of infections (McFarlane et al. 2008: Bliss et al. 1999), activation of mast cells, basophils, and eosinophils is intimately associated with a Th2 response. This response is an essential component of protective immunity against large pathogens such as helminths, but also is responsible for Th1 hypersensitivity. Females typically are more resistant to helminth infections and more prone to conditions such as asthma than males (de Marco et al. 2000; Klein 2004). It is hardly surprising therefore, that physiological levels of 17β-estradiol stimulate the release of preformed mediators from mast cells and basophil granules as well as synthesis of new mediators (Zaitsu et al. 2007). In contrast, testosterone generally inhibits mast cell/basophil activity (Menendez-Pelaez et al. 1992; Vliagoftis et al. 1992). High doses of 17α-estradiol increase the degranulation of human blood eosinophils both *in vivo* and *in vitro* (Silva et al. 1997). Similarly, a combination of 17β-estradiol and progesterone enhances human eosinophil degranulation (Hamano et al. 1998). The extent to which sex hormones affect the activity of mast cells, basophils, and eosinophils during parasitic infection requires further investigation.

6.3.3 Macrophages

Macrophages play crucial roles in immunity to parasite infections as inflammatory mediators, antigen presenting cells, and host cells for many intracellular organisms, such as *Leishmania*. The effect of sex hormones on these cells is extremely complex with both estrogens and androgens having pro- and anti-inflammatory activities. This is probably in part a reflection of the massive heterogeneity of macrophage populations as well as the individual experimental systems utilized. Estrogen upregulates phagocytosis, but downregulates the production of reactive oxygen intermediates, but not reactive nitrogen intermediates (Chao et al. 1994). Nevertheless, low doses of estrogen stimulate nitric oxide synthase (NOS) 2 production (Verthelyi, 2006) and enhance nitric oxide (NO)-mediated killing of *Leishmania* parasites (Lezama-Davila et al. 2007; Osorio et al. 2008). Overall, the indications are that testosterone downregulates the innate and inflammatory activity of macrophages by inhibiting NO production (Friedl et al. 2000) and toll-like receptor 4 (TLR4) expression (reviewed in Chapter 5 and Rettew et al. 2008). More recent studies using ERα and ERβ deficient severe combined immunodeficiency mice and microbial stimuli have demonstrated TNF-α production and antimicrobial activity are significantly enhanced in murine macrophages deficient in ERα, but not ERβ (Lambert et al. 2004). Similarly, in a TcR T cell transgenic model system, ERα-deficient murine splenic macrophages have enhanced ability to induce T cell proliferation and IFN-γ production (Lambert et al. 2004). This indicates that estrogen enhances Th2 activity and explains the generally increased resistance of females to helminth infections and the increased incidence of asthma in this sex in animal models.

6.3.4 Dendritic Cells

Dendritic cells (DCs) are professional antigen presenting cells, act as the major bridge between innate and adaptive immune responses and play a pivotal role in directing the magnitude and nature of the acquired response. A significant amount of work has identified estrogen receptor ligands on DCs, how these receptors influence DC differentiation from precursor cells, and subsequently modulate function (reviewed Chapters 2, 3, and Kovats and Carreras, 2008). Estradiol acts on DC progenitor cells to regulate GM-CSF- and Flt3Ligand (Flt3L)-mediated differentiation (Paharkova-Vatchkova et al. 2004; Carreras et al. 2008). By contrast, testosterone does not promote the differentiation of DCs from bone marrow precursors (Paharkova-Vatchkova et al. 2004) while studies in mice and humans suggest that testosterone limits the density of Langerhan cells (Koyama et al. 1989; Galasso et al. 1996). As yet, the direct and indirect effects of estrogens and androgens on DC differentiation and function have not been correlated with the outcome of parasite diseases. However, future studies will, undoubtedly, identify these cells as major targets of sex hormone influences in such diseases.

6.3.5 NK Cells

NK cells play significant roles in innate defense against parasites via their cytolytic activity and ability to produce IFN-γ. Numerous studies, though not all, demonstrate that 17β-estradiol downregulates NK cell activity in a dose-dependent manner (Baral et al. 1995; Nilsson and Carlsten 1994). Recent data illustrate that testosterone can increase the cytotoxicity and migration of NK cells isolated from peripheral blood (Lang et al. 2003). Consequently, it would be anticipated that where a potent Th1 innate response and NK cells are required for protection against a parasite infection being a male would be a distinct advantage. Interestingly, males are more resistant than females to both *T. gondii* (Roberts et al. 2001) and *B. microti* (Aguilar-Delfin et al. 2001, 2003) where NK cell responses mediate significant protection. During *Plasmodium chabaudi* infection of C57BL/6 mice, production of IFN-γ by NK cells is critical for protection (Mohan et al. 1997). Male C57BL/6 mice are more susceptible to *P. chabaudi* infection and have lower IFN-γ concentrations than females; ovariectomy of females significantly reduces IFN-γ concentrations and increases susceptibility to infection as compared with intact female mice (Cernetich et al. 2006; Mohan et al. 1997). Further, exogenous replacement of 17β-estradiol (with or without progesterone) in ovariectomized females increases IFN-γ concentrations and reduces the development of disease (Klein et al. in press).

6.3.6 B Cells

It is well documented that females produce stronger humoral responses than males. Consequently, while estrogen promotes IgM and IgG production from B cells in human peripheral blood monocytes (Kanda and Tamaki 1999), testosterone inhibits this activity (Kanda et al. 1996). However, whether these hormones directly influence B cells awaits clarification as the antibody responses may be the indirect result of hormones influencing cytokine production from other cell populations. Examination of comparative humoral immune responses to parasites (Table 6.3) indicates that an absolute correlation between antibody production, sex, and susceptibility to infection does not exist and that a more clinical dissection of how sex hormones influence B cell antibody class and subclass switching and production is required. Consequently, sex differences in response to *P. chabaudi* infection are not mediated by B cells because female μMT mice (i.e., mice devoid of B cells) are still more resistant to *P. chabaudi* than male μMT mice (Cernetich et al. 2006).

6.3.7 T Cells

Sex and pregnancy-associated hormones have been shown to be key players in the regulation of T cell activity, as detailed in Chapters 2, 8, and 9. Functional receptors for both testosterone and estrogen have been identified in $CD4^+$ and $CD8^+$ lymphocyte populations and a plethora of diverse effects have been attributed to hormonal influence (Benten et al. 1999 and 1998). As detailed in Chapter 2, there is a substantial body of evidence that supports an important role for sex hormones in the development and function of T cells. Furthermore, these effects are likely to influence sex differences in prevalence, severity, and susceptibility to a wide range of parasitic infections. Overall, estrogen depending on conditions of study, has the ability paradoxically to promote either a Th1 or a Th2 response whereas androgens downregulate adaptive immunity (reviewed in Roberts et al. 1996). This would reflect the overall greater resistance of females to both parasites requiring a protective Th2 response as well as those requiring a Th1 response.

6.4 Role of Sex Steroids in Response to Specific Parasites

A comprehensive list of how sex hormones and gonadectomy modulate susceptibility to parasite infection as measured by parasite burdens, pathology, and mortality is given in Table 6.4.

Table 6.4 Role of sex hormones, orchidectomy and ovariectomy in determining susceptibility to parasitic infection

Parasite	Host species	Effect of sex hormones	References
Babesia microti	*Mus musculus*	$E_2\uparrow$	Aguilar-Delfin et al. (2001)
L. major	*Mus musculus*	T $\uparrow$	Mock and Nacy (1988)
L. mexicana	*Mus musculus*	$E_2\downarrow O\uparrow G\downarrow$	Alexander (1988), Khamis et al. (unpublished), Lezama-Davila et al. (2008)
L. panamensis	*Mesocricetus auratus*	$T\uparrow$	Travi et al. (2002)
L. donovani	*Mus musculus*	$P\uparrow T\uparrow$	Jones et al. (2008), Liu et al. (2006)
Plasmodium berghei	*Mus musculus*	$T\uparrow G\downarrow$	Kamis and Ibrahim (1989)
P. chabaudi	*Mus musculus*	$T\uparrow G\downarrow$	Benten et al. (1992), Benten et al. (1997)
P. falciparum	*Mus musculus*	$T\uparrow E_2\uparrow P\uparrow$	Escobedo et al. (2005)
Toxoplasma gondii	*Mus musculus*	$T\downarrow E_2\uparrow P\uparrow G\downarrow$	Kittas and Henry (1980), Walker et al. (1997), Liesenfeld et al. (2001)
Trypanosoma cruzi	*Mus musculus*	$E_2\downarrow$	de Souza et al. (2001)
	Calomys callosus	$T\uparrow E_2\downarrow$	do Prado et al. (1998, 1999)
Angiostrongylus malaysiensis	*Rattus norvegicus*	$T\uparrow G\downarrow$	Kamis et al. (1992)
Brugia malayi	*Mus musculus*	$T\uparrow$	Rajan et al. (1994)
Heterakis spumosa	*Mus musculus*	$T\uparrow G\downarrow$	Harder et al. (1992)
Nippostrongylus brasiliensis	*Millardia meltada*	$G\downarrow$	Tiuria et al. (1994)
Rhabdias bufonis	*Rana temporaria*	$E_2\downarrow$	Lees and Bass (1960)
Strongyloides ratti	*Mus musculus*	$T\uparrow G\downarrow$	Kiyota et al. (1984)
S. venezuelensis	*Rattus norvegicus*	$T\uparrow E_2\downarrow G\downarrow O\uparrow$	Rivero et al. (2002)
Trichinella spiralis	*Rattus norvegicus*	$T\uparrow E_2\downarrow G\downarrow$	Mankau and Hamilton (1972)
Trichostrongylus retortaeformis	*Oryctolagus cuniculus*	$O\downarrow$	Hobbs et al. (1999b)
Gorgoderina vitelliloba	*Rana temporaria*	$E_2\downarrow$	Lees and Bass (1960)
Polystoma integerrimum	*Rana temporaria*	$E_2\downarrow$	Lees and Bass (1960)
Schistosoma haematobium	*Mus musculus*	$T\downarrow$	Escobedo et al. (2005)
Schistosoma mansoni	*Mus musculus*	$T\downarrow C\uparrow$	Nakazawa et al. (1997)
Taenia crassiceps	*Mus musculus*	$T\downarrow E_2\uparrow G\uparrow P\uparrow$	Escobedo et al. (2005)
T. solium	*Sus scrofa*	$G\uparrow$	Morales et al. (2002)
Hypoderma tarandi	*Rangifer tarandus tarandus*	$G\downarrow$	Folstad et al. (1989)

$\uparrow$ Increase in susceptibility $\downarrow$ Decrease in susceptibility

Testosterone (T); Oestradiol (E_2); Progesterone (P); Orchidectomy of males (G); Ovariectomy of females (O)

(Modified from Klein, 2004)

6.4.1 Sexual Dimorphism in the Response to Leishmania infection: A Not So Simple Paradigm

There has been a tendency amongst immunologists, in particular, when addressing questions relating to pathology and protection to use *Leishmania major* infections of mice as the overarching experimental model of leishmaniasis. As the paradigm for the CD4$^+$ Th1–Th2 dichotomy is largely based upon the curing/noncuring responses, respectively, to *L. major* infection, the importance of this model is hard to dispute and any observations that male or female sex hormones can influence the Th1/Th2 bias have significance. Consequently, in apparent agreement with the traditional view that males have more robust cellular immune responses than females, male DBA/2 and B10/129 mice are more resistant to *L. major* than females (Giannini 1986; Alexander 1988). In addition, a vaccine trial in humans utilizing killed *L. major* was more effective in males than females (Modabber 2000). Conversely, females rather than males tend to develop persistent lesions when infected with another species causing cutaneous leishmaniasis in the Old World, *L. tropica* (Greenblatt 1980). However, numerous *Leishmania* species, in addition to *L. major,* cause disease in humans as well as other mammals. Indeed the Old World species causing cutaneous leishmaniasis diverged from the New World species (e.g., *L. mexicana* and *L. amazonensis*) some 40–80 million years ago and are as distant phylogenetically from each other as they are from *L. donovani* which causes visceral leishmaniasis. Consequently, it is hardly surprising that these parasites have evolved different strategies to survive not only within different hosts but also different tissues within the host. Therefore, not only different virulence factors have been identified for different parasite species but disease outcome in the host is also often under different host genetic controls, even within the same tissue site. As a result, there are differences in the immunobiology of *Leishmania* species, and age and sex can modulate disease outcomes. Thus, unlike the epidemiological and experimental reports on *L. major* and *L. tropica*, females are more resistant than males to cutaneous infection with *L. mexicana* in both humans and mice (Alexander 1988; Roberts et al. 1990; Satoskar and Alexander 1995; Satoskar et al. 1998) and visceral leishmaniasis caused by *L. donovani* in humans (de Beer et al. 1991; Jahn et al. 1986) or *L. infantum* in dogs (Zaffaroni et al. 1999). Tissue site may, in part, be important as male BALB/c and DBA/2 mice are more susceptible than females if infected systemically rather than cutaneously (Mock and Nacy 1988). Nevertheless, protective immunity, against all species studied to date, requires IL-12-driven production of IFN-γ, via NK cells and Th1 cells to activate macrophages to kill the intracellular amastigote stage of the life cycle via the generation of toxic oxygen and nitrogen intermediates. Consequently, the simplistic generalization that males have more pronounced cell-mediated responses than females does not stand close scrutiny with this archetypal model of Th1-mediated protection. Thus, female DBA/2 mice infected with *L. mexicana* develop much stronger Th1 responses as measured by IFN-γ production, delayed-type hypersensitivity (DTH) responses, and IgG2a antibody than similarly infected male mice (Satoskar and Alexander

1995; Satoskar et al. 1998). In humans infected with *L. mexicana*, females generally have increased Th1 responses as measured by DTH reactions and decreased Th2 responses as measured by IgE production than males (Lynch et al. 1982). Resolution of infection in women also is faster. That sex hormones can mediate the disease phenotype during leishmaniasis (Mock and Nacy 1988) has been demonstrated in systemic *L. major* infections where gonadectomy increases the resistance of males while testosterone implants increases the susceptibility of females. Both testosterone and estrogen may influence the response of DBA/2 mice to *L. mexicana* as ovariectomy increases the susceptibility of females to cutaneous infection whereas gonadectomy increases the resistance of males (Khamis et al. unpublished data). Similarly, estrogen and testosterone may influence immunity at the level of the macrophage. Estrogen, for example, at physiological levels, can promote macrophage-mediated killing of *L. mexicana* via increased NO production independent of proinflammatory cytokine production (Lezama-Davila et al. 2008). Testosterone negatively regulates murine bone marrow-derived macrophage p38 MAPK activation upon infection with *L. donovani* and, consequently, promotes parasite survival (Liu et al. 2006). Sex hormones influence the cross-talk in signaling pathways in macrophages, which is of particular significance given that resistance of DBA/2 mice to *L. mexicana* is under single gene control that is mapped to a region of mouse chromosome 4, with homology to human 9p, for which Jak1 and Jak2 kinase are candidates (Roberts et al. 1990). These observations are significant as IFN-γ and IL-4 signaling via Jak1 and Jak2 are pivotal in regulating the immune response and influencing the activation of macrophages to control intracellular parasitism (Muller et al. 1993). *L. mexicana* is particularly potent as compared with *L. major* at downregulating macrophage IFN-γ-induced Jak1, Jak2, and STAT1 activation (Rosas et al. 2005), whereas *L. donovani* has similar activity to *L. mexicana* (Nandan and Reiner 1995). Estrogen upregulates the expression of IFN-γ mRNA by T cells (Grasso and Muscettola 1990), whereas testosterone inhibits production of this cytokine (Araneo et al. 1991). Indeed, our recent studies using CD4+ T cell-specific IL-4R$\alpha^{-/-}$ BALB/c mice have revealed a previously undetected underlying male susceptibility to *L. mexicana* involving T cells. CD4$^+$ T cell specific IL-4R$\alpha^{-/-}$ male mice, unlike their female counterparts, were unable to resolve infection and had a less polarized Th1 response (Bryson et al. unpublished data). Although it is accepted that protective immunity against *Leishmania* spp. is mediated by a Th1 response, the identification of a precise role for Th2 responses in nonhealing disease remains more elusive. While IL-4 and IL-13 can enhance disease progression in appropriate models of both Old and New World leishmaniasis, a disease-promoting role for these cytokines in visceral leishmaniasis caused by *L. donovani* is difficult to ascertain because infected individuals display a mixed Th1/Th2 profile such that IL-4 and IL-13 can be readily detected alongside IFN-γ (reviewed McMahon-Pratt and Alexander 2004). Indeed, even in the BALB/c*L. major* model, IL-4 administered early during infection can prime BALB/c DCs for IL-12 production and facilitate a protective Th1 response (Biedermann et al. 2001). IL-10, which is produced from macrophages, DCs, Th2 cells, regulatory T cells or a combination of cells, is the major immunosuppressive cytokine promoting visceral disease and maintaining chronic *L. major* infections (Suffia et al. 2006). The cellular sources of IL-10 are potentially subject to sex hormone control. *L. mexicana* and *L. amazonensis* can also subvert the development of a Th1 response by IL-4-independent as well as IL-4-dependent mechanisms (McMahon-Pratt and Alexander 2004). Thus, the Th1/Th2 paradigm of resistance/susceptibility to intracellular parasites is a gross oversimplification of a far more complicated network of regulatory/counter-regulatory interactions, which are also subject to further modulation by sex hormones. Consequently, the outcome of disease will differ according to the *Leishmania* species being studied, the tissue site examined, and the age or sex of the host. Nevertheless, studies using these organisms have provided fascinating insights into how sex hormones can influence basic immunological mechanisms controlling the disease outcomes. Further knowledge of this area will provide key insights as to the future rational development of appropriate strategies for therapeutic intervention or vaccination.

6.4.2 Plasmodium *Infections and Th1/Th2 Immunity: Are Females Better Balanced?*

Malaria represents one of the world's major public health problems, with 40% of the world's population at risk, and a child dying every 30 seconds of disease (Global Malaria 2008). Although effector mechanisms of the immune system are required to control disease, excessive responses can result in severe immunopathology and mortality (reviewed by Lamb et al. 2006). Individuals may develop severe and often life-threatening complications such as: cerebral malaria, placental malaria, severe anemia, metabolic acidosis, and shock-like responses that lead to multiorgan failure. Pathology often results from the sequestration of parasitized RBCs (pRBCs) in tissue sites and subsequent inflammatory immune responses (Schofield and Grau 2005). *P. falciparum* encodes a family of cell-surface proteins that allow pRBCs to bind to deep microvascular endothelial beds within organs. These localized pRBCs can lead to organ pathology by directly blocking blood flow and by attracting proinflammatory immune cells into the tissue (Engwerda et al. 2005a and b). Many studies indicate that while Th1 responses and the production of proinflammatory Th1 cytokines, such as IFN-γ and IL-12, are involved in the control of malaria (Batchelder et al. 2003; Favre et al. 1997; Stevenson et al. 2001), overproduction of these cytokines and IL-18 can lead to disease pathogenesis. Therefore, a balance exists between appropriate proinflammatory responses and uncontrolled inflammation leading to pathogenesis. The early production of IL-12 by DCs and IFN-γ by NK and natural killer T (NKT)-cells plays an important role by directing downstream immune responses, including the differentiation of CD4$^+$ T cells into Th1 and Th2 subsets, and the stimulation of protective antibody production by B cells (Stevenson and Urban 2006; Hansen et al. 2007; Hansen and Schofield 2004; Hansen et al. 2003). Regulatory cytokines, such as IL-10 and TGF-β, are important in dampening "runaway" proinflammatory responses and aid in the prevention of disease pathology (Li and O'Malley 2003; Omer and Riley 1998; Omer et al. 2003). The literature contains conflicting reports concerning the effects of sex on clinical disease in humans. While there seems to be a consensus that the incidence of clinical *Plasmodium* infection is similar between males and females (Venugopalan et al. 1997; Wildling et al. 1995), studies have suggested that males have higher levels of parasitemia (Wildling et al. 1995; Landgraf et al. 1994), but females have higher mortality rates (Kochar et al. 1999). On the other hand, the vast majority of animal models of disease suggest that males are more susceptible than females. Rodent models of disease using *P. chabaudi* show that female C57BL/10 (Benten et al. 1992a and b) and C57BL/6 (Cernetich et al. 2006) mice have less infection-induced mortality compared with male mice. In the latter study, females also recovered from anaemia and infection-associated weight loss faster, and exhibited higher expression of IL-10, IL-12Rβ, and IFN-γ and protective antibodies than males. Testosterone seems to be a key modulator in the sexual dimorphism exhibited during *P. chabaudi* infection, as testosterone treatment of female mice can prevent self-healing, whereas castration of male mice leads to self-healing (Benten et al. 1992a,b). Interestingly, in this disease model, testosterone is not signaling through classical androgen receptors or estrogen receptors (Benten et al. 1992a, 1992b ; Benten et al. 1993; Wunderlich et al. 2002). Even testosterone administered to female C57BL/10 mice orally is capable of increasing mortality rates and is associated with a decreased capacity of peritoneal cells to generate reactive oxygen species (Mossman et al. 1997). Testosterone treatment of female mice results in increases of CD8$^+$ T cells in the spleen and decreases in overall splenocyte cell numbers (Benten et al. 1993). Interestingly, female C57BL/10 mice treated with testosterone for 3 weeks and then left untreated for 12 weeks exhibit similar susceptibility to disease as mice treated with testosterone for 3 weeks followed by immediate infection (Benten et al. 1997; Wunderlich et al. 1992). Mortality is seen in the mice that were left untreated for 12 weeks, despite the fact that testosterone-induced alterations in spleen cell composition and phenotype had been reversed. These findings indicate that genetically controlled host resistance undergoes hormonal imprinting and, following initial induction by testosterone, suppressive mechanisms remain intact, even after its withdrawal (Benten et al. 1997). Once mice have acquired immunity against *P. chaubaudi*, they become unresponsive to testosterone and do not undergo immune suppression in response to testosterone treatment (Wunderlich et al. 1992). The immunosuppressive effect of testosterone has important implications in therapeutic approaches as male mice and testosterone-treated female mice have markedly reduced vaccine protection compared with untreated female mice (Wunderlich et al. 1993). The use of gene-deficient mice has further served to dissect how testosterone could exacerbate disease. The lymphotoxin β receptor is involved in the signaling pathway of the LT$\alpha\beta_2$ heterotrimer as well as several members of the TNF family (reviewed in Tumanov et al. 2007). Alterations to these pathways can result in major defects in spleen architecture as well as the loss of some secondary lymphoid tissues, resulting in increased susceptibility to infection (Futterer et al. 1998; Fu and Chaplin 1999). However, while

$Lt\beta R^{-/-}$ males are more susceptible than WT males to disease (Krucken et al. 2005), $Lt\beta R^{-/-}$ females are even more resistant than WT females, although this resistance is still suppressible by testosterone (Krucken et al. 2005). Treatment of resistant female $Lt\beta R^{-/-}$ mice with testosterone causes severe pathology in the liver, indicating that the liver is intimately involved in testosterone-mediated immune dysregulation (previous studies focused mainly on the immune modulation seen in the spleen) (Wunderlich et al. 2005; Krucken et al. 2005). Susceptibility is associated with downregulation of *Dusp1* and *Sult2a* gene expression which are upstream of ERK1/2 activation and DHEA sulfate production, respectively both of which are associated with protection against *Plasmodium* infection. Interestingly, estrogen appears to also have an immunosuppressive effect in murine *P. chabaudi* infections, although similar to testosterone, estradiol is not capable of suppressing responses in mice already immune to *P. chabaudi* (Benten et al. 1992a and b). Estradiol suppresses the normally self-healing infections of female and castrated male C57BL/6 mice. This suppressive effect is abrogated by the estrogen receptor blockers, tamoxifen and clomifene. In contrast to the decrease in $CD8^+$ T-cells and overall splenic cell numbers seen in testosterone-treated mice, estradiol-treated mice exhibit reduced numbers of $CD4^+$ T-cells. The immunosuppressive effects of testosterone, but not estradiol, can be adoptively transferred via splenocytes into syngeneic mice, suggesting that T cells may be directly modulated by testosterone (Benten et al. 1992a and b; Benten et al. 1993). To further support this notion, studies by Zhang et al. (2000) found that while male $IL-4^{-/-}$ and $IFN-\gamma^{-/-}$ mice were more susceptible to disease than wild-type (WT) male counterparts, no difference was seen between WT and $IFN-\gamma^{-/-}$ females, but when male WT and $IFN-\gamma^{-/-}$ mice were castrated, they exhibited a resistant phenotype similar to that seen in females (Zhang et al. 2000). Understanding how sex hormones mediate resistance and susceptibility to malaria will provide a foundation that can serve in the development of more efficacious therapies that will have a major impact on global health.

6.4.3 Toxoplasma gondii *Infections and Th1/Th2 Immunity: The Males Have It*

T. gondii is one of the exceptions to the female host supremacy paradigm, with females being more susceptible to disease than males. *T. gondii* is one of the most prevalent human parasites, with a worldwide distribution and seropositivity of about 30% (Lang et al. 2007). Although *T. gondii* is capable of infecting almost any mammalian nucleated host cell, its definitive hosts are members of the *Felidae* family. Intermediate hosts can become infected via a number of different mechanisms, including ingestion of undercooked meat containing *T. gondii* cysts, fecal–oral transmission of oocysts shed directly from cats, blood transfusions, and organ transplantation from infected donors, and vertical transmission from an infected mother to her fetus. Most information concerning immunity to *T. gondii* infection has been established using murine models. Control of *T. gondii* infection is dependent on the ability to mount a Th1 response, relying on IFN-γ production by both the innate and adaptive arms of the immune system. During the early innate immune response, IL-12 production by DCs, neutrophils, and macrophages is important in driving IFN-γ production by NK cells (reviewed by Denkers 2003). IL-12 production by innate immune cells is most likely downstream of parasite recognition by toll-like receptors and a MyD88-dependent signaling cascade. This is supported by the fact that MyD88-deficient mice have a significant defect in IL-12 production in response to soluble tachyzoite antigen and higher mortality resulting from insufficient IFN-γ production (Scanga et al. 2002). In addition to its role in innate responses, IL-12 also drives the development and proliferation of CD4$^+$ Th1 cells and CD8$^+$ T cells, both of which are important sources of IFN-γ. IFN-γ plays a key role during infection by inducing the expression of several enzymes involved in mediating anti-toxoplasma immunity. These enzymes include several members of the P27 GTPase family, including interferon inducible GTPase (Taylor et al. 2000), LRG-47, and IRG-47 (Collazo et al. 2001), which could result in lysosomal degradation of the parasite or induction of NOS2 (Yap and Sher 1999). Despite the fact that IFN-γ responses have potent microbicidal capabilities, *T. gondii* is able to evade these responses in hosts and survive latently as bradyzoites within cysts. Following potent IFN-γ-dependent responses, hosts must dampen potentially toxic pro-inflammatory immune sequelae by secreting anti-inflammatory proteins, including IL-10 and lipoxin A$_4$. *T. gondii* may play a role in lipoxin secretion, potentially allowing for its escape from host proinflammatory defenses (reviewed by Aliberti 2005). Although most infections in immunocompetent humans are asymptomatic, lymphadenopathy can be a clinical symptom. In sexually mature adults infected with toxoplasmosis, lymphadenopathy is more commonly seen in females than in males (Beverley et al. 1976). Additionally, toxoplasmic encephalitis is more common in female AIDS patients (Phillips et al. 1994). In mouse models of disease, female mice are more susceptible to disease than are males (Roberts et al. 1995; Johnson et al. 2002; Walker et al. 1997; Liesenfeld et al. 2001). A study by (Roberts et al. 1995) found that BALB/K male and female mice exhibit temporal differences in cytokine production, in which male mice produce more IFN-γ earlier during infection. The inability of female mice to mount an early IFN-γ response may be responsible for higher mortality and parasite burdens. Male BALB/c SCID mice also have longer survival times, lower parasite burdens, and less severe immunopathology than do female SCID mice. Male SCID mice produce IL-12 more quickly than females and also produce higher levels of IFN-γ more rapidly postinfection (Walker et al. 1997). Another study found that female mice are more susceptible to peroral infection with *T. gondii*, with female mice dying earlier, having higher numbers of tachyzoites, and exhibiting severe necrosis of the small intestine than males. Treatment of female mice with testosterone reduces intestinal parasite numbers and disease pathology (Liesenfeld et al. 2001). Although testosterone abrogates disease, many studies report that female sex hormones exacerbate disease. Gonadectomized guinea pigs and mice treated with the synthetic female hormone hexoestrol exhibit overwhelming disease compared with untreated controls (Kittas and Henry 1979; Kittas and Henry 1980). Additionally, treatment of ovariectomized female mice with pharmacological doses of potent estrogen compounds, including 17β-estradiol, diethylstilbestrol, or α-dienestrol, renders mice more susceptible to disease, as measured by brain cyst formation. Treatment of ovariectomized female mice with physiological concentrations of estrogen, however, had no effect on disease outcome, indicating that pharmacological, but not physiological, concentrations of estrogen mediate susceptibility (Pung and Luster 1986). Treatment of ovariectomized female mice with pharmacological doses of potent estrogen compounds, including 17β-estradiol, diethylstilbestrol, or alpha-dienestrol, renders mice more susceptible to disease, as measured by brain cyst formation (Pung and Luster 1986). During pregnancy, both estrogen and progesterone levels are high. These hormones are associated with a dampening of Th1 immune responses in order to prevent the mother's immune response from attacking the fetus (Raghupathy 1997). Therefore, it is not surprising that pregnant female mice are even more susceptible to disease than nonpregnant females, exhibiting a decreased capacity to produce IFN-γ (Shirahata et al. 1992). Administration of Th1-associated cytokines such as IL-12 and IFN-γ can increase survival of *T. gondii*-infected pregnant mice (Shirahata et al. 1993). This is discussed in significant detail in Chapter 9 of this book. There is the school of thought that males have a more robust innate immune response whereas females develop stronger adaptive immune responses leading very often to pathologies in females associated with immune hyperactivity. Certainly, the earlier inflammatory response of male SCID mice to this parasite would be consistent with this view point as would the enhanced pathology in chronically infected intact female mice (Roberts et al. 1995). Furthermore, it has been postulated (Yap et al. 2006) that as part of its life history strategy, *T. gondii* has evolved potent mechanisms of stimulating the innate immune system to not only control initial parasite growth, but also to direct the developing adaptive immune response so that a chronic, but asymptomatic, disease state can be maintained to the benefit of both host and parasite survival. To achieve this outcome, *T. gondii* has a number of TLR ligands, including GPI-anchors (Debierre-Grockiego et al. 2007), HSP70 (Mun et al. 2005; Aosai et al. 2006), and profilin (Lauw et al. 2005), as well as *T. gondii* cyclophilin 18 which induces IL-12 production through ligation of the CCR5 receptor (Aliberti et al. 2003). These processes activate macrophages, DCs, and NK cells resulting in the production of IL-12, IFN-γ, TNF-α, and NOS2 which in turn control parasite growth, induce Th1 cell expansion, and facilitate the development of cytolytic CD8$^+$ T cells (Parker et al. 1991; Johnson et al. 2002; Combe et al. 2005). It is significant in this context, therefore, that anti-microbial activity is significantly enhanced in murine macrophages deficient in ERα which also have enhanced ability to induce T cell proliferation and IFN-γ production (Lambert et al. 2004).

6.4.4   Trypanosomes

6.4.4.1   American Trypanosomiasis

Trypanosomes cause two major forms of disease in humans: American trypanosomiasis or Chagas' disease, and African trypanosomiasis, also known as sleeping sickness. Chagas' disease, caused by *Trypanosoma cruzi* and carried by the triatome bug, is responsible for approximately 50,000 deaths per year, primarily in Latin America. In humans, Chagas' disease has an early acute phase characterized by circulating parasites, followed by a chronic phase characterized by tropism for the myocardium and enteric nervous system (Morel and Lazdins 2003). Chronic disease can occur as late as 30 years post-infection (Andrade and Andrews 2005), although most patients never proceed to the chronic stage of disease and remain asymptomatic carriers (Dutra et al. 2005). Deaths are mainly due to chronic chagasic cardiomyopathy (CCC), which is characterized by cardiomegaly, severe arrhythmias, and heart failure (Higuchi et al. 2003), although digestive dysfunctions in patients with enteric disease may occur and result in death (Dutra et al. 2005). Both innate and adaptive arms of the immune system are critical in controlling *T. cruzi* infection during the acute phase. TLR signaling in macrophages is critical in the activation of CD4$^+$ T cells (Tarleton 2007). These CD4$^+$ T cells produce IFN-γ, which is critical in the activation of macrophages and CD8$^+$ T cells, both of which play a role in parasite clearance (Silva et al. 2003; Tarleton 1990). During acute disease, however, massive thymocyte apoptosis occurs which may contribute to the prevention of hosts from effectively clearing disease (reviewed in Savino et al. 2007). During chronic disease, T cells appear to play a role in the inflammatory pathogenesis associated with CCC (Cunha-Neto and Kalil 2001; Gomes et al. 2003). Two major hypotheses exist to explain the pathogenesis of chagasic myocarditis: the autoimmune hypothesis and the parasite persistence hypothesis. The autoimmune hypothesis states that disease pathology is due to autoantibodies or autoreactive T cells that result from molecular mimicry between parasite and host antigens. The parasite persistence hypothesis posits that inflammatory damage is due to immune responses directed at clearing parasites, and is based on the fact that *T. cruzi* does remain in heart tissue and treatments that reduce parasite burdens in the heart do result in disease abrogation (reviewed in Girones and Fresno 2003). Girones and Fresno suggest that aspects of both hypotheses may be at work in the pathogenesis seen in clinical cases. Although a great deal remains to be understood about the role of the immune system during disease clearance and pathogenesis, research has shown that sex hormones do play a role in its modulation. It is well documented that in experimental infections, male mice and rats are more susceptible to disease than females (Chapman et al. 1975; de Souza et al. 2001; dos Santos et al. 2005; D'Ambrosio et al. 2008). In addition, males are clinically more likely to develop severe cardiomyopathies (Basquiera et al. 2003) and exhibit abnormal electrocardiograms more often than females (Brabin and Brabin 1992). Male mice also are more susceptible to thymocyte apoptosis than females (Mucci et al. 2005).

Resistance appears to be directly linked to the presence of female sex hormones, as ovariectomized females are more susceptible to disease (Chapman et al. 1975; Santos et al. 2007; D'Ambrosio et al. 2008). A role for male sex hormones in mediating susceptibility is less clear, as gonadectomized males are reported to have reduced parasite burdens (D'Ambrosio et al. 2008) and unaltered parasitemia and mortality (Chapman et al. 1975) when compared with gonadally intact males, which likely reflects differences in mouse and parasite strains. Interestingly, treatment of female mice with high pharmacological doses of 17β-estradiol increases parasitemia and mortality, whereas low physiological doses have either no effect or reduced parasite burden and death (de Souza et al. 2001). Limited research has been performed to study differences in the levels of immune factors between the sexes, although male mice produce higher levels of TNF-α and IL-10 compared with female mice (do Prado et al. 1998; Schuster and Schaub 2001). More research needs to be performed to further understand how sex hormones modulate immune cells and their responses during *T. cruzi* infection.

6.4.4.2 African Trypanosomiasis

African trypanosomiasis represents a particularly devastating disease in Africa, with 60 million people at risk in 36 countries (Barrett et al. 2003) and the number of cases rising each year (Simarro et al. 2008). African trypanosomiasis is responsible for 50,000 deaths a year, and is almost always fatal in the absence of treatment (Barrett et al. 2003). Human African trypanosomiasis is caused by *T. brucei gambiense* and *T. brucei rhodesiense*, while experimental models of disease often employ the animal parasite *T. brucei brucei*. African trypanosomes are distinct from most protozoan parasites in that they exist extracellularly within their hosts. Disease is characterized by two distinct stages – an early hemolytic stage followed by a late meningoencephalitic stage. The late stage is characterized by leukocyte infiltration into the CNS, astrocyte, and microglial activation, and acute neuroinflammation ultimately resulting in coma and death (Bisser et al. 2006). African trypanosomes have developed many mechanisms to evade host immune responses, but they are most well known for their ability to switch their immunodominant variant surface glycoprotein (VSG), thus always keeping one step ahead of B-cell-mediated clearance (Barry and McCulloch 2001). VSG is also responsible for polyclonal B cell activation, which can lead to the generation of autoantibodies (Kazyumba et al. 1986) and immune complex disease (Lambert et al. 1981) in humans. VSG also drives Th1 immune responses in hosts (Schleifer et al. 1993), and while IFN-γ does aid in host defense via macrophage activation, paradoxically it also serves as a parasite growth factor (Bakhiet et al. 1993) and mediates parasite passage across the blood–brain barrier (Masocha et al. 2004). Furthermore, while activated macrophages are responsible for trypanocidal activity (Mansfield et al. 2002), they also are responsible for immunosuppression via NO, prostaglandins, and TNF-α (reviewed in Mansfield and Paulnock 2005). Therefore, immune responses during African trypanosomiasis represent a complex balance in which parasites have successfully manipulated host defenses to their own advantage. Sex hormones play a role in modulating immunity to African trypanosomes, as researchers have found that female mice have lower parasitemia and survive longer than their male counterparts (Greenblatt and Rosenstreich 1984). Additionally, during human disease, males have trypanosomes in their cerebral spinal fluid more often than females, and have more relapses following treatment (Pepin et al. 1994). Unfortunately, no further research has taken place into determining the roles of specific sex hormones during disease, although it has been shown that infection can result in hypogonadism and decreased testosterone and estradiol levels, both clinically and experimentally (Soudan et al. 1993, Soudan et al. 1992, Hublart et al. 1990, Boersma et al. 1989). Hopefully, future research will further elucidate the roles of sex hormones in modulating immunity to disease so that more efficacious treatments can be developed.

6.4.5   Other Protozoans – a Mixed Bag

6.4.5.1   Giardiasis and Amebiasis

While not as extensively studied, other parasites for which females are more resistant than males include *Giardia* and *Entamoeba* species. While *Giardia* and *Entamoeba* are both intestinal protozoans transmitted fecal–orally, amoebiasis is a much more severe and clinically significant disease, causing 40–100,000 deaths each year (reviewed in Stanley 2003). Giardiasis most often presents as a self-limiting, non-invasive disease causing diarrhoea and intestinal malabsorption, (reviewed by Roxstrom-Lindquist et al. 2006), whereas *Entamoeba* causes ameobic colitis and ulcers, and is capable of invading other organs such as the liver, resulting in amoebic liver abscesses (ALA) (reviewed by Garcia-Zepeda et al. 2007). Although host defenses against both parasites are not well characterized, research indicates that both innate and adaptive immunity play a role in parasite clearance. Neutrophils play an important role in both diseases, as human neutrophil defensins have anti-giardial properties *in vitro* (Aley et al. 1994), and several studies have shown that neutrophil-depleted mice have more severe hepatic and intestinal amoebiasis (reviewed by Guo et al. 2007). An *in vivo* role for macrophages and NO-mediated killing has only been demonstrated in invasive amebiasis (Seydel et al. 2000; reviewed by Faubert 2000 and Roxstrom-Lindquist et al. 2006). Although mast cells play a critical role in controlling giardiasis (Li et al. 2004), their role in amebiasis has yet to be determined (Guo et al. 2007). Both B cell responses and secretory IgA production (Langford et al. 2002; Haque et al. 2006), and Th1 responses and IFN-γ production (Venkatesan et al. 1997; Seydel et al. 2000; Haque et al. 2007) are key to protective immunity in both diseases. Although there is a great deal more to understand about the fundamental immune responses against giardiasis and amebiasis, researchers have found that the host sex can influence the outcome of both diseases. Most human clinical studies find no difference in disease prevalence between the sexes; one study performed in Brazil, however, found that male dogs have a higher prevalence of giardiasis than their female counterparts (Oliveira-Sequeira et al. 2002). Studies performed using C57BL/6 mice infected with *G. muris* found that while males and females shed similar numbers of cysts during acute stages of disease (Daniels and Belosevic 1995b), males begin passing cysts earlier (Daniels and Belosevic 1995a) and continue to pass cysts for a significantly longer time (day 60 for males; day 18–20 for females) than females (Daniels and Belosevic 1994; Daniels and Belosevic 1995a, 1995b). In addition, male mice have higher parasite burdens in their gut (Daniels and Belosevic 1995a and 1995b), and more prolonged disaccharidase deficiency (Daniels and Belosevic 1995a) than female mice. Few studies have investigated the mechanisms behind female resistance to giardiasis, but one study using C57BL/6 mice showed that females have elevated levels of parasite-specific IgG and stronger IgG2b and IgG3 responses than males (Daniels and Belosevic 1994). Similar to giardiasis, the prevalence of infection with *Entamoeba* in humans is similar between males and females, although invasive disease predominates in males (Acuna-Soto et al. 2000). Studies in mice parallel what is seen in human disease, with female mice clearing ALA within 3 days and male mice recovering after day 14. In mice, females have significantly more early IFN-γ-producing cells compared with male mice that have higher levels of IL-4-producing cells. In females, the IFN-γ is produced, in part, by NKT cells, as NKT-deficient female mice have exacerbated ALA compared with their WT counterparts (Lotter et al. 2006). In addition, other studies have shown that female mice tend to mount stronger antiamebic antibody responses than do male mice (Moreno-Fierros et al. 1992). Although no current studies have been performed to date, research using gonadectomized mice will aid in understanding the role of specific sex hormones in modulating immunity to *Entamoeba* and *Giardia*.

6.4.5.2 Babesiosis

Babesiosis is an emerging zoonotic pathogen carried by ixodid ticks, existing in an erythrocytic form within mammalian hosts. Babesiosis occurs primarily in temperate regions of the United States and Europe. In Europe, disease is caused by the cattle parasite *Babesia divergens*, whereas cases in the US are primarily due to infection with the rodent parasite *B. microti* and occasionally the recently discovered species WA1 (reviewed by Homer et al. 2000). *B. divergens* infections are very rare, but when they do occur they usually afflict asplenic individuals, and are very serious with a 42% mortality rate (Gorenflot et al. 1998). In the US, babesiosis affects both spleen-intact and asplenic individuals, and is usually self-limiting and sometimes asymptomatic (Homer et al. 2000), although coinfection with *B. microti* may exacerbate Lyme disease and complicate its treatment (Krause et al. 1996). Both innate and adaptive immune responses play a role in immunity against babesiosis, although the roles of specific immune effectors are still unclear. Both NK cells and macrophages may play a role in innate responses, as high levels of NK cell activity are seen in resistant strains of mice infected with *B. microti* (Eugui and Allison 1980), and macrophage depletion is associated with increased susceptibility to disease (Mzembe et al. 1984; Saeki and Ishii 1996). IFN-γ producing CD4$^+$ T cells also are important in parasite clearance (Igarashi et al. 1999), although an absolute role for T-cell-mediated immunity is unclear, as C57BL/6 SCID mice are highly resistant to disease, despite their lack of T cells (Aguilar-Delfin et al. 2001). These discrepancies may be due to the fact that different *Babesia* species and strains of mice have been used in experimental studies. A similar discrepancy is seen in research focusing on sex differences during infection. Researchers studying *B. microti* have found that male mice are more susceptible to disease (Goble 1966); whereas those studying WA1 report that female mice are more susceptible (Aguilar-Delfin et al. 2001). While the mechanisms behind female susceptibility to WA1 remain to be studied, research into male susceptibility to *B. microti* indicates that testosterone causes longer and more severe infections in mice that have been castrated and implanted with testosterone compared to those castrated and implanted with inert oil (Hughes and Randolph 2001a). Additionally, higher-ranking mice within a home cage have higher levels of testosterone, which is associated with depressed levels of serum immunoglobulin and reduced resistance to infection with *B. microti* (Barnard et al. 1994). Interestingly, ixodid ticks preferentially attach to rodents with high testosterone levels (Hughes and Randolph 2001b). Although testosterone exacerbates disease, estrogens do not confer resistance to disease to *B. microti*, as one study found that mice pretreated with 17β-estradiol have similar patterns of infection as control mice (Wood and Clark 1982). More research needs to examine both the basic immune responses against *Babesia* parasites, and the role that sex hormones play during disease, with special attention to differences seen between parasite species and mouse strains.

6.4.5.3 Trichomoniasis

Trichomoniasis is a sexually transmitted disease (STD) caused by the flagellated protozoan *Trichomonas vaginalis*. Over 170 million people are infected worldwide (World Health 1995), and in inner city areas of the United States the prevalence can approach 25% in STD clinics (reviewed in Schwebke and Burgess 2004). *T. vaginalis* causes vaginitis in women that can persist for as long as 3–5 years (Bowden and Garnett 2000); although it is estimated that as many as 50% of cases are asymptomatic (Fouts and Krause 1980). The vast majority of cases in men are asymptomatic, with the course of infection lasting 10 days or less (Krieger 1990), although trichomoniasis may be emerging as an important cause of non-gonococcal urethritis (Krieger et al. 1993). Trichomoniasis is of particular concern, as resistance to standard metronidazole treatment is emerging (Schmid et al. 2001). Infection in pregnant women also is associated with low birth weights and preterm deliveries (Cotch et al. 1997). Infection also is correlated with an increased frequency of HIV contraction in women (Laga et al. 1993; Leroy et al. 1995) and increased transmission by men (Hobbs et al. 1999a). Very little is known about the immunology of trichomoniasis because there is no suitable animal model to study the disease. However, it is known that natural immunity to infection is only partially protective, with up to 30% of patients experiencing reinfection following treatment (Niccolai et al. 2000). Antibodies may mediate partial protection (Abraham et al. 1996), especially those directed against putative parasite adhesion molecules (Arroyo et al. 1992). Studies using euthymic and athymic BALB/c mice also suggest that T cells may play a role in clearing infection (Martinotti et al. 1988). As mentioned above, females may have very prolonged infections with *T. vaginalis*, whereas clinical cases in men are rare (Van Andel et al. 1996) and of a very short duration, indicating that sex plays a role during infection, with women being more susceptible to disease than men. This is not surprising, as studies have found that *T. vaginalis* has androgen and estrogen receptors on its cell surface (Ford et al. 1987). Interestingly, in order to study *T. vaginalis* in laboratory rodents, female mice must receive estrogen treatments to establish disease (Cappuccinelli et al. 1974; Van Andel et al. 1996). Similarly, in clinical studies, women volunteers also require estrogenization to establish the disease (Azuma 1968). Furthermore, conditions associated with high levels of estrogen, such as menses and pregnancy, exacerbate *T. vaginalis* infections (Brown 1972). 17β-estradiol enhances the *in vitro* growth of parasites, whereas testosterone and progesterone inhibit growth at early time points (Stein and Cope 1933; reviewed by Garber et al. 1991). Other studies using euthymic and athymic BALB/c mice found that females are more susceptible than males to developing abscesses following subcutaneous injections of *T. vaginalis* (Landolfo et al. 1981; Martinotti et al. 1988). Conversely, some *in vitro* studies have found that estrogen inhibits *T. vaginalis* growth (Sugarman and Mummaw 1988) and inhibits a virulence factor, cell-detaching factor, which is correlated with disease severity (Garber et al. 1991). This may explain why intravaginal estrogen pellets alleviate disease symptoms (Garber et al. 1991). Developing a better understanding of how estrogen and other sex hormones modulate immunity to trichomoniasis will lead to better treatment and prevention options, although a lack of a good animal model will make this a difficult endeavor.

6.4.6 Sexual Dimorphism in Response to Helminth Parasites: Female Th2 Dominance

Protective immunity against helminth parasites is generally dependent on the development of a strong Th2 response involving IL-4 and IL-13, in particular (reviewed by Finkelman et al. 2004), and for the most part females are more resistant than males. This is, particularly, true for gut parasitic nematode infections. For example, in a C57BL/6 experimental model, significantly higher numbers of *Strongyloides ratti* are found in males than in similarly infected females. Ovariectomy of females has no effect, but injection of testosterone into females or males increases gut worm burdens (Kiyota et al. 1984). *Nippostrongylus brasiliensis* infection of Indian soft-furred rats reveals that males are less able than females to expel intestinal worms and egg release through feces also is prolonged in male rats (Tiuria et al. 1994). Gonadectomy significantly reduces worm burdens in male rats whereas ovariectomy has no effect on parasite burdens in females (Tiuria et al. 1994). Thus, increased susceptibility of males to gut parasitic nematodes may be direct result of androgenic as opposed to estrogenic influences on immunity, as evident in a majority of parasitic infections (Table 6.4). How androgens mediate these effects awaits clarification. However, estrogens can increase the resistance of male CD1 mice against *T. spiralis* as measured both by adult worm burdens and tissue larvae (Reddington et al. 1981). Gonadectomy also increases the resistance of males while testosterone treatment increases susceptibility of females. Consequently, both testosterone and estradiol influence control of *T. spiralis* in a reciprocal counter-regulatory fashion. Control of *T. spiralis* is Th2 dependent, but mast cells also play a crucial role in worm expulsion (Knight et al. 2008). Estrogen at physiological concentrations not only activates mast cells via a nongenomic ERα to release a substantial amount of preformed mediators but also could stimulate synthesizes of new mediators (Zaitsu et al. 2006). Conversely, testosterone inhibits mast cell secretion of histamine and serotonin following stimulation by compound 48/80 or neuropeptide substance P, a process promoted by estrogen (Vliagoftis et al. 1992). Although differences in how sex hormones influence mast cell activity may explain why females can expel some gut parasitic nematodes better than males, these cells do not play a significant role in control of all gut parasitic nematodes, *Trichuris muris* (Bancroft et al. 1998) being a significant example. Here studies utilizing cytokine deficient mice might reveal underlying mechanisms modulating sex hormone influences on immunity and outcome of parasite infection (Bancroft et al. 2000; Hayes et al. 2007). Both WT female BALB/c and C57BL/6 mice are resistant to *T. muris* infection as are female BALB/c IL-4$^{-/-}$ (Bancroft et al. 2000). Male IL-4$^{-/-}$ of both BALB/c and C57BL/6 strains and female C57BL/6 IL-4$^{-/-}$ mice are susceptible to *T. muris* (Bancroft et al. 2000). The observation that female BALB/c IL-4$^{-/-}$ mice are still resistant to infection may reflect their continued ability to generate IL-13, which is not evident in C57BL/6 mice or male BALB/c IL-4$^{-/-}$ animals (Bancroft et al. 2000). Administration of rIL-13 to normally susceptible male BALB/c IL-4$^{-/-}$ mice induces worm expulsion while neutralization of IL-13 in BALB/c IL-4$^{-/-}$ mice inhibits expulsion (Bancroft et al. 2000). The authors suggest a link in this model between estrogen and the ability to produce IL-13 which is the main mediator of resistance against this parasite (Bancroft et al. 1998). Further studies utilizing the TNF-α receptor p55$^{-/-}$ and p75$^{-/-}$ mice on the C57 BL/6 background reveal further sexual dimorphism in response to this parasite (Hayes et al. 2007). Although not classically a Th2 cytokine, TNF-α protects against infection with *T. muris*. While female p55$^{-/-}$ and p75$^{-/-}$ mice fully expel the parasites by day 35, male mice harbor chronic infections and p55$^{-/-}$ male mice are more susceptible than p75$^{-/-}$ animals. The observed sex difference is reversed by treating female mice with neutralizing IL-13 and male mice with rIL-13. While IL-13 levels in TNF-αR deficient male mice are similar to male WT mice, p55$^{-/-}$ male mice have increased IFN-γ production which is not evident in the equivalent female mice at day 21. These studies clearly demonstrate the influence as well as complexity of sex hormones in modulating the Th2 balance in favor of females during gut parasitic infections. While females are invariably more resistant to infection with helminthic parasites including filarial nematodes, in which Th2 responses promote resistance to infection, a notable exception has recently been described. During *Litomosoides sigmodontis*, filarial survival is reduced in male compared with female BALB/c mice, whereas the prevalence and density of micro-filariae are higher in female than male mice (Graham et al. 2005). C57BL/6 mice are more resistant than BALB/c mice to *L. sigmodontis* and resistance in C57BL/6 mice is Th2-dependent, despite the fact that BALB/c mice develop potent Th2 responses (Le Goff et al. 2002). Susceptibility of BALB/c mice appears to be caused by the genera-tion of potent regulatory T cell responses that overcome Th2 effector functions that permit survival of the adult parasite (Taylor et al. 2005 and 2007). Of particular significance is that estradiol at physiological levels expands $CD4^{+}CD25^{+}FoxP3^{+}$ regulatory T cells expressing IL-10 (Tai et al. 2008) and IL-10 is essential for microfilariae persistence (Hoffman et al. 2001). Laboratory models do not always mimic natural human infection. Thus, while human males are more susceptible to *Schistosoma mansoni* infection than females (Degu et al. 2002), males are more resistant to experimental murine infections and this is mediated by testosterone (Eloi-Santos et al. 1992; Nakazawa et al. 1997). Nevertheless, while murine studies examined the early stages of infection with this parasite, human studies concentrated on chronic disease and consequently the comparison is not ideal. Thus, sex differences in immune responses to parasites at different stages of infection is a further complicat-ing dimension that has to be considered when examining the influence of sex on disease outcome. Indeed, the ebb and flow of sex-mediated influences over the course of an infection may result in no apparent difference in disease phenotype during chronic disease. For example, no sex-related difference in disease phenotype is noted during chronic *S. haemotobium* infections and yet at that time females have a comparatively weaker Th1 response and produce significantly more IL-10 and TGF-β than males (Remoue et al. 2001) indicating a more pronounced regulatory T cell response at that stage.

6.5 Implications for Human Disease/Consideration for Drug and Vaccine Development

This chapter has presented data to show that males and females develop distinct immune responses to various parasitic diseases. While females are usually less susceptible to males, there are notable differences that further complicate the understanding of an already complex interaction between sex hormones and the immune system. A great deal of the research in this field focuses on *in vitro* systems and animal models of disease, which do not always translate into the clinical picture seen during human disease. This issue is very important during the design and development of novel therapeutic agents. The literature demonstrates that hormonal differences between the sexes can influence both the innate and adaptive arms of immunity. Differences in both T and B cell responses have been noted, which has critical implications in studies focusing on vaccine development and efficacy. The research in this field has demonstrated a clear need for treatment options that are tailored to individual patients, taking into account age, sex, and genetics.

6.6   Future Direction for the Field

Overall, the immune response that dominates in one or the other sex is largely dependent on the parasite being investigated. With regard to inflammatory disease, investigators working with autoimmunity would suggest that the higher incidence in females is a result of a Th1-biased response in this sex, while those working with allergic diseases would highlight the higher female incidence is as a result of a Th2-dominant response. Studies of parasite infections also demonstrate that females mount a more vigorous response to control the disease whether it is Th2 dominant to control infection with helminth pathogens or Th1-biased response to control intracellular parasites. However, there are numerous examples that contradict this simple model and we have to acknowledge the complexity and potential chronicity of the host–parasite interplay. Several tissue sites and immune cells may be involved at any one time, all of which may be responsive to sex hormone influences. We currently have new molecular tools available which allow us to at last dissect and characterize these various interactions. Hormone receptor knockout mice have, for example, allowed us to differentiate the roles of ERα and ERβ globally in vivo and at the single cell level in vitro. The technology to create second generation temporal (switch on/switch off) and spatial knockout (tissue specific) as well as gene reporter mice will enable us to identify how and when cells are affected by hormones to impact sex differences in responses to parasites.

References

Abraham M, Desjardins MC, Filion LG, Garber GE (1996) Inducible immunity to *Trichomonas vaginalis* in a mouse model of vaginal infection. Infect Immun 64:3571–3575

Acuna-Soto R, Maguire JH, Wirth DF (2000) Gender distribution in asymptomatic and invasive amebiasis. Am J Gastroenterol 95:1277–1283

Aguilar-Delfin I, Homer MJ, Wettstein PJ, Persing DH (2001) Innate resistance to *Babesia* infection is influenced by genetic background and gender. Infect Immun 69:7955–7958

Aguilar-Delfin I, Wettstein PJ, Persing DH (2003) Resistance to acute babesiosis is associated with interleukin-12- and gamma interferon-mediated responses and requires macrophages and natural killer cells. Infect Immun 71:2002–2008

Alexander J (1988) Sex differences and cross-immunity in DBA/2 mice infected with *L. mexicana* and *L. major*. Parasitology 96:297–302

Alexander J, Stimson WH (1988) Sex hormones and the course of parasitic infection. Parasitol Today 4:189–193

Aley S, Zimmerman M, Hetsko M, Selstedt ME, Gillin FD (1994) Killing of *Giardia lamblia* by cryptins and cationic neutrophil peptides. Infect Immun 62:5397–5403

Aliberti J (2005) Host persistence: exploitation of anti-inflammatory pathways by *Toxoplasma gondii*. Nat Rev Immunol 5:162–170

Aliberti J, Valenzuela JG, Carruthers VB, Hieny S, Andersenn JM, Charest H, Reis-Sousa C, Fairlamb A, Ribeiro JM, Sher A (2003) Molecular mimicry of a CCR5 binding-domain in the microbial activation of dendritic cells. Nat Immunol 4:485–490

Andrade LO, Andrews NW (2005) The *Trypanosoma cruzi*-host-cell interplay: location, invasion, retention. Nat Rev Microbiol 3:819–823

Aosai F, Rodriguez Pena MS, Mun HS, Fang H, Mitsunaga T, Norose K, Kang HK, Bae YS, Yano A (2006) *Toxoplasma gondii*-derived heat shock protein 70 stimulates maturation of murine bone marrow-derived dendritic cells via Toll-like receptor 4. Cell Stress Chaperones 11:13–22

Araneo BA, Dowell T, Diegel M, Daynes RA (1991) Dihydrotestosterone exerts a depressive influence on the production of interleukin-4 (IL-4), IL-5, and gamma-interferon, but not IL-2 by activated murine T-cells. Blood 78:688–699

Arroyo R, Engbring J, Alderete J (1992) Molecular basis of host epithelial cell recognition by *Trichomonas vaginalis*. Mol Microbiol 6:853–862

Azuma T (1968) A study of the parasitizing condition of *Trichomonas vaginalis* with special reference to the relationship between estrogen and the growth of *Trichomonas vaginalis*. J Jpn Obstet Gynecol 15:168–172

Bakhiet M, Olsson T, Mhlanga J, Buscher P, Lycke N, Van der Meide P, Kristensson K (1993) Human and rodent interferon-gamma as a growth factor for *Trypanosoma brucei*. Eur J Immunol 26:1359–1364

Bancroft AJ, McKenzie AN, Grencis RK (1998) A critical role for IL-13 in resistance to intestinal nematode infection. J Immunol 160:3453–3461

Bancroft AJ, Artis D, Donaldson DD, Sypek JP, Grencis RK (2000) Intestinal nematode expulsion in IL-4 KO mice is IL-13 dependent. Eur J Immunol 30:2083–2091

Baral E, Nagy E, Berczi I (1995) Modulation of natural killer cell-mediated cytotoxicity by tamoxifen and estradiol. Cancer 75:591–599

Barnard CJ, Behnke JM, Sewell J (1994) Social behaviour and susceptibility to infection in house mice (*Mus musculus*): effects of group size, aggressive behaviour and status-related hormonal responses prior to infection on resistance to *Babesia microti*. Parasitology 108:487–496

Barrett MP, Burchmore RJ, Stich A, Lazzari JO, Frasch AC, Cazzulo JJ, Krishna S (2003) The trypanosomiases. Lancet 362:1469–1480

Barry JD, McCulloch R (2001) Antigenic variation in trypanosomes: enhanced phenotypic variation in a eukaryotic parasite. Adv Parasitol 49:1–70

Basquiera AL, Sembaj A, Aguerri AM, Omelianiuk M, Guzman S, Moreno Barral J, Caeiro TF, Madoery RJ, Salomone OA (2003) Risk progression to chronic Chagas cardiomyopathy: influence of male sex and of parasitaemia detected by polymerase chain reaction. Heart 89:1186–1190

Batchelder JM, Burns JM Jr, Cigel FK, Lieberg H, Manning DD, Pepper BJ, Yanez DM, van der Heyde H, Weidanz WP (2003) *Plasmodium chabaudi adami*: interferon-gamma but not IL-2 is essential for the expression of cell-mediated immunity against blood-state parasites in mice. Exp Parasitol 105:159–166

Behnke JM, De Clercq D, Sacko M, Gilbert FS, Ouattara DB, Vercruysse J (2000) The epidemiology of human hookworm infection in the southern region of Mali. Trop Med Int Health 5:343–354

Benten WP, Wunderlich F, Mossman H (1992a) *Plasmodium chaubadi*: estradiol suppresses acquiring, but not once-acquired immunity. Exp Parasitol 75:240–247

Benten WP, Wunderlich F, Mossman H (1992b) Testosterone-induced suppression of self-healing *Plasmodium chabaudi*: an effect not mediated by androgen receptors? J Endocrinol 135:407–413

Benten WP, Wunderlich F, Herrmann R, Kuhn-Velten WN (1993) Testosterone-induced compared with oestradiol-induced immunosuppression against *Plasmodium chabaudi* malaria. J Endocrinol 139:487–494

Benten WP, Ulrich P, Kuhn-Velten WN, Vohr HW, Wunderlich F (1997) Testosterone-induced susceptibility to *Plasmodium chaubadi* malaria: persistence after withdrawal of testosterone. J Endocrinol 153:275–281

Benten WP, Benten M, Lieberherr M, Giese G, Wunderlich F (1998) Estradiol binding to cell surface raises cytosolic free calcium in T cells. FEBS Lett 442:349–353

Benten WP, Benten M, Lieberherr M, Giese G, Wrehlke C, Stamm O, Sekeris CE, Mossmann H, Wunderlich F (1999) Functional testosterone receptors in plasma membranes of T cells. FASEB J 13:123–133

Beverley JK, Fleck DG, Kwantes W, Ludlam GB (1976) Age-sex distribution of various diseases with particular reference to toxoplasmic lymphadenopathy. J Hyg 76:215–228

Biedermann T, Zimmermann S, Himmelrich H, Gumy A, Egeter O, Sakrauski AK, Seegmüller I, Voigt H, Launois P, Levine AD, Wagner H, Heeg K, Louis JA, Röcken M (2001) IL-4 instructs TH1 responses and resistance to *Leishmania major* in susceptible BALB/c mice. Nat Immunol 2:1054–1060

Bisser S, Ouwe-Missi-Oukem-Boyer ON, Toure FS, Taoufiq Z, Bouteille B, Buguet A, Mazier D (2006) Harbouring in the brain: A focus on immune evasion mechanisms and their deleterious effects in malaria and human African trypanosomiasis. Int J Parasitol 36:529–540

Bliss SK, Zhang Y, Denkers EY (1999) Murine neutrophil stimulation by *Toxoplasma gondii* antigen drives high level production of IFN-gamma-independent IL-12. J Immunol 63:2081–2088

Boersma A, Noireau F, Hublart M, Boutignon F, Lemesre JL, Racadot A, Degand P (1989) Gonadatropic axis and *Trypanosoma brucei gambiense* infection. Ann Soc Belg Med Trop 69:127–135

Bowden FJ, Garnett GP (2000) *Trichomonas vaginalis* epidemiology: parameterising and analysing a model of treatment interventions. Sex Transm Inf 76:248–256

Brabin L, Brabin BJ (1992) Parasitic infection in woman and their consequences. Adv Parasitol 31:1–81

Brown MT (1972) Trichomoniasis. Practitioner 209:639–644

Buxton D (1998) Protozoan infections (Toxoplasma gondii, Neospora caninum and Sarcocystis spp.) in sheep and goats: recent advances. Vet Res 29:289–310

Cappuccinelli P, Latters C, Cagliani J, Ponzi AN (1974) Features of intravaginal *Trichomonas vaginalis* infection in the mouse and the effect of estrogen treatment and immuno depression. G Batteriol Virol Immunol 61:31–40

Carreras E, Turner S, Paharkova-Vatchkova V, Mao A, Dascher C, Kovats S (2008) Estradiol acts directly on bone marrow myeloid progenitors to differentially regulate GM-CSF or Flt3 Ligand mediated dendritic cell differentiation. J Immunol 180:727–738

Cernetich A, Garver LS, Jedlicka AE, Klein PW, Kumar N, Scott AL, Klein SL (2006) Involvement of gonadal steroids and gamma interferon in sex differences in response to blood-stage malaria infection. Infect Immun 74:3190–3203

Chao TC, Vanalten PJ, Walter RJ (1994) Steroid sex-hormones and macrophage function-modulation of reactive oxygen intermediates and nitrite release. Am J Reprod Immunol 32:43–52

Chapman WL Jr, Hanson WL, Waits VB (1975) The influence of gonadectomy of host on parasitemia and mortality of mice infected with *Trypanosoma cruzi*. J Parasitol 61:213–216

Chen D, Copeman DB, Burnell J, Hutchinson GW (2000) Helper T cell and antibody responses to infection of CBA mice with *Babesia microti*. Parasite Immunol 22:81–88

Collazo CM, Yap GS, Sempowski GD, Lusby KC, Tessarollo L, Vande Woude GF, Sher A, Taylor GA (2001) Inactivation of LRG-47 and IRG-47 reveals a family of interferon γ-inducible genes with essential, pathogen-specific roles in resistance to infection. J Exp Med 194:181–187

Combe CL, Curiel TJ, Moretto MM, Khan IA (2005) NK cells help to induce CD8 (+)-T-cell immunity against *Toxoplasma gondii* in the absence of CD4 (+) T cells. Infect Immun 73:4913–4921

Cotch MF, Pastorek JG, Nugent RP, Hillier SL, Gibbs RS, Martin DH, Eschenbach DA, Edelman R, Carey JC, Reegan JA, Krohn MA, Klebanoff MA, Rao AV, Rhoads GG (1997) *Trichomonas vaginalis* associated with low birth weight and preterm delivery. Sex Transm Dis 24: 361–362

Cunha-Neto E, Kalil J (2001) Heart-infiltrating and peripheral T cells in the pathogenesis of human Chagas' disease cardiomyopathy. Autoimmunity 34:187–192

D'Ambrosio FR, Caetano LC, dos Santos CD, Abrahao AA, Pinto AC, do Prado JC (2008) Alterations triggered by steroid gonadal hormones in triglycerides and the cellular immune response of *Calomys callosus* infected with the Y strain of *Trypanosoma cruzi*. Vet Parasitol 152:21–27

Daniels CW, Belosevic M (1994) Serum antibody responses by male and female C57BL/6 mice infected with *Giardia muris*. Clin Exp Immunol 97:424–429

Daniels CW, Belosevic M (1995a) Comparison of the course of infection with *Giardia muris* in male and female mice. Int J Parasitol 25:131–135

Daniels CW, Belosevic M (1995b) Disaccharidase activity in male and female C57BL/6 mice infected with *Giardia muris*. Parasitol Res 81:143–147

de Beer P, el Harith A, Deng LL, Semiao-Santos SJ, Chantal B, van Grootheest M (1991) A killing disease epidemic among displaced Sudanese population identified as visceral leishmaniasis. Am J Trop Med Hyg 44:283–289

De Marco R, Locatelli F, Sunyer J, Burney P (2000) Differences in incidence of reported asthma related to age in men and women. A retrospective analysis of the data of the European Respiratory Health Survey. Am J Respir Crit Med 162:68–74

de Souza EM, Rivera MT, Araujo-Jorge TC, de Castro SL (2001) Modulation induced by estradiol in the acute phase of *Trypanosoma cruzi* infection in mice. Parasitol Res 87:513–520

Debierre-Grockiego F, Campos MA, Azzouz N, Schmidt J, Bieker U, Resende MG, Mansur DS, Weingart R, Schmidt RR, Golenbock DT, Gazzinelli RT, Schwarz RT (2007) Activation of TLR2 and TLR4 by Glycosylphosphatidylinositols Derived from *Toxoplasma gondii*. J Immunol 179:1129–1137

Degu G, Mengistu G, Jones J (2002) Some factors affecting prevalence of and immune responses to *Schistosoma mansoni* in schoolchildren in Gorgora, northwest Ethiopia. Ethiop Med J 40:345–352

Denkers EY (2003) From cells to signaling cascades: manipulation of innate immunity by *Toxoplasma gondii*. FEMS Immunol Med Microbiol 39:193–203

Desmonts G, Courvreur J (1974) Congenital toxoplasmosis: a prospective study of 378 pregnancies. N Engl J Med 290:1110–1116

do Prado JC, Leal MP, Anselmo-Franci JA, De Andrade HF, Kloetzel JK (1998) Influence of female gonadal hormones on the parasitaemia of female *Calomys callosus* infected with the 'Y' strain of *Trypanosoma cruzi*. Parasitol Res 84:100–105

Dos Santos CD, Toldo MP, do Prado JC (2005) *Trypanosoma cruzi*: the effects of dehydroepiandrosterone (DHEA) treatment during experimental infection. Acta Trop 95:109–115

Dutra WO, Rocha MO, Teixeira MM (2005) The clinical immunology of human chagas disease. Trends Parasitol 21:581–587

Eloi-Santos S, Olsen NJ, Correa-Oliveira R, Colley DG (1992) *Schistosoma mansoni*: mortality, pathophysiology, and susceptibility differences in male and female mice. Exp Parasitol 75:168–175

Elton C (1931) Health and parasites of a wild mouse population. Proc Zool Soc London 4:651–721

Engwerda CR, Beattie L, Amante FH (2005a) The importance of the spleen in malaria. Trends Parasitol 21:75–80

Engwerda CR, Belnoue E, Gruner AC, Renia L (2005b) Experimental models of cerebral malaria. Curr Top Microbiol Immunol 297:103–143

Eugui EM, Allison AC (1980) Differences in susceptibility of various mouse strains to haemo-protozoan infections: possible correlation with natural killer activity. Parasitol Immunol 2:277–292

Evans RH (2001) Baylisascaris procyonis (Nematoda; Ascaridoidea) in raccoon (Procyon lotol) in Orange County, California. Vector Borne Zoonotic Dis 1:239–242

Faubert G (2000) Immune response to Giardia duodenalis. Clin Microbiol Rev 13:35–54

Favre N, Ryffel B, Bordmann G, Rudin W (1997) The course of Plasmodium chabaudi chabaudi infections in interferon-gamma receptor deficient mice. Parasite Immunol 19:375–383

Finkelman FD, Shea-Donohue T, Morris SC, Gildea L, Strait R, Madden KB, Schopf L, Urban JF Jr (2004) Interleukin-4- and interleukin-13-mediated host protection against intestinal nematode parasites. Immunol Rev 201:139–155

Ford LC, Hammill HA, Delange RJ, Bruchner DA, Suzuki-Charez F, Mickus HL, Lebberz TB (1987) Determination of estrogen and androgen receptors in Trichomonas vaginalis and the effects of anti-hormones. Am J Obstet Gynecol 156:1119–1121

Fouts AC, Krause SJ (1980) Trichomonas vaginalis: re-evaluation of its clinical presentation and laboratory diagnosis. J Infect Dis 141:137–143

Friedl R, Brunner M, Moeslinger T, Spieckermann PG (2000) Testosterone inhibits expression of inducible nitric oxide synthase in murine macrophages. Life Sci 68:417–429

Fu Y-X, Chaplin DD (1999) Development and maturation of secondary lymphoid tissues. Annu Rev Immunol 17:399–433

Futterer A, Mink K, Luz A, Kosco-Vilbois MH, Pfeffer K (1998) The lymphotoxin β receptor controls organogenesis and affinity maturation in peripheral lymphoid tissues. Immunity 9:59–70

Galasso F, Altamura V, Sbano E (1996) Effects of topical testosterone propionate on the positive nickel patch test. J Dermatol Sci 13:76–82

Garber GE, Lemchuk-Favel LT, Rousseau G (1991) Effect of beta-estradiol on production of the cell-detaching factor of Trichomonas vaginalis. J Clin Microbiol 29:1847–1849

Garcia-Zepeda EA, Rojas-Lopez A, Esquivel-Velazquez M, Ostoa-Saloma P (2007) Regulation of the inflammatory immune response by the cytokine/chemokine network in amoebiasis. Parasite Immunol 29:679–684

Giannini MS (1986) Sex-influenced response in the pathogenesis of cutaneous leishmaniasis in mice. Parasite Immunol 8:31–37

Girones N, Fresno N (2003) Etiology of Chagas disease myocarditis: autoimmunity, parasite persistence, or both? Trends Parasitol 19:19–22

Global Malaria Programme (2008) WHO. http://malaria.who.int. Cited 16 May 2008

Goble FC (1966) Pathogenesis of blood protozoa. In: Soulsby EJL (ed) Biology of parasites. Academic Press, New York and London, pp 237–254

Gomes JA, Bahia-Oliveira LM, Rocha MO, Martins-Filho OA, Gazzinelli G, Correa-Oliveira R (2003) Evidence that development of severe cardiomyopathy in human Chagas' disease is due to a Th1-specific immune response. Infect Immun 71:1185–1193

Gorenflot A, Moubri K, Precigout E, Carcy B, Schetters TP (1998) Human babesiosis. Ann Trop Med Parasitol 92:489–501

Graham AL, Taylor MD, Le Goff L, Lamb TJ, Magennis M, Allen JE (2005) Quantitative appraisal of murine filariasis confirms host strain differences but reveals that BALB/c females are more susceptible than males to Litomosoides sigmodontis. Microbes Infect 7:612–618

Grasso G, Muscettola M (1990) The influence of beta-estradiol and progesterone on interferon gamma production in vitro. Int J Neurosci 51:315–317

Greenblatt CL (1980) The present and future of vaccination for cutaneous leishmaniasis. Prog Clin Biol Res 47:259–285

Greenblatt HC, Rosenstreich DL (1984) Trypanosoma rhodesiense infection in mice: sex dependence of resistance. Infect Immun 43:337–340

Guo X, Houpt E, Petri WA Jr (2007) Crosstalk at the initial encounter: interplay between host defense and ameba survival strategies. Curr Opin Immunol 19:376–384

Halvorsen O, Skorping A, Stokkan KA (1984) Humoral immunity and output of first-stage larvae of Elaphostrongylus rangiferi (Nematoda, Metastrongyloidea) by infected reindeer, Rangifer tarandus tarandus. J Helminthol. 58:13–18

Hamano N, Terada N, Maesako K, Numata T, Konno A (1998) Effect of sex hormones on eosinophilic inflammation in nasal mucosa. Allergy Asthma Proc 19:263–269

Hansen DS, Schofield L (2004) Regulation of immunity and pathogenesis in infectious diseases by CD1d-restricted NKT cells. Int J Parasitol 34:15–25

Hansen DS, Siomos MA, De Koning-Ward T, Buckingham L, Crabb BS, Schofield L (2003) CD1d-restricted NKT cells contribute to malarial splenomegaly and enhance parasite-specific antibody responses. Eur J Immunol 33:2588–2598

Hansen DS, D'Ombrain MC, Schofield L (2007) The role of leukocytes bearing Natural Killer Complex receptors and Killer Immunoglobulin-like Receptors in the immunology of malaria. Curr Opin Immunol 19:416–423

Haque R, Mondal D, Duggal P, Kabir M, Roy S, Farr BM, Sack RB, Petri WA Jr (2006) Entamoeba histolytica infection in children and protection from subsequent amebiasis. Infect Immun 74:904–909

Haque R, Mondal D, Shu J, Roy S, Kabir M, Davis AN, Duggal P, Petri WA Jr (2007) Correlation of interferon-gamma production by peripheral blood mononuclear cells with childhood malnutrition and susceptibility to amebiasis. Am J Trop Med Hyg 76:340–344

Hayes KS, Bancroft AJ, Grencis RK (2007) The role of TNF-α in Trichuris muris infection I: influence of TNF-α receptor usage, gender and IL-13. Parasite Immunol 29:575–582

Higuchi ML, Benvenuti LA, Reis MM, Metzger M (2003) Pathophysiology of the heart in Chagas' disease: current status and new developments. Cardiovasc Res 60:96–107

Hoffman WH, Pfaff AW, Schulz-Key H, Soboslay PT (2001) Determinants for resistance and susceptibility to microfilaraemia in Litomosoides sigmondontis filariasis. Parasitology 122:641–649

Homer MJ, Aguilar-Delfin I, Telford SR 3rd, Krause PJ, Persing DH (2000) Babesiosis. Clin Microbiol Rev 13:451–469

Hublart M, Tetaert D, Croix D, Boutignon F, Degand P, Boersma A (1990) Gonadotropic dysfunction produced by Trypanosoma brucei brucei in the rat. Acta Trop 47:177–184

Hughes VL, Randolph SE (2001a) Testosterone increses the transmission potential of tick-borne parasites. Parasitology 123:365–371

Hughes VL, Randolph SE (2001b) Testosterone depresses innate and acquired resistance to ticks in natural rodent hosts: a force for aggregated distributions of parasites. J Parasitol 87:49–54

Igarashi I, Suzuki R, Waki S, Tagawa Y, Seng S, Tum S, Omata Y, Saito A, Nagasawa H, Iwakura Y, Suzuki N, Mirami T, Toyoda Y (1999) Roles of CD4$^+$ T cells and gamma interferon in protective immunity against Babesia microti infection in mice. Infect Immun 67:4143–4148

Jahn A, Lelmett JM, Diesfeld HJ (1986) Seroepidemiological study on kala-azar in Barlingo District, Kenya. J Trop Med Hyg 89:91–104

Johnson J, Suzuki Y, Mack D, Mui E, Estes R, David C, Skamene E, Forman J, McLeod R (2002) Genetic analysis of influences on survival following Toxoplasma gondii infection. Int J Parasitol 32:179–185

Kanda N, Tamaki K (1999) Estrogen enhances immunoglobulin production by human PBMCs. J Allergy Clin Immunol 103:282–288

Kanda N, Tsuchida T, Tamaki K (1996) Testosterone inhibits immunoglobulin production by human peripheral blood mononuclear cells. Clin Exp Immunol 106 410–415

Kazyumba G, Berney M, Brighouse G, Cruchaud A, Lambert PH (1986) Expression of the B cell repertoire and autoantibodies in human African trypanosomiasis. Clin Exp Immunol 65:10–18

Kittas C, Henry L (1979) Effect of sex hormones on the immune system of guinea-pigs and on the development of toxoplasmic lesions in non-lymphoid organs. Clin Exp Immunol 36:16–23

Kittas C, Henry L (1980) Effect of sex hormones on the response of mice to infection with *Toxoplasma gondii*. Br J Exp Pathol 61:590–600

Kiyota M, Korenaga M, Nawa Y, Koyani M (1984) Effect of androgen on the expression of the sex differences in susceptibility to infection with *Strongyloides ratti* in C57BL/6 mice. Aust L Exp Biol Med Sci 62:607–618

Klein SL (2004) Hormonal and immunological mechanisms mediating sex differences in parasite infection. Parasite Immunol 26:247–264

Klein PW, Easterbrook JD, Lalime EN, Klein SL (2008) Estrogen and progesterone affect responses to *Plasmodium chabaudi* in female C57BL/6 mice. Gender Medicine 5:423–433

Knight PA, Brown JK, Pemberton D (2008) Innate immune response mechanisms in the intestinal epithelium: potential roles for mast cells and goblet cells in the expulsion of adult *Trichinella spiralis*. Parasitology 135:655–670

Kochar DK, Thanvi I, Joshi A, Shubhakaran A, Agarwal N, Jain N (1999) Mortality trends in falciparum malaria–effect of gender difference and pregnancy. J Assoc Physicians India 47:774–778

Kovats S, Carreras E (2008) Regulation of dendritic cell differentiation and function by estrogen receptor ligands. Cell Immunol 252:81–90

Koyama Y, Nagao S, Ohashi K, Takahashi H, Marunouchi T (1989) Effect of systemic and topical application of testosterone propionate on the density of epidermal Langerhans cells in the mouse. J Invest Dermatol 92:86–90

Krause PJ, Telford SR 3rd, Spielman A, Sikand V, Ryan R, Christianson D, Burke G, Brassard P, Pollack R, Peck J, Persing DH (1996) Concurrent Lyme disease and babesiosis. Evidence for increased severity and duration of illness. JAMA 275:1657–1660

Krieger JN (1990) Epidemiology and clinical manifestations of urogenital trichomoniasis in men. In: Honigberg BM (ed) Trichomonads parasitic in humans. Springer, New York, pp 235–245

Krieger JN, Verdon M, Siegel N, Holmes KK (1993) Natural history of urogenital trichomoniasis in men. J Urol 149:1455–1458

Krucken J, Dkhil MA, Braun JV, Schroetel RM, El-Khadragy M, Carmeliet P, Mossmann H, Wunderlich F (2005) Testosterone suppresses protective responses of the liver to blood-stage malaria. Infect Immun 73:436–443

Laga M, Manoka A, Kivuvu M, Malele B, Tuliza M, Nzila N, Goeman J, Behets F, Batter V, Alary M (1993) Non-ulcerative sexually transmitted diseases as risk factors for HIV-1 transmission in women: results from a cohort study. AIDS 7:95–102

Lamb TJ, Brown DE, Potocnik AJ, Langhorne J (2006) Insights into the immunopathogenesis of malaria using mouse models. Expert Rev Mol Med 6:1–22

Lambert PH, Berney M, Kazyumba G (1981) Immune complexes in serum and in cerebrospinal fluid in African trypanosomiasis. Correlation with polyclonal B cell activation and with intracerebral immunoglobulin synthesis. J Clin Invest 67:77–85

Lambert KC, Curran EM, Judy BM, Lubahn DB, Estes DM (2004) Estrogen receptor-α deficiency promotes increased TNF-α secretion and bacterial killing by murine macrophages in response to microbial stimuli *in vitro*. J Leuk Biol 75:1116–1172

Landgraf B, Kollaritsch H, Wiedermann G, Wernsdorfer WH (1994) Parasite density of *Plasmodium falciparum* malaria in Ghanaian schoolchildren: evidence for influence of sex hormones? Trans R Soc Trop Med Hyg 88:73–74

Landolfo S, Martinotti MG, Martinetto P, Forni G, Rabagliati AM (1981) *Trichomonas vaginalis*: dependence of resistance among different mouse strains upon the non-H-2 gene haplotype, sex, and age of recipient hosts. Exp Parasitol 52:312–318

Lang F, Drell TL, Niggemann B, Zanker KS, Entschladen F (2003) Neurotransmitters regulate the migration and cytotoxicity in natural killer cells. Immunol Lett 90:165–172

Lang C, Gross U, Luder CGK (2007) Subversion of innate and adaptive immune responses by *Toxoplasma gondii*. Parasitol Res 100:191–203

Langford TD, Housley MP, Boes M, Chen J, Kagnoff MF, Gillin FD, Eckmann L (2002) Central importance of immunoglobulin A in host defense against *Giardia* spp. Infect Immun 70:11–18

Lauw FN, Caffrey DR, Golenbock DT (2005) Of mice and man: TLR11 (finally) finds profilin. Trends Immunol 26:509–511

Le Goff L, Lamb TJ, Graham AL, Harcus Y, Allen JE (2002) IL-4 is required to prevent filarial nematode development in resistant but not susceptible strains of mice. Int J Parasitol 32:1277–1284

Leroy V, De Clercq A, Ladner J, Bogaerts J, Van de Perre P, Dabis F (1995) Should screening of genital infections be part of antenatal care in areas of high HIV prevalence? A prospective cohort study from Kigali, Rwanda, 1992–1993. Genitourin Med 71:207–211

Lezama-Davila CM, Issac-Marquez AP, Barbi J, Oghumu S, Satoskar AR (2007) 17β-Estradiol increases *Leishmania mexicana* killing in macrophages from DBA/2 mice by enhancing production of nitric oxide but not pro-inflammatory cytokines. Am J Trop Med Hyg 76:1125–1127

Lezama-Davila CM, Issac-Marquez AP, Barbi J, Cummings HE, Lu B, Satoskar AR (2008) Role of phosphatidylinositol-3-kinase-gamma (PI3Kgamma)-mediated pathway in 17beta-estradiol-induced killing of *L. mexicana* in macrophages from C57BL/6 mice. Immunol Cell Biol 86(6): 539–543

Li E, Zhou P, Petrin Z, Singer SM (2004) Mast cell-dependent control of *Giardia lamblia* infections in mice. Infect Immun 72:6642–6649

Liesenfeld O, Nguyen TA, Pharke C, Suzuki Y (2001) Importance of gender and sex hormones in regulation of susceptibility of the small intestine to peroral infection with *Toxoplasma gondii* tissue cysts. J Parasitol 87:1491–1493

Liu L, Wang L, Zhao Y, Wang Y, Qiao Z (2006) Testosterone attenuates p38 MAPK pathway during *Leishmania donovani* infection of macrophages. Parasitol Res 99:189–193

Lopez A, Dietz VJ, Wilson M, Navim TR, Jones JL (2000) Preventing congenital toxoplasmosis. Morb Mort Week Report 49:57–71

Lotter H, Jacobs T, Gaworski I, Tannich E (2006) Sexual dimorphism in the control of amebic liver abscess in a mouse model of disease. Infect Immun 74:118–124

Lynch NR, Yarzabal L, Verde O, Avila JL, Monzon H, Convit J (1982) Delayed-type hypersensitivity and immunoglobulin E in American cutaneous leishmaniasis. Infect Immun 38:877–881

Mansfield JM, Paulnock DM (2005) Regulation of innate and acquired immunity in African trypanosomiasis. Parasite Immunol 27:361–371

Mansfield JM, Davis TH, Dubois ME (2002) Immunobiology of African trypanosomes: new paradigms, newer questions. In: Black SJ, Seed JR (eds) The African trypanosomes. Kluwer, Boston, pp 79–86

Martinotti MG, Musso T, Savoia D (1988) Influence of gender in pathogenesis of trichomoniasis in congenitally athymic (nude) mice. Genitourin Med 64:18–21

Masocha W, Robertson B, Rottenberg ME, Mhlanga J, Sorokin L, Kristensson K (2004) Cerebral vessel laminins and IFN-gamma define *Trypanosoma brucei brucei* penetration of the blood-brain barrier. J Clin Invest 114:689–694

McFarlane E, Perez C, Charmoy M, Allenbach C, Carter KC, Alexander J, Tacchini-Cottier F (2008) Neutrophils contribute to development of a protective immune response during onset of infection with *Leishmania donovani*. Infect Immun 76:532–541

McMahon-Pratt D, Alexander J (2004) Does the Leishmania major paradigm of pathogenesis and protection hold for New World cutaneous leishmaniases or the visceral disease? Immunol Rev 201:206–224

Menendez-Pelaez A, Mayo JC, Sainz RM, Perez M, Antolin I, Tolivia D (1992) Development and hormonal regulation of mast cells in the Harderian gland of Syrian hamsters. Anat Embryol (Berl) 186:91–97

Mock BA, Nacy CA (1988) Hormonal modulation of sex differences in resistance to *Leishmania major* systemic infections. Infect Immun 56:3316–3319

Modabber F (2000) First generation leishmaniasis vaccines in clinical development: moving but what next? In EC Euroleish/WHO-TDR Workshop. Second generation vaccines against leishmaniasis: from antigen discovery to clinical trial. World Health Organization, Geneva, Switzerland, pp 35–39

Mohan K, Moulin P, Stevenson MM (1997) Natural killer cell cytokine production, not cytotoxicity, contributes to resistance against blood-stage *Plasmodium chabaudi* AS infection. J Immunol 159:4990–4998

Morel CM, Lazdins J (2003) Chagas' disease. Nat Rev Microbiol 1:14–15

Moreno-Fierros L, Campos-Rodriguez R, Enriquez-Rincon F (1992) Sex differences in systemic and local immune responses to *Entamoeba histolytica* after intraperitoneal and rectal immunization in Balb/c mice. Arch Med Res 23:153–155

Mossman H, Benten WP, Galanos C, Freudenberg M, Kuhn-Velten WN, Reinauer H, Wunderlich F (1997) Dietary testosterone suppresses protective responsiveness to *Plasmodium chabaudi* malaria. Life Sci 60:839–848

Mucci J, Mocetti E, Leguizamon MS, Campetella O (2005) A sexual dimorphism in thrathymic sialylation survey is revealed by the trans-sialidase from *Trypanosoma cruzi*. J Immunol 174:4545–4550

Mun HS, Aosai F, Norose K, Piao LX, Fang H, Akira S, Yano A (2005) Toll-like receptor 4 mediates tolerance in macrophages stimulated with *Toxoplasma gondii*-derived heat shock protein 70. Infect Immun 73:4634–4642

Murray CJL, Salomon JA, Mathers CD, Lopez AD (eds) (2002) Summary measures of population health: concepts, ethics, measurement and applications. WHO, Geneva

Mzembe SA, Lloyd S, Soulsby EJ (1984) Macrophage mediated resistance to *Babesia microti* in *Nematospiroides dubius*-infected mice. Z Parasitenkd 70:753–761

Nakanishi H, Horii Y, Terashima K, Fujita K (1990) Age-related changes of the susceptibility to infection with *Brugia pahangi* in male and female BALB/c mice. J Parasitol 76:283–285

Nakazawa M, Fantappie MR, Freeman GL, Eloi-Santos S, Olsen NJ, Kovacs WJ, Secor WE, Colley DG (1997) *Schistosoma mansoni*: susceptibility difference sbetween male and female mice can be mediated by testosterone during early infection. Exp Parasitol 85:233–240

Nandan D, Reiner NE (1995) Attenuation of gamma interferon-induced tyrosine phosphorylation in mononuclear phagocytes infected with *Leishmania donovani*: selective inhibition of signaling through Janus kinases and Stat1. Infect Immun 63:4495–4500

Niccolai L, Kopicko JJ, Kassie A, Petros H, Clark RA, Kissinger P (2000) Incidence and predictors of reinfection with *Trichomonas vaginalis* in HIV-infected women. Sex Transm Dis 27:284–288

Nilsson N, Carlsten H (1994) Estrogen induces suppression of natural killer cell cytotoxicity and augmentation of polyclonal B cell activation. Cell Immunol 158:131–139

Oliveira-Sequeira TC, Amarante AF, Ferrari TB, Nunes LC (2002) Prevalence of intestinal parasites in dogs from Sao Paulo State, Brazil. Vet Parasitol 103:19–27

Omer FM, Riley EM (1998) Transforming growth factor beta production is inversely correlated with severity of murine malaria infection. J Exp Med 188:39–48

Omer FM, de Souza JB, Riley EM (2003) Differential induction of TGF-beta regulates proinflammatory cytokine production and determines the outcome of lethal and nonlethal *Plasmodium yoelii* infections. J Immunol 171:5430–5436

Osorio Y, Bonilla DL, Peniche AG, Melby PC, Travi BL (2008) Pregnancy enhances the innate immune response in experimental cutaneous leishmaniasis through hormone-modulated nitric oxide production. J Leukoc Biol 83:1413–1422

Paharkova-Vatchkova V, Maldonado R, Kovats S (2004) Estrogen preferentially promotes the differentiation of CD11c$^+$ CD11b (intermediate) dendritic cells from bone marrow precursors. J Immunol 172:1426–1436

Parker SJ, Roberts CW, Alexander J (1991) CD8$^+$ T cells are the major lymphocyte subpopulation involved in the protective immune response to *Toxoplasma gondii* in mice. Clin Exp Immunol 84:207–212

Pepin J, Milford F, Khonde A, Niyonsenga T, Loko L, Mpia B (1994) Gambiense trpanosomiasis: frequency of, and risk factors for, failure of melarsoprol therapy. Trans R Soc Trop Med Hyg 88:447–452

Pereira-Bueno J, Quintanilla-Gozalo A, Perez-Perez V, Alvarez-Garcia G, Collantes-Fernandez E, Ortega-Mora LM (2004) Evaluation of ovine abortion associated with Toxoplasma gondii in Spain by different diagnostic techniques. Vet Parasitol 7:33–43

Phillips AN, Anunes F, Stergious G, Ranki A, Jensen F, Bentwich Z, Sacks T, Pedersen C, Lundgren JD, Johnson AM (1994) A sex comparison of rates of new AIDS-defining disease and death in 2554 AIDS cases. AIDS Europe study group. AIDS 8:331–835

Pung OJ, Luster MI (1986) *Toxoplasma gondii*: decreased resistance to infection in mice due to estrogen. Exp Parasitol 61:48–56

Raghupathy R (1997) Th1-type immunity is incompatible with successful pregnancy. Immunol Today 18:478–482

Reddington JJ, Stewart GL, Kramar GW, Kramar MA (1981) The effects of host sex and hormones on *Trichinella spiralis* in the mouse. J Parasitol 67:548–555

Remoue F, Van To D, Schacht AM, Picquet M, Garraud O, Vercruysse J, Ly A, Capron A, Riveau G (2001) Gender-dependent specific immune response during chronic human *Schistosomiasis haematobia*. Clin Exp Immunol 124:62–68

Rettew JA, Huet-Hudson YM, Marriot I (2008) Testosterone reduces macrophage expression in the mouse of toll-like receptor 4, a trigger for inflammation and innate immunity. Biol Reprod 78:432–437

Roberts M, Alexander J, Blackwell JM (1990) Genetic analysis of *Leishmania mexicana* infection in mice: single gene (Scl-2) controlled predisposition to cutaneous lesion development. J Immunogenet 17:89–100

Roberts CW, Cruickshank SM, Alexander J (1995) Sex-determined resistance to *Toxoplasma gondii* is associated with temporal differences in cytokine production. Infec Immun 63:2549–2555

Roberts CW, Satoskar A, Alexander J (1996) Sex steroids, pregnancy-associated hormones and immunity to parasitic infection. Parasitol Today 12:382–387

Roberts CW, Walker W, Alexander J (2001) Sex-asociated hormones and immunity to protozoan parasites. Clin Microbiol Rev 14:476–488

Rosas LE, Keiser T, Barbi J, Satoskar AA, Septer A, Kaczmarek J, Lezama-Davila CM, Satoskar AR (2005) Genetic background influences immune responses and disease outcome of cutaneous L. mexicana infection in mice. Int Immunol 17:1347–1357

Roxstrom-Lindquist K, Palm D, Reiner D, Ringqvist E, Svard SG (2006) Giardia immunity–an update. Trends Parasitol 22:26–31

Saeki H, Ishii T (1996) Effect of silica treatment on resistance to *Babesia rodhaini* infection in immunized mice. Vet Parasitol 61:201–210

Santos CD, Levy AM, Toldo MP, Azevedo AP, Prado JC Jr (2007) Haematological and histopathological findings after ovariectomy in *Trypanosoma cruzi* infected mice. Vet Parasitol 143:222–228

Satoskar A, Alexander J (1995) Sex-determined susceptibility and differential IFN-γ and TNF-α mRNA expression in DBA/2 mice infected with *Leishmania Mexicana*. Immunology 84:1–4

Satoskar A, Al-Quassi HH, Alexander J (1998) Sex-determined resistance against *Leishmania mexicana* is associated with the preferential induction of a Th1-like response and IFN-γ production by female but not male DBA/2 mice. Immunol Cell Biol 76:159–166

Savino W, Villa-Verde DM, Mendes-da-Cruz DA, Silva-Monteiro E, Perez AR, Del Pilar Aoki M, Bottasso O, Guinazu N, Silva-Barbosa SD, Gea S (2007) Cytokines and cell adhesion receptors in the regulation of immunity to *Trypanosoma cruzi*. Cytokine Growth Factor Rev 18:107–124

Scanga CA, Aliberti J, Jankovic D, Tilloy F, Bennouna S, Denkers EY, Medzhitov R, Sher A (2002) Cutting edge: MyD88 is required for resistance to *Toxoplasma gondii* infection and regulates parasite-induced IL-12 production by dendritic cells. J Immunol 168:5997–6001

Schleifer KW, Filutowicz H, Schopf LR, Mansfield JM (1993) Characterization of T helper cell responses to the trypanosome variant surface glycoprotein. J Immunol 150:2910–2919

Schmid G, Narcisi E, Mosure D, Secor E (2001) Prevalence of metronidazole resistant *Trichomonas vaginalis* in a gynecology clinic. J Reprod Med 46:545

Schofield L, Grau GE (2005) Immunological processes in malaria pathogenesis. Nat Rev Immunol 5:722–735

Schuster JP, Schaub GA (2001) Experimental Chagas disease: The influence of sex and psychoneuroimmunological factors. Parasitol Res 87:994–1000

Schwebke JR, Burgess D (2004) Trichomoniasis. Clin Microbiol Rev 17:794–803

Seydel KB, Smith SJ, Stanley SL Jr (2000) Innate immunity to amebic liver abscess is dependent on gamma interferon and nitric oxide in a murine model of disease. Infect Immun 68:400–402

Shirahata T, Muroya N, Ohta C, Goto H, Nakane A (1992) Correlation between increased susceptibility to primary *Toxoplasma gondii* infection and depressed production of gamma interferon in pregnant mice. Microbiol Immunol 36:81–91

Shirahata T, Muroya N, Ohta C, Goto H, Nakane A (1993) Enhancement by recombinant human interleukin 2 of host resistance to *Toxoplasma gondii* infection in pregnant mice. Microbiol Immunol 37:583–590

Silva H, Tchernitchin AN, Tchernitchin NN (1997) Low doses of estradiol-17 alpha degranulate blood eosinophil leucocytes and high doses alter homeostatic mechanisms. Med Sci Res 25:201–204

Silva JS, Machado FS, Martins GA (2003) The role of nitric oxide in the pathogenesis of Chagas disease. Front Biosci 8:s314–s325

Simarro PP, Jannin J, Cattand P (2008) Eliminating human African trypanosomiasis: where do we stand and what comes next? PLoS Med 5:174–180

Soudan B, Tetaert D, Racadot A, Degand P, Boersma A (1992) Decrease of testosterone level during an experimental African trypanosomiasis: involvement of a testicular LH receptor desensitization. Acta Endocrinol 127:86–92

Soudan B, Tetaert D, Hublart M, Racadot A, Croix D, Boersma A (1993) Experimental "chronic" African trypanosomiasis: endocrine dysfunctions generated by parasitic components released during the trypanolytic phase in rats. Exp Clin Endocrinol 101:166–172

Stanley SL Jr (2003) Amoebiasis. Lancet 361:1025–1034

Stein I, Cope E (1933) *Trichomonas vaginalis*. Am J Obstet Gynecol 25:819–825

Stevenson MM, Urban BC (2006) Antigen presentation and dendritic cell biology in malaria. Parasite Immunol 28:5–14

Stevenson MM, Su Z, Sam H, Mohan K (2001) Modulation of host responses to blood-stage malaria by interleukin-12: from therapy to adjuvant activity. Microbes Infect 3:49–59

Suffia IJ, Reckling SK, Piccirillo CA, Goldzmid RS, Belkaid Y (2006) Infected site-restricted Foxp3+ natural regulatory T cells are specific for microbial antigens. J Exp Med 203:777–788

Sugarman B, Mummaw N (1988) The effect of hormones on *Trichomonas vaginalis*. J Gen Microbiol 134:1623–1628

Tai P, Wang J, Jin H, Song X, Yan J, Kang Y, Zhao L, Xiaojin A, Du X, Chen X, Wang S, Xia G, Wang B (2008) Induction of regulatory T cells by physiological level estrogen. J Cell Physiol 214:456–464

Tarleton RL (1990) Depletion of CD8+ T cells increases susceptibility and reverses vaccine-induced immunity in mice infected with *Trypanosoma cruzi*. J Immunol 144:717–724

Tarleton RL (2007) Immune system recognition of *Trypanosoma cruzi*. Curr Opin Immunol 19:430–434

Taylor GA, Collazo CM, Yap GS, Nguyen K, Gregorio T, Taylor LS, Eagleson B, Secrest L, Southon E, Reid SW, Tessarollo L, Bray M, McVicar DW, Komschlies KL, Young HA, Biron

CA, Sher A, Vande Woude GF (2000) Pathogen specific loss of host resistance in mice lacking the IFN-γ-inducible gene IGTP. Proc Natl Acad USA 97:751–755

Taylor MD, Le Goff L, Harris A, Malone E, Allen JE, Maizels RM (2005) Removal of regulatory T cell activity reverses hyporesponsiveness and leads to filarial parasite clearance in vivo. J Immunol 174:4924–4933

Taylor MD, Harris A, Babayan SA, Bain O, Culshaw A, Allen JE, Maizels RM (2007) CTLA-4 and CD4⁺CD25⁺ Regulatory T Cells Inhibit Protective Immunity to Filarial Parasites *in vivo*. J Immunol 179:4626–4634

Tiuria R, Horrii Y, Tateyama S, Tsuchiya K, Nawa Y (1994) The Indian soft-furred rat, *Millardia meltada*, a new host for *Nippostrongylus brasiliensis*, showing androgen-dependent sex difference in intestinal mucosal defence. Int J Parasitol 24:1055–1057

Tumanov AV, Christiansen PA, Fu Y-X (2007) The role of lymphotoxin receptor signaling in diseases. Curr Mol Med 7:567–578

US Department of Agriculture "Charting the cost of food safety", Food Review, May-August, 1994

Van Andel RA, Kendall LV, Franklin CV, Riley LK, Besch-Williford CL, Hook RR Jr (1996) Sustained estrogenization is insufficient to support long-term exper mentally induced genital *Trichomonas vaginalis* infection in BALB/c mice. Lab Anim Sci 46:689–690

Venkatesan P, Finch RG, Wakelin D (1997) A comparison of mucosal inflammatory responses to *Giardia muris* in resistant B10 and susceptible BALB/c mice. Parasite Immunol 19:137–143

Verthelyi D (2006) Female's Heightened Immune Status: Estrogen, T Cells, and Inducible Nitric Oxide Synthase in the Balance. Endocrinology 147:659–661

Veugopalan PP, Shenoy DD, Kamath A, Rajeev A (1997) Distribution of malarial parasites: effect of gender of construction workers. Indian J Med Sci 51:89–92

Vliagoftis H, Dimitriadou V, Boucher W, Rozniecki JJ, Correia I, Raam S, Theoharides TC (1992) Estradiol augments while tamoxifen inhibits rat mast cell secretion. Int Arch Allergy Immunol 98:398–409

Walker W, Roberts CW, Ferguson DJ, Jabbari H, Alexander J (1997) Innate immunity to *Toxoplasma gondii* is influenced by gender and is associated with differences in interleukin-12 and gamma interferon production. Infect Immun 65:1119–1121

Wildling E, Winkler S, Kremsner PG, Brandts C, Jenne L, Wernsdorfer WH (1995) Malaria epidemiology in the province of Moyen Ogoov, Gabon. Trop Med Parasitol 46:77–82

Williams KAB, Scott JM, MacFarlane DE, Williamson JMW, Elias Jones TF, Williams H (1981) Congenital toxoplasmosis: a prospective survey in the west of Scotland. J Infect 3:219–229

Wood PR, Clark IA (1982) Apparent irrelevance of NK cells to resolution of infections with *Babesia microti* and *Plasmodium vinckei petteri* in mice. Parasite Immunol 4:319–327

World Health Organization (1995) An overview of selected curable sexually transmitted diseases. Global programs on Aids. World Health Organization, Geneva, Switzerland, pp 2–27

Wunderlich F, Benten WP, Bettenhaeuser U, Schmitt-Wrede HP, Mossman H (1992) Testosterone-unresponsiveness of existing immunity against *Plasmodium chaubadi* malaria. Parasite Immunol 14:307–320

Wunderlich F, Maurin W, Benten WP, Schmitt-Wrede HP (1993) Testosterone impairs efficacy of protective vaccination against *P. chaubadi* malaria. Vaccine 11:1097–1099

Wunderlich F, Benten WP, Lieberherr M, Guo Z, Stemm O, Wrehlke C, Sekeris CE, Mossman H (2002) Testosterone signaling in T cells and macrophages. Steroids 67:535–538

Wunderlich F, Dkhil MA, Mehnert LI, Braun JV, El-Khadragy M, Borsch E, Hermsen D, Benten WP, Pfeffer K, Mossmann H, Krücken J (2005) Testosterone responsiveness of spleen and liver in female lymphotoxin beta receptor-deficient mice resistant to blood-stage malaria. Microbes Infect 7:399–409

Yap GS, Sher A (1999) Cell-mediated immunity to *Toxoplasma gondii*: initiation, regulation and effector function. Immunobiology 210:240–247

Yap GS, Shaw MH, Ling Y, Sher A (2006) Genetic analysis of host resistance to intracellular pathogens: lessons from studies of *Toxoplasma gondii* infection. Microbes Infect 8:1174–1178

Zaffaroni E, Rubaudo L, Lanfranchi P, Mignone W (1999) Epidemiological patterns of canine leishmaniasis in Western Liguria (Italy). Vet Parasitol 81:11–19

Zaitsu M, Yamasaki F, Ishii E, Midoro-Horiuti T, Goldblum RM, Hamasaki Y (2006) Interleukin-18 primes human basophilic KU812 cells for higher leukotriene synthesis. Prostaglandins Leukot Essent Fatty Acids 74:61–66

Zaitsu M, Narita S, Lambert KC, Grady JJ, Estes DM, Curran EM, Brooks EG, Watson CS, Goldblum RM, Midoro-Horiuti T (2007) Estradiol activates mast cells via a non-genomic estrogen receptor-alpha and calcium influx. Mol Immunol 44:1977–1985

Zhang ZH, Chen L, Saito S, Kanagawa O, Sendo F (2000) Possible Modulation by Male Sex Hormone of Th1/Th2 Function in Protection against *Plasmodium chabaudi chabaudi* AS Infection in Mice. Exp Parasitol 96:121–129

Chapter 7
Sex Differences in Parasitic Infections: Beyond the Dogma of Female-Biased Resistance

Galileo Escobedo, Marco A. De León-Nava, and Jorge Morales-Montor

Abstract Sex differences in parasitic infections are a biological phenomenon of considerable significance for individual health and disease as well as for the evolution of a species. The general rule of thumb is that females are more resistant to infectious diseases than males. There are, however, many notable exceptions to this rule that illustrate a female bias in susceptibility to infection. By studying sex differences to cysticercosis infection, it has been demonstrated that females are more likely than males to become infected, to carry larger parasite loads, to be more severely affected, and more reticent to developing protective immunity. Our animal studies illustrate that female-biased susceptibility to parasitic infection is influenced by hormones, reproductive status, age, and genetic background. The mechanisms underlying the sexual dimorphism in murine cysticercosis involve the effects of sex steroids on both the host immune and central nervous systems as well as directly on the parasite. In this chapter, the causes of female-biased susceptibility to parasitic, and possibly other, infections are examined.

7.1 Introduction

It is widely reported in the literature that female mammals are more resistant to parasitic infections than males because of sex-associated differences in exposure to parasites and the immunosuppressive properties of testosterone (T). This paradigm implies that the sexual dimorphism in response to parasites is mediated primarily by the immune system of the host, which disregards the fact that some parasites can directly respond to the distinct sex steroid hormone profiles of their female and male hosts (Escobedo et al. 2005).

J. Morales-Montor (✉)
Departamento de Inmunología, Instituto de Investigaciones Biomédicas, Universidad Nacional Autónoma de México, AP 70228, México D.F. 04510, México
e-mail: jmontor66@biomedicas.unam.mx

S.L. Klein and C.W. Roberts (eds.), *Sex Hormones and Immunity to Infection*,
DOI 10.1007/978-3-642-02155-8_7, © Springer-Verlag Berlin Heidelberg 2010

The hypothesis that females are more resistant to parasites than males has been termed the "female supremacy hypothesis," which has prevailed for over half a century (Addis 1946), despite thoughtful recommendations against simplification (Zuk and McKean 1996). This hypothesis is strengthened by reports of elevated mortality rates in males, including humans, compared with females, as a result of exposure to infectious agents (Moore and Wilson 2002; Owens 2002; Zuk and McKean 1996; do Prado et al. 1998; Watanabe et al. 1999; Klein 2000; Ganley and Rajan 2001; Hughes and Randolph 2001; Roberts et al. 2001; Verthelyi 2001). The relative resistance of females to infection has been associated with the differences between the sexes in life history strategies, including attracting mates (Zuk 1994), establishing social hierarchy (Barnard et al. 1998), engaging in reproductive behavior (Kavaliers and Colwell 1993; Morales et al. 1996; Willis and Poulin 2000), and expending energy mounting an immune response to a replicating parasite (Hanssen et al. 2003).

7.2   Sex Differences in Immune Responses to Parasites

As noted in Chaps. 2 and 3 of this book, immune responses in males and females differ when confronted with microbes and this is largely caused by sex steroids, including estrogen, progesterone (P4), and T, binding to ligand-specific receptors on or in the cytoplasm of immune cells (Benten et al. 2002; Grossman et al. 1979, 1991). Despite the consistent observation that sex steroids affect the functioning of immune cells, these sex differences are highly dependent on the parasite-host system and discrepancies do exist. For example, although concentrations of IFN-γ are higher in female than male mice during malaria infection (Cernetich et al. 2006), male mice produce more IFN-γ and TNF-α than females in response to *Toxoplasma gondii* (Roberts et al. 1995; Walker et al. 1997). In response to *Taenia crassiceps*, a parasite to which females are more susceptible than males, males have higher IFN-γ responses than females, females produce more IL-10 than males, and concentrations of IL-6 and IL-12 are equivalent between the sexes (Terrazas et al. 1998). Finally, in hamsters, males develop more profound lesions than females following infection with *Leishmania panamensis* or *L. guyanensis* (Travis et al. 2002). Male hamsters also have higher concentrations of IL-4, IL-10, and TGF-β than females; conversely, concentrations of IL-12 and IFN-γ do not differ between the sexes.

We interpret these data to reveal that the host immune response is not sex specific, but rather male and female hosts have evolved immunological responses that are specific to the invading parasite, responsive to parasite strategies at different times during infection, the number of parasites, and the location of parasites in the host. For example, extracellular, but not intracellular, stages of parasites are vulnerable to antibodies and complement because of direct interactions with the external surface of the parasite (Philipp et al. 1980). Such antibody-mediated damage has been documented to affect some parasite life cycle stages but certainly not all, i.e., tachyzoites in toxoplasmosis (Johnson and Sayles 2002), early larvae in cysticercosis (Restrepo et al. 2001), merozoites in malaria (Daly and Long 1995), trophozoites in amoebiasis (Ghosh et al. 1998), and promastigotes in leishmaniasis (Rafati et al. 2001). Some of the extracellular stages of parasites manage to escape from circulation and become sequestered inside their target cells, apparently unscathed by antibodies or other harmful immune effectors. The immune response against intracellular parasites is largely dependent on the expression of parasite antigens on the membrane of the infected cell or peptides presented in MHC molecules (Kyes et al. 2001), In theory, T helper 2 (Th2) cell dominated immune responses, that favor high levels of antibody production, would be most appropriate against extracellular stages of parasites. In contrast, T helper 1 (Th1) responses, which induce inflammatory mediators and CD8+ T cells and eliminate infected cells, would be most effective against intracellular parasites (Sher et al. 1992). If female hosts favor Th2 responses (Roberts et al. 2001), they should be more resistant than males to extracellular parasites and to intracellular infections in their initial stages when the parasites are migrating toward their protected intracellular locations. In contrast, if androgens favor Th1 responses (Roberts et al. 2001), then males should control intracellular parasites more effectively than females, especially during the late stages of infection. Lack of congruity between expected and observed results most likely comes from oversimplistic expectations regarding the interactions of the endocrine and immune systems. The majority of Chap. 6 discusses parasitic diseases to which males are more susceptible than females. The focus of this chapter, however, will be on those parasitic infections to which females show increased susceptibility compared with males (please see Table 6.2 for a summary).

7.3 Examples of Female-Biased Susceptibility to Parasites

7.3.1 Protozoan Parasites

One of the most well-studied parasites to which females are more susceptible than males is *Toxoplasma gondii*. In mouse models, females develop more severe brain inflammation and are more likely to die following infection than males (Walker et al. 1997). Male mice produce higher concentrations of TNF-α, IL-12, and IFN-γ than females during acute infection (Roberts et al. 1995; Walker et al. 1997). Ovariectomy of female mice reduces, whereas administration of 17β-estradiol (E2) exacerbates, the development of tissue cysts caused by *T. gondii* infection (Liesenfeld et al. 2001; Pung and Luster 1986). Further evidence of hormonal modulation of immune responses to *T. gondii* is provided by the observed differential susceptibility to infection during pregnancy, which is detailed in Chap. 9 of this book. Human studies of sex differences in *T. gondii* infection are scarce because most healthy adults are asymptomatic. Among immunocompromised individuals, however, *T. gondii*-induced encephalitis is more prevalent among women than men (Phillips et al. 1994).

Infection of rodents with *T. gondii* results in pronounced behavioral alterations, including increased exploratory behavior and aggression, which may make them more conspicuous to and less fearful of the definitive host, the cat (Arnott et al. 1990; Berdoy et al. 1995; Webster et al. 1994). Unfortunately, behavioral studies of *T. gondii* in rodents have been conducted irrespective of sex; thus, whether there is a sexual dimorphism in the behavioral consequences of infection requires exploration. In humans, however, infection with *T. gondii* is associated with an increased perceived dominance as well as increased concentrations of T in men (Flegr et al. 2008).

7.3.2   Helminth Parasites

7.3.2.1   Cestode parasites

Taenia crassiceps is an intestinal cestode of canines (definitive host) and of various extraintestinal tissues of rodents (intermediate host) in its larval (cysticercus) stage (Freeman 1962). Experimental cysticercosis caused by *T. crassiceps* in mice simply requires the intraperitoneal injection of live cysticerci (Culbreth et al. 1972). Intraperitoneal cysticerci reproduce asexually by exogenous budding, developing massive parasite loads in a few months (Esch and Smith, 1976) that may even approximate the body weight of the host, without causing obvious signs of sickness (Larralde et al. 1995). The cysticerci also survive and reproduce in vitro under culture conditions in media free of fetal calf serum. These features of experimental murine cysticercosis have made it a convenient model for studying the immuno-logical, genetic, and reproductive factors involved in susceptibility to infection and parasite proliferation (Sciutto et al. 2002).

During the acute stage of *T. crassiceps* infection, females are more susceptible to infection than males (Larralde et al. 1995). Experimental findings, to date, have shown that in different congenic and syngenic strains of mice, females become infected more often than males and carry more cysticerci than males, with signifi-cant between-strain variations (Fragoso et al. 1996, 1998; Sciutto et al. 1991; Huerta et al. 1992; Larralde et al. 1995; Terrazas et al. 1998; Morales-Montor et al., 2001b, 2002a,b). Specifically, sex differences are pronounced among BALBc/AnN strains of mice as well as in wild-caught house mice (Fragoso et al. 2008). Our data show that in 16 of 17 different mouse strains tested, female mice carry larger parasite loads than males. The coefficient of variation for female mice is significantly smaller than that for males, illustrating that males show a greater variability in their ability to restrain cysticercus growth. Female-biased susceptibil-ity to murine cysticercosis are notable in response to diverse parasite strains, including HYG (a slow growth parasites), WFU (a slow growth parasite), and ORF (a rapid growth parasite) (Fragoso et al. 2008). Genes linked with the major histocompatibility complex (H-2) influence parasite growth. H-2 congenic mice on a BALB background differ in their susceptibility to *T. crassiceps*: BALB/c (H-2d) are the most susceptible, whereas BALB/K (H-2k) and BALB/B (H-2b) mice are comparatively resistant. Non-H-2 genes have no significant effect on susceptibility in H-2d strains, as reflected by the similar parasite loads in BALB/c and DBA/2. Recombinant mouse strain alleles (Kk, Ik, Sd, Dd) also influence susceptible, indicating that S regions, D regions, or both of the H-2d complex are involved in the control of resistance to murine cysticercosis (Sciutto et al. 1991). Differential susceptibility to *T. crassiceps* in BALB/c substrains of mice also may be associated with differences in serum T levels and Qa-2 protein expression (Fragoso et al. 1996).

Estrogens favor, whereas androgens and P4 inhibit, parasite reproduction (Bojalil et al. 1993; Terrazas et al. 1994; Morales-Montor et al. 2002b; Vargas-Villavicencio et al. 2006). Gonadectomy and thymectomy equalize parasite loads between the sexes by greatly increasing the susceptibility of males to infection and modestly decreasing the susceptibility of females to infection (Huerta et al. 1992; Terrazas et al. 1994; Morales-Montor et al. 2002b). Male mice also are better protected by vaccination against *T. crassiceps* than are females (Cruz-Revilla et al. 2000). The exogenous administration of E2 to gonadectomized females increases parasite load, but exposure of gonadectomized males to dihydrotestosterone (DHT) reduces parasite loads (Morales-Montor et al. 2002b). Treatment of female mice with the estrogen receptor antagonist Tamoxifen reduces parasite load (Vargas-Villavicencio et al. 2007). Adaptive transfer of T cells to thymectomized male mice can reinstate immunity against *T. crassiceps* (Bojalil et al. 1993). The early Th1 response (e.g., production of IL-2 and IFN-γ) hinders *T. crassiceps* growth during acute infection (Terrazas et al. 1999; Toenjes et al. 1999; Rodriguez-Sosa et al. 2002), whereas the Th2 response (e.g., production of IL-4 and IL-10) prevails at later times of infection, but is incapable of slowing *T. crassiceps* growth (Terrazas et al. 1998; Toenjes et al. 1999). In summary, these findings illustrate that during murine cysticersosis, androgens favor a Th1 response that limits parasite growth and estrogens favor a Th2 response that permits parasite reproduction (Huerta et al. 1992; Bojalil et al. 1993; Terrazas et al. 1994; Morales-Montor et al., 2002a, 2002b, 2002c).

Sex differences in susceptibility to cysticercosis have been reported for other host-taeniid species systems. For example, sex steroids have been implicated in porcine cysticercosis caused by *Taenia solium* because both castration and pregnancy nearly double the prevalence of naturally acquired cysticercosis in rural pigs (Morales et al. 2002). In humans, women are more commonly afflicted with severe neurocysticercosis and show higher proinflammatory cytokine profiles than men (Del Brutto et al. 1988; Fleury et al. 2003). We hypothesize that sex steroids modulate natural infections of humans and pigs with *T. solium*, because sex steroids affect experimental *T. crassiceps* infections in laboratory mice.

Several studies indicate that *T. crassiceps* infection alters the concentrations of sex steroids as well as other physiological systems of the host that presumably increase successful reproduction of the parasite during chronic infection. In fact, during chronic infection, male mice become "feminized," with increased concentrations of estrogens and reduced concentrations of androgens compared with noninfected male mice. These hormonal changes are accompanied by a reduced expression of male reproductive behavior in response to stimulus females (Morales et al. 1996). The reduced mating behavior in infected male rodents can be restored to normal levels by exposure to exogenous T suggesting that suppressed T, and not pathology caused by infection, underlies the reduced sexual behavior (Morales et al. 1996). Further, the administration of the aromatase inhibitor Fadrozole reduces feminization of male mice and decreases parasite loads (Morales-Montor et al. 2002c). We hypothesize that the hormonal changes are triggered by high concentrations of IL-6 during late stages of infection (Morales-Montor et al. 2001b).

Serum, splenic, and testicular IL-6 are augmented in infected compared with uninfected male mice (Morales-Montor et al, 2001a,b, 2002b). When infected male mice have an intact immune system, there is an increase in serum E2 and a decrease in T and DHT. Conversely, when the immune system is knocked down by irradiation or neonatal thymectomy, there is no effect of chronic infection on serum sex hormone concentrations (i.e., levels remain similar to those from uninfected male mice) (Morales-Montor et al, 2001b). IL-6 is critical in this puzzle: IL-6$^{-/-}$ (KO) mice do not exhibit the "feminization" that wild-type (WT) male mice do during *T. crassiceps* infection and reconstitution of IL-6 in KO mice results in subsequent feminization (Morales-Montor et al, 2001b;2002b). IL-6 activates aromatase expression in the testes of the cysticercotic mice, which causes aromatization of androgens to estrogens. The increased serum levels of follicle stimulating hormone (FSH) in chronically infected male mice supports the hypothesis that FSH also may be a factor involved in the feminization process in infected male mice (Morales-Montor et al, 2002b). To further study the role of IL-6 and macrophage-migration inhibitory factor (MIF), we infected IL-6 and MIF KO mice and measured the number of parasites and serum sex steroid levels. IL-6 and MIF KO mice of both sexes harbored similar numbers of parasites, with no change in sex-hormone levels. However, in WT mice, the sex-associated susceptibility to infection is observed concomitantly with the feminization of chronically infected mice (Morales-Montor et al, 2002b). These results suggest a role for both IL-6 and MIF in sex-associated susceptibility in murine *T. crassiceps* cysticercosis.

We have also demonstrated that the central nervous system of infected, feminized male mice responds to peripheral infection by overexpressing *c-fos* mRNA in the hypothalamus, hippocampus, and preoptic area (Morales-Montor et al. 2004). One hypothesis that we would like to test is that hormonal changes induced in the host promote the overexpression of the *c-fos* gene in the parasite as well as the host cells (Escobedo et al. 2004), in a manner similar to what has been demonstrated after stress or other immune challenges (Pacheco-Lopez et al. 2002). Thus, in murine cysticercosis, parasite proliferation is both a cause and consequence of the interactions among the host nervous, immune, and endocrine systems.

7.3.2.2   Nematode parasites

Another parasite disease for which sex differences in intensity and prevalence are reported is schistosomiasis. In this case, the intensity and prevalence of *Schistosoma* infection in endemic areas are higher in males than females, among both children and adults (Degu et al. 2002; Marguerite et al. 1999). In mouse models of *S. mansoni* infection, however, the sex difference is reversed. Female mice are more susceptible to infection and develop higher inflammatory responses, as measured by organ weights and delayed type hypersensitivity responses, than males (Boissier et al. 2003; Eloi-Santos et al. 1992). Administration of T protects, whereas castration exacerbates disease as measured by worm burden and host mortality following inoculation with *S. mansoni* parasites (Nakazawa et al. 1997). Clarification about why the sex difference in response to *Schistosoma* parasites is reversed in mice compared with humans is necessary to establish the utility of this mouse model for study of human schistosomiasis.

A detailed understanding of the mechanisms underlying female-biased susceptibility to these parasites is critical to predicting sex-biased responses to other infectious microbes. Also, the consideration that both hosts and parasites have coevolved to survive and reproduce may assist with interpreting discrepancies in the field. Finally, an organismal or systems level of analysis may be more informative about the causes of sex differences than more reductionist approaches (Oltvai and Barabási, 2002; Strohman 2003).

7.4 Evidence that Hormones Contribute to Female-Biased Susceptibility

7.4.1 Host Hormones Directly Affect Parasite Growth and Reproduction

Hormones regulate a variety of cellular and physiological functions of organisms, including immune function (Grossman et al. 1991; Derijk and Berkenbosch 1991; Hughes and Randolph 2001; Vertheleyi 2001; Barnard et al. 1998; Sternberg 1997). The immunological responsiveness to hormones is clearly evident during various parasitic diseases including malaria, schistosomiasis, toxoplasmosis, cysticercosis, trypanosomiasis, leishmaniasis (Cernetich et al. 2006; Roberts et al. 2001; Remoue et al. 2001; Zhang et al. 2000; do Prado et al. 1999; Larralde et al. 1995). The specific effects of hormones on host immunity, in general, and in response to parasites, in particular, are detailed elsewhere in this book (Chaps. 2 and 6). Thus, we would like to present compelling data illustrating that host hormones have direct effects on parasite growth and reproduction, which may make males and females differentially susceptible to infection.

Not only can hormones affect host responses to infection, but parasites can exploit the host endocrine system for their own benefit. Recent experimental evidence (Ghansah et al. 2002; Remoue et al. 2002; Morales-Montor et al. 2001a; Dissanayake 2000; Freilich et al. 2000; Townson and Tagboto 1996; Lingnau et al. 1993; Charlab et al. 1990; Maswoswe et al. 1985; Escobedo et al. 2004) has led us to suggest a mechanism of host exploitation by the parasite. In this system of "trans-regulation," the parasite benefits directly from host-derived hormones or growth factors, to allow rapid establishment and increased growth and reproduction. Trans-regulation phenomena in parasites have been scarcely explored. Evidence from our laboratory illustrate that treatment of *T. crassiceps* (ORF strain) with E2 and P4 in vitro stimulates reproduction, whereas in vitro treatment with T or DHT inhibits and even exerts a slight toxic effect on the parasite (Escobedo et al, 2004). The effect of E2 on parasite growth can be interrupted by the administration of tamoxifen, a compound that is well known for its antiestrogenic effects (Escobedo et al, 2004). On the other hand, T and DHT directly affect parasitic DNA integrity probably by activating apoptotic mechanism in the cysticercus cells. This experimental finding is not dependent on a nuclear receptor because flutamide (a well-studied and used antiandrogen) does not affect parasite reproduction in vitro (Escobedo et al, 2004).

Trans-regulation has been reported in other parasitic infections. In *P. falciparum*, for example, the in vitro cortisol treatment of merozoites increases the number and size of the gametocytes produced (Maswoswe et al. 1985). Treatment of merozoites with insulin, E2, P4, or T also increases the number of gametocytes produced in vitro (Lingnau et al. 1993). In contrast, when these parasites are treated with 16-α-bromoepiandrosterone, a DHEA analog, growth is diminished by up to 25% (Freilich et al. 2000). In vitro experiments with *S. mansoni* reveal that DHEA or cortisol treatment of cercaria and schistosomules inhibits parasite viability and oviposition up to 100% (Morales-Montor et al. 2001a). Moreover, treatment of the adult *S. haematobium* with T diminishes fertility and, thus, the reproductive capacity of this parasite (Remoue et al. 2002).

Synthesis of DNA as well as metabolic activity (inducing receptors with tyrosine kinase, protein C kinase and MAPK activity) is increased in *Trypanososma cruzi* amastigotes treated in vitro with murine epidermal growth factor (EGF) (Ghansah et al. 2002). Similarly, murine EGF stimulates the in vitro development and maturation of *Brugia malayi* microfilariae (Dissanayake 2000). Moreover, the in vitro treatment of *Onchocerca volvulus* and *O. lienalis* microfilariae with 20-hydroxyecdisone induces differentiation to the infective stage (Townson and Tagboto 1996). Granulocyte macrophage-colony stimulating factor (GM-CSF) augments in vitro growth of *L. mexicana* (Charlab et al. 1990). Treatment of *T. crassiceps* cysticerci with E2 increases, whereas treatment with T or DHT diminishes, the reproductive capacity of the parasite (Escobedo et al. 2004). In addition, viability, growth, and infective capacity of cysticerci are also increased up to 200% following treatment with estrogen, but are inhibited by androgen treatment (Escobedo et al. 2004).

Direct exposure of *Entamoeba histolytica* trophozoites to varying concentrations of E2, P4, T, or DHT has little effect on parasite viability or proliferation. Conversely, treatment of *E. histolytica* trophozoites with cortisol stimulates parasite proliferation in a dose-dependent manner (Carrero et al. 2006). In contrast, exposure of *E. histolytica* trophozoites to dehydroepiandrosterone (DHEA) inhibits proliferation, reduces adherence and motility, and causes lysis of trophozoites in a dose-dependent manner. Moreover, cortisol increases and DHEA decreases

E. histolytica DNA synthesis as determined by ^{3}H-thymidine incorporation. Lysis of trophozoites by DHEA appears to be caused by a necrotic rather than an apoptotic process, as determined through patterns of DNA fragmentation and TUNEL assays (Carrero et al, 2006). The activity of glucose-6-phosphate dehydrogenase in trophozoite extracts is partially inhibited in presence of DHEA. Although DHEA inhibits *E. histolytica* growth, administration of DHEA to infected hamsters exacerbates amebic liver abscesses (Carrero et al, 2006) illustrating that steroids act both directly on *E. histolytica* and the host.

The hormonal microenvironment may favor or inhibit survival of parasites differentially between the sexes. This may represent a highly evolved host-parasite relationship in which certain hormones appear to serve as proliferation or death factors that influence the establishment of infection, independent of the host immune response. All of this amounts to the parasite exploiting endocrine mechanisms developed by the host for its own advantage.

7.4.2 Parasite-Derived Hormonal Mechanisms Affect Parasite Physiology

7.4.2.1 Synthesis of hormones by parasites

Parasites, such as *S. mansoni*, synthesize, incorporate, and metabolize lipids (Meyer et al. 1970; Rumjanek and Simpson 1980). These parasites also synthesize isoprenoic compounds (Foster et al. 1993) and produce several molecular species of phospholipids (Brouwers et al. 1998). Helminths also produce ecdysteroids (Mendis et al. 1983, 1984; Mercer et al. 1987, 1990; Cleator et al. 1987; Fleming 1985; Evershed et al. 1987). The production and physiological significance of sex steroid-like hormones in parasites have received little attention (Romano et al. 2003). The synthesis of steroid hormones by *S. mansoni* was investigated by Briggs (1972), who found that these parasites had the capacity to perform several steps in the steroidogenic pathway. Particularly, he showed that only adult schistosomes, but not schistosome eggs or miracidia, have the ability to convert cholesterol, corticosteroids, and sex hormones to various metabolites. For instance, when 4-Androstene-3, 17-dione was added to cultured parasites, it was bioconverted to 5α-androstane-3,17-dione and androsterone, cortisone was metabolized to cortisol, and estrone (E1) to E2. Pregnenolone treatment yielded P4 and 17α-hydroxyprogesterone, while T was bioconverted to 5α-dihydrotestosterone, 17α-hydroxyprogesterone to 4-androstene-3, 17-dione, and 17α-hydroxypregnenolone was bioconverted to 4-androstene-3, 17-dione. When cholesterol was added to the cultured worms, pregnenolone was produced.

T. crassiceps cysticerci also can synthesize steroid hormones in vitro (Gomez et al. 2000). The parasites can transform [^{3}H] androstenedione to [^{3}H] T. Also *T. solium* cysticerci have the capacity to synthesize the same androgens.

Conversely, isolated scoleces incubated in the presence of [(3)H]androstenedione yield small quantities of [(3)H] E2 (Valdéz et al., 2006). Furthermore, the incubation of trypomastigotes of the *T. cruzi* (Tulahuén strain) with 3H-androstenedione results in synthesis 3H-T, 3H-E2 and 3H-E1. Metabolism of 3H-DHEA by the parasites yields 3H-androstendione and 3H-androstendiol. These results indicate that *T. cruzi* trypomastigotes produce androgens and estrogens when incubated in the presence of steroid precursors and suggest the presence of active parasite steroidogenic enzymes (Vacchina et al. 2008).

7.4.2.2   Synthesis of hormone receptors by parasites

The observation that parasites can respond directly to host hormones suggests that parasites may synthesize receptors and have the capacity to bind host hormones to direct downstream transcriptional events. In this context, Sani et al. (1985) characterized specific retinol and retinoic acid-binding proteins in *O. volvulus, O. gibsoni, Dipetalonema vitae, Brugia pahangi*, and *Dirofilaria immitis*. The role ascribed to these proteins was to bind to host hormones and mediate the biological effects of retinoids on parasite growth, reproduction, and differentiation. The genome of *O. volvulus* encodes for at least three classical nuclear receptors, two of which (i.e., OvNR-1 and OvNR-2) have been characterized and are similar to the retinoid receptors in vertebrates and to the *Drosophila melanogaster* protein EiP78c (Unnasch et al. 1999). Computer modeling of these receptors suggests that these molecules possess a ligand binding-cavity of a size and form capable of binding to vertebrate steroids (Unnasch et al. 1999; Yates et al. 1995).

S. *mansoni* have receptors capable of binding to E2, which provides a likely mechanism for the protective effect of E2 in infected mice and hamsters (Barrabes et al. 1979, 1986). Classical nuclear receptors for sex steroids, thyroid hormones, and ecdisteroids have been characterized in *S. mansoni* (de Mendoca et al. 2000). Homology between these receptors and those described in *Drosophila*, mice, and humans ranges from 70 to 95% (de Mendoca et al. 2000). The presence of a receptor (SmRTK-1) with tyrosine kinase activity in *S. mansoni* has been determined. Preferential localization of SmRTK-1 in sporocysts and oocysts may favor differentiation and growth processes in this parasite (Vicogne et al. 2003). *S. haematobium* also synthesizes a protein of 28 kDa (Sh28GST) capable of binding to T and facilitating transport, metabolism, and physiological action of this hormone in the parasite (Remoue et al. 2002).

We have shown that *T. crassiceps* cysticerci express an androgen receptor, both isoforms of the classic estrogen receptor (ERα and β), but not P4 receptors A or B (Escobedo et al. 2004). Thus, the direct effects of estrogens and androgens on *T. crassiceps* reproduction may be due to the binding of E2 and T to their respective receptors on the parasite. Conversely, P4 may affect *T. crassiceps* cysticercosis through metabolism into E2 (Vargas-Villavicencio et al. 2005). Binding of the ER to classic estrogen response elements may be responsible for the activation of AP-1 complex genes in the normal metabolism of *T. crassiceps* (Escobedo et al. 2004).

7.5   Conclusions and Future Directions

There are many exceptions to the general rule of thumb that males are more susceptible to parasites than females. Our work and the work of others reveal that hormones, including androgens, estrogens, and glucocorticoids, affect both the parasite and host responses to infection. Future studies must continue to consider the interactions of hormones with cytokines in a network that result in sex differences in response to infection (Besedovsky and del Rey 2002). An area of fruitful research will be further examination into the responses of parasites to the hormonal environment of each host sex and in particular how this directly contributes to sex differences in parasitic infections. Further, the observation that parasites, such as *T. crassiceps*, can feminize the hormonal milieu of male mice illustrates the complexities of these host-parasite interactions and underscores why simplistic rules of thumb can result in misinterpretation of important findings.

As noted previously in this Chapter, the immune system of males and females are not dimorphic; rather, the immune responses generated against both self and nonself antigen are different between the sexes. Immunological sensitivity to hormones, and possibly neurotransmitters, may have evolved as an adaptation for individuals of either sex to confront infection successfully. even if by different mechanisms. Sex differences in immune function also may have evolved to solve sex-specific challenges, like pregnancy or competition for territories or mates (Grossman 1979; Zuk 1994; Kavaliers et al. 2001). Selective pressures may drive the evolution of dimorphic immunity to infection and create a need to balance the defense of the host against infection with successful reproduction and survival of the host (Grossman et al. 1979; Gaillard and Spinedi 1998; Agrawal and Lively 2001; Charles et al. 2002; Moore and Wilson 2002; Owens 2002; Potti et al. 2002; Tella et al. 2002). A compromise may be achieved by a transient enhancement or suppression of immunity, which could allow pregnancy in females with offspring that are partially nonself and to allow males to utilize high androgen concentrations to develop secondary sex traits, defend territories, and obtain mates (Martal et al. 1995; Matzinger 2002; Medzhitov and Janeway 2002; Zuk 1994). Pregnancy, in particular, demands immunological allowance that would be beneficial to offspring survival and transient immunoendocrine regulation by hormones and cytokines may ensure that the fetus is not damaged (Barnea 2001). Similarly, males that can better cope with the stress of competition and can balance this with susceptibility to infection may be more likely to achieve successful matings and reproductive success as discussed in Chapter 1 of this book.

Parasite responsiveness to host hormones also should be considered from an evolutionary perspective. We hypothesize that the differential effects of parasites on male and female hosts may serve to increase the probability of transmission of parasites to definitive hosts. Further, feminization of males during chronic murine cysticercosis may be regarded as a parasite-mediated behavior that increases transmission of the parasite to its definitive hosts, carnivores, and completion of the parasite's life cycle (Gourbal et al. 2001). Similarly, the increased susceptibility of females to acute *T. crassiceps* infection may represent the optimal host for increased reproduction of worms (Poulin and Thurn 1996; Zuk and McKean 1996; Panhuis et al. 2001).

The ability of a parasite to differentially affect a female or a male of the same species could potentially be due to hormonal regulation of the immune response or direct hormonal affects on the parasite. Understanding the contribution of each of these and characterization of the parasite molecules involved may facilitate the development of drugs that counteract the effects of hormones on the immune system or the parasite. We propose that immunologists, microbiologists, and biomedical scientists must consider the sex of the host when examining responses to infection. Our observation of female-biased susceptibility to *T. crassiceps* extends beyond parasites and has been documented for viruses and bacteria, as noted in Chapters 4, 5, and 11 of this book. The only way to fully tease apart the mechanisms underlying female-biased susceptibility to infection will be to conduct both in vivo and in vitro studies. It also is premature to assume that all sex differences in response to infection are mediated by sex steroids. Our work with glucocorticoid-mediated effects on *S. mansoni* highlights this point. Thus, we propose a network that has numerous connections and that may establish simple cause–effect relationships in parasitism. Identification of the most connected nodes within the network would be a way to examine the principal participants in sex differences in response to infection. One could hypothesize that important neuro-endocrine system connections with the immune system are established during embryonic development, when gonadal differentiation occurs (Klein et al. 2002; Martin 2000; Sinisi et al. 2003) and principal criteria for immunological self- and danger signal recognition also appear to be set (Matzinger 2002; Medzhitov and Janeway 2002).

Acknowledgments Financial support was provided by Dirección General de Asuntos del Personal Académico, Universidad Nacional Autonoma de México (UNAM), IN213108, by The National Council of Science and Technology of Mexico (CONACyT, 58283) and by a special grant from Fundación Miguel Alemán-SSA, the three of them to J.M.-M. G.E., and MADL were CONACyT Ph.D. fellows.

References

Addis CJ Jr (1946) Experiments on the relation between sex hormones and the growth of tape-worms. J Parasitol 32:229–236

Agrawal AF, Lively CM (2001) Parasites and the evolution of self-fertilization. Int J Organic Evol 55:869–879

Arnott MA, Cassella JP, Aitken PP, Hay J (1990) Social interactions of mice with congenital Toxoplasma infection. Ann Trop Med Parasitol 84:149–156

Barnard CJ, Behnke JM, Gage AR, Brown H, Smithurst PR (1998) The role of parasite-induced immunodepression, rank and social environment in the modulation of behaviour and hormone concentration in male laboratory mice (*Mus musculus*). Proc R Soc Lond B Biol Sci 22:693–701

Barnea ER (2001) Embryo maternal dialogue: From pregnancy recognition to proliferation control. Early Preg 5:65–66

Barrabes A, Duong TH, Combescot C (1979) Effect of testosterone or progesterone implants on the intensity of experimental infestation with *Schistosoma mansoni* in the female golden hamster. C R Seances Soc Biol Fil 173:153–6

Barrabes A, Goma-Mouanda J, Reynouard F, Combescot C (1986) 17 beta-estradiol receptors in *Schistosoma mansoni*. Contribution to the explanation of the protective power of this hormone in Schistosoma mansoni bilharziasis in the mouse. Preliminary study. Ann Parasitol Hum Comp 61:637–41

Benten WP, Becker A, Schmitt-Wrede HP, Wunderlich F (2002) Developmental regulation of intracellular and surface androgen receptors in T cells. Steroids 67:925–931

Berdoy M, Webster JP, MacDonald DW (1995) The manipulation of rat behaviour by *Toxoplasma gondii*. Mammalia 59:605–613

Besedovsky HO, del Rey A (2002) Introduction: Immune-neuroendocrine network. Front Hormone Res 29:1–14

Boissier J, Chlichlia K, Digon Y, Ruppel A, Mone H (2003) Preliminary study on sex-related inflammatory reactions in mice infected with Schistosoma mansoni. Parasitol Res 91(2):144–50

Bojalil R, Terrazas LI, Govezensky T, Sciutto E, Larralde C (1993) Thymus-related cellular immune mechanisms in sex associated resistance to experimental murine cysticercosis (*Taenia crassiceps*). J Parasitol 79:384–389

Briggs MH (1972) Metabolism of steroid hormones by schistosomes. Biochim Biophys Acta 280:481–485

Brouwers JFHM, Van Hellemond JJ, van Golde LMG, Tielens AG (1998) Ether lipids and their possible physiological function in adult *Schistosoma mansoni*. Mol Biochem Parasitol 96:49–58

Carrero JC, Cervantes C, Moreno Mendoza N, Saavedra E, Morales Montor J, Laclette JP (2006) Dehydroepiandrosterone decreases while cortisol increases in vitro growth and viability of *Entamoeba histolytica*. Microbes Infect 8:323–31

Cernetich A, Garver LS, Jedlicka AE, Klein PW, Kumar N, Scott AL, Klein SL (2006) Involvement of gonadal steroids and gamma interferon in sex differences in response to blood-stage malaria infection. Infect Immun 74:3190–3203

Charlab R, Blaineau C, Schechtman D, Barcinski MA (1990) Granulocyte-macrophage colony-stimulating factor is a growth-factor for promastigotes of *Leishmania mexicana* amazonensis. J Protozool 37:352–7

Charles S, Morand S, Chasse JL, Auger DP (2002) Host patch selection induced by parasitism: Basic reproduction ratio r(0) and optimal virulence. Theor Pop Biol 62:97–109

Cleator M, Delves CJ, Howells RE, Rees HH (1987) Identity and tissue localization of free and conjugated ecdysteroids in adults of *Dirofilaria immitis* and *Ascaris suum*. Mol Biochem Parasitol 25:93–105

Cruz-Revilla C, Rosas G, Fragoso G, Lopez-Casillas F, Toledo A, Larralde C, Sciutto E (2000) *Taenia crassiceps* cysticercosis: Protective effect and immune response elicited by DNA immunization. J Parasitol 86:67–74

Culbreth KL, Esch GW, Kuhn RE (1972) Growth and development of larval *Taenia crassiceps* (Cestoda) III. The relationships between larval biomass and the uptake and incorporation of C-leucine. Exp Parasitol 32:272–281

Daly TM, Long CA (1995) Humoral response to a carboxyterminal region of the merozoite surface protein-1 plays a predominant role in controlling blood-stage infection in rodent malaria. J Immunol 55:236–243

Degu G, Mengistu G, Jones J (2002) Some factors affecting prevalence of and immune responses to Schistosoma mansoni in schoolchildren in Gorgora, northwest Ethiopia. Ethiop Med J 40:345–52

Del Brutto OH, Garcia E, Talamas O, Sotelo J (1988) Sex related severity of inflammation in parenchymal brain cysticercosis. Arch Int Med 148:544–546

Derijk R, Berkenbosch F (1991) The immune-hypothalamo-pituitary-adrenal axis and autoimmunity. Int J Neurosci 59:91–100

Dissanayake S (2000) Upregulation of a raf kinase and a DP-1 family transcription factor in epidermal growth factor (EGF) stimulated filarial parasites. Int J Parasitol 30:1089–97

Do Prado JC, Leal MP, Anselmo-Franci JA, de Andrade HF, Kloetzel JK (1998) Influence of female gonadal hormones on the parasitemia of female *Calomys callosus* infected with the "Y" strain of *Trypanosoma cruzi*. Parasitol Res 84:100–105

Do Prado JC, Jr LAM, Leal MP, Bernard E, Kloetzel JK (1999) Influence of male gonadal hormones on the parasitemia and humoral response of male Calomys callosus infected with the Y strain of *Trypanosoma cruzi*. Parasitol Res 85:826–9

Eloi-Santos S, Olsen NJ, Correa-Oliveira R, Colley DG (1992) Schistosoma mansoni: mortality, pathophysiology, and susceptibility differences in male and female mice. Exp Parasitol 75:168–75

Esch GW, Smith JD (1976) Studies on the in vitro culture of *Taenia crassiceps*. Int J Parasitol 6:143–149

Escobedo G, Larralde C, Chavarria A, Cerbón MA, Morales-Montor J (2004) Molecular mechanism involved in the differential effects of sex steroids on the reproduction and infectivity of *Taenia crassiceps*. J Parasitol 90:1235–1244

Escobedo G, Roberts CW, Carrero JC, Morales-Montor J (2005) Parasite regulation by host hormones: an old mechanism of host exploitation? Trends Parasitol 21:588–593

Evershed RP, Mercer JG, Rees HH (1987) Capillary gas chromatography/mass spectrometry of ecdysteroids. J Chromatogr 390:357–369

Flegr J, Lindova J, Kodym P (2008) Sex-dependent toxoplasmosis-associated differences in testosterone concentrations in humans. Parasitology 135:427–431

Fleming MW (1985) *Ascaris suum*: role of ecdysteroids in molting. Exp Parasitol 60:207–210

Fleury A, Gomez T, Alvarez I, Meza D, Huerta M, Chavarria A, Carrillo-Mezo RA, Lloyd C, Dessein A, Preux PM, Dumas M, Larralde C, Sciutto E, Fragoso G (2003) High prevalence of calcified silent neurocysticercosis in a rural village of Mexico. Neuroepidemiology 22:139–145

Foster JM, Pennock JF, Marshall I, Rees HH (1993) Biosynthesis of isoprenoid compounds in *Schistosoma mansoni*. Mol Biochem Parasitol 61:275–284

Fragoso G, Lamoyi E, Mellor A, Lomeli C, Govezensky T, Sciutto E (1996) Genetic control of susceptibility to *Taenia crassiceps* cysticercosis. Parasitology 112:119–124

Fragoso G, Lamoyi E, Mellor A, Lomeli C, Hernandez M, Sciutto E (1998) Increased resistance to *Taenia crassiceps* murine cysticercosis in Qa-2 transgenic mice. Infect Immun 66:760–764

Fragoso G, Meneses G, Sciutto E, Fleury A, Larralde C (2008) Preferential growth of Taenia crassiceps cysticerci in female mice holds across several laboratory mice strains and parasite lines. J Parasitol 94(2):551–3

Freeman RS (1962) Studies on the biology of *Taenia crassiceps* (Zeder 1800) Rudolphi, 1810 (Cestoda). Can J Zool 40:969–990

Freilich D, Ferris S, Wallace M, Leach L, Kallen A, Frincke J, Ahlem C, Hacker M, Nelson D, Hebert J (2000) 16-alpha-bromoepiandrosterone, a dehydroepiandrosterone (DHEA) analogue, inhibits *Plasmodium falciparum* and *Plasmodium berghei* growth. Amer J Trop Med Hyg 63:280–283

Gaillard RC, Spinedi E (1998) Sex- and stress-steroids interactions and the immune system: Evidence for a neuroendocrineimmunological sexual dimorphism. Dom An Endocrinol 15:345–352

Ganley LS, Rajan TV (2001) Endogenous testosterone levels do not affect filarial worm burdens in mice. Exp Parasitol 98:29–34

Ghansah TJ, Ager EC, Freeman-Junior P, Villalta F, Lima MF (2002) Epidermal growth factor binds to a receptor on *Trypanosoma cruzi* amastigotes inducing signal transduction events and cell proliferation. J Eukaryot Microbiol 49:383–90

Ghosh PK, Gupta S, Leon LR, Ghosh R, Ruiz-Ordaz BH, Ortiz-Ortiz L (1998) Intestinal amoebiasis: Antibody-secreting cells and humoral antibodies. J Diarrhoeal Dis Res 1:1–7

Gomez Y, Valdez RA, Larralde C, Romano MC (2000) Sex steroics and parasitism: *Taenia crassiceps* cisticercus metabolizes exogenous androstenedione to testosterone in vitro. J Steroid Biochem Mol Biol 74:143–7

Gourbal BE, Righi M, Petit G, Gabrion C (2001) Parasite-altered host behavior in the face of a predator: Manipulation or not? Parasitol Res 87:186–192

Grossman CJ, Sholiton LJ, Nathan P (1979) Rat thymic estrogen receptor. I. Preparation, location, and physiochemical properties. J Steroid Biochem 11:1233–40

Grossman CJ, Roselle GA, Mendenhall CL (1991) Sex steroid regulation of autoimmunity. J Steroid Biochem Molec Biol 40:649–659

Hanssen SA, Folstad I, Erikstad KE (2003) Reduced immunocompetence and cost of reproduction in common eiders. Oecologia 136:457–464

Huerta L, Terrazas LI, Sciutto E, Larralde C (1992) Immunological mediation of gonadal effects on experimental murine cysticercosis caused by *Taenia crassiceps* metacestodes. J Parasitol 78:471–476

Hughes VL, Randolph SE (2001) Testosterone increases the transmission potential of tick-borne parasites. Parasitol 123:365–71

Johnson LL, Sayles PC (2002) Deficient humoral responses underlie susceptibility to *Toxoplasma gondii* in CD4-deficient mice. Infect Immunity 70:185–191

Kavaliers M, Colwell DD (1993) Multiple opioid system involvement in the mediation of parasitic-infection induced analgesia. Brain Res 623:316–320

Kavaliers M, Choleris E, Colwell DD (2001) Brief exposure to female odors "emboldens" male mice by reducing predatorinduced behavioral and hormonal responses. Hormones Behav 40:497–509

Klein SL (2000) The effects of hormones on sex differences in infection: From genes to behavior. Neuro Biobehavior Rev 24:627–638

Klein SL, Marson AL, Scott AL, Ketner G, Glass GE (2002) Neonatal sex steroids affect responses to Seoul virus infection in male but not female Norway rats. Brain Behav Immunol 16:736–746

Kyes S, Horrocks P, Newbold C (2001) Antigenic variation at the infected red cell surface in malaria. Ann Rev Microbiol 55:673–707

Larralde C, Morales J, Terrazas I, Govezensky T, Romano MC (1995) Sex hormone changes induced by the parasite lead to feminization of the male host in murine *Taenia crassiceps* cysticercosis. J Steroid Biochem Mol Biol 52:575–580

Liesenfeld O, Nguyen TA, Pharke C, Suzuki Y (2001) Importance of gender and sex hormones in regulation of susceptibility of the small intestine to peroral infection with Toxoplasma gondii tissue cysts. J Parasitol 87:1491–3

Lingnau A, Margos G, Maier WA, Seitz HM (1993) The effects of hormones on the gameto-cytogenesis of *Plasmodium falciparum* in vitro. Appl Parasitol 34: 53–60

Marguerite M, Gallissot MC, Diagne M, Moreau C, Diakkhate MM, Roberts M et al (1999) Cellular immune responses of a Senegalese community recently exposed to Schistosoma mansoni: correlations of infection level with age and inflammatory cytokine production by soluble egg antigen-specific cells. Trop Med Int Health 4:530–43

Martal J, de la Llosa-Permier MP, Chene N, Huynh L, Millet S, Haridon RL, Assal A, Roignant N, Chaouat G (1995) Trophoblastic interferons and embryonal immune tolerance. Contracep Fert Sexual 23:562–572

Martin JT (2000) Sexual dimorphism in immune function: the role of prenatal exposure to androgens and estrogens. Eur J Pharmacol 405:251–261

Maswoswe SM, Peters W, Warhurst DC (1985) Corticosteroid stimulation of the growth of *Plasmodium falciparum* gametocytes *in vitro*. Ann Trop Med Parasitol 79:607–16

Matzinger P (2002) The danger model: A renewed sense of self. Science 296:301–305

Medzhitov R, Janeway CA (2002) Decoding the patterns of self and nonself by the innate immune system. Sci 296:298–300

Mendis AHW, Rose ME, Rees HH, Goodwin TW (1983) Ecdysteroids in adults of the filarial nematode *Dirofilaria immitis*. Mol Biochem Parasitol 9:209–226

Mendis AHW, Rees HH, Goodwin TW (1984) The occurrence of ecdysteroids in the cestode *Monienza expansa*. Mol Biochem Parasitol 10:123–138

Mercer JG, Munn AE, Rees HH (1987) *Echinococcus granulosus*: occurrence of ecdysteroids in protoscoleces and hydatic cyst fluid. Mol Biochem Parasitol 24:203–214

Mercer JG, Barker GC, Howells RE, Rees HH (1990) Investigation of ecdysteroid excretion by adult *Dirofilaria immitis* and *Brugia pahangi*. Mol Biochem Parasitol 39:89–96

Meyer F, Meyer H, Bueding E (1970) Lipid metabolism in the parasitic and free-living flatworms, *Schistosoma mansoni* and *Dugesia dorotocephala*. Biochem Biophys Acta 210:257–266

Moore SL, Wilson K (2002) Parasites as a viability cost of sexual selection in natural populations of mammals. Science 297:2015–2018

Morales J, Larralde C, Arteaga M, Govezensky T, Romano MC, Morali DG (1996) Inhibition of sexual behavior in male mice infected with *Taenia crassiceps* cysticerci. J Parasitol 82:689–693

Morales J, Velasco T, Tovar V, Fragoso G, Fleury A, Beltran C, Villalobos N, Aluja A, Rodarte LF, Sciutto E, Larralde C (2002) Castration and pregnancy of rural porcines significantly increase the prevalence of naturally acquired *Taenia solium* cysticercosis. Vet Parasitol 30:41–48

Morales-Montor J, Baig S, Mitchell R, Deway K, Hallal-Calleros C, Damian RT (2001a) Immunoendocrine interactions during chronic cysticercosis determine male mouse feminization: Role of IL-6. J Immunol 167:4527–4533

Morales-Montor J, Mohamed F, Ghaleb AM, Baig S, Hallal-Calleros C, Damian RT (2001b) In vitro effects of hypothalamic-pituitary-adrenal axis (HPA) hormones *on Schistosoma mansoni*. J Parasitol 87:1132–1139

Morales-Montor J, Baig S, Kabbani A, Damian RT (2002a) Do interleukin- 6 and macrophage-migration inhibitory factor play a role during sex-associated susceptibility in murine cysticercosis? Parasitol Res 88:901–904

Morales-Montor J, Baig S, Hallal-Calleros C, Damian RT (2002b) Taenia crassiceps: androgen reconstitution of the host leads to protection during cysticercosis. Exp Parasitol 100:209–216

Morales-Montor J, Hallal-Calleros C, Romano C, Damian RT (2002c) Inhibition of P-450 aromatase prevents feminisation and induces protection during cysticercosis. Int J Parasitol 32:1379–1387

Morales-Montor J, Arrieta I, del castillo LI, Rodríguez-Dorantes M, Cerbón MA, Larralde C (2004) Remote sensing of intraperitoneal parasitism by the host's brain: Regionalchanges of c-*fos* gene expression in the brain of feminised cysticercotic male mice. Parasitol 128:1–9

Nakazawa M, Fantappie MR, Freeman GL Jr, Eloi-Santos S, Olsen NJ, Kovacs WJ et al (1997) Schistosoma mansoni: susceptibility differences between male and female mice can be mediated by testosterone during early infection. Exp Parasitol 85:233–40

Oltvai ZN, Barabási AL (2002) Systems biology Life's complexity pyramid. Science 298 (5594):763–4

Owens IP (2002) Ecology and evolution: Sex differences in mortality rate. Science 297:2008–2009

Panhuis TM, Butlin R, Zuk M, Tregenza T (2001) Sexual selection and speciation. Trends Ecol Evol 1:364–371

Philipp M, Parkhouse RM, Ogilvie BM (1980) Changing proteins on the surface of a parasitic nematode. Nature 287:538–540

Phillips AN, Antunes F, Stergious G, Ranki A, Jensen GF, Bentwich Z et al (1994) A sex comparison of rates of new AIDS-defining disease and death in 2554 AIDS cases AIDS in Europe Study Group. Aids 8:831–835

Potti J, Davila JA, Tella JL, Frias O, Villar S (2002) Gender and viability selection on morphology in fledgling pied flycatchers. Mol Ecol 11:1317–1326

Poulin R, Thurn JR (1996) Sexual inequalities in helminth infections: A cost of being a male? Am Nat 147:287–295

Pung OJ, Luster MI (1986) *Toxoplasma gondii*: decreased resistance to infection in mice due to estrogen. Exp Parasitol 61:48–56

Rafati S, Salmanian AH, Hashemi K, Schaff C, Belli S, Fasel N (2001) Identification of *Leishmania major* cysteine proteinases as targets of the immune response in humans. Mol Biochem Parasitol 113:35–43

Remoue F, Van To D, Schacht AM, Picquet M, Garraud O, Vercruysse J, Ly A, Capron A, Riveau G (2001) Gender-dependent specific immune response during chronic human *Schistosomiasis haematobia*. Clin Exp Immunol 124:62–8

Remoue F, Mani JC, Pugniere M, Schacht AM, Capron A, Riveau G (2002) Functional specific binding of testosterone to *Schistosoma haematobium* 28-Kilodalton glutathione S-transferase. Infect Immun 70:601–5

Restrepo BI, Aguilar MI, Melby PC, Teale JM (2001) Analysis of the peripheral immune response in patients with neurocysticercosis: Evidence for T cell reactivity to parasite glycoprotein and vesicular fluid antigens. Am J Trop Med Hyg 65:366–370

Roberts CW, Cruickshank SM, Alexander J (1995) Sex-determined resistance to Toxoplasma gondii is associated with temporal differences in cytokine production. Infect Immun 63:2549–2555

Roberts CW, Walker W, Alexander J (2001) Sex-associated hormones and immunity to protozoan parasites. Clin Microbiol Rev 14:476–488

Romano MC, Valdéz RA, Cartas AL, Gómez Y, Larralde C (2003) Steroid hormone production by parasites: the case of *Taenia crassiceps* and *Taenia solium* cysticerci. J Steroid Biochem Mol Biol 85:221–225

Rumjanek FD, Simpson AJG (1980) The incorporation and utilization of radiolabelled lipids by adult *Schistosoma mansoni*. Mol Biochem Parasitol 1:31–44

Sani BP, Vaid A, Comley JC, Montgomery JA (1985) Novel retinoid-binding proteins from filarial parasites. Biochem J 232:577–83

Sciutto E, Fragoso G, Diaz ML, Valdez F, Montoya RM, Govezensky T, Lomeli C, Larralde C (1991) Murine *Taenia crassiceps* cysticercosis: H-2 complex and sex influence on susceptibility. Parasitol Res 77:243–246

Sciutto E, Fragoso G, Manoutcharian K, Gevorkian G, Rosas-Salgado G, Hernandez-Gonzalez M, Herrera-Estrella L, Cabrera-Ponce L, López-Casillas F, Gonzalez-Bonilla C, Santiago-Machuca A, Ruiz-Pérez F, Sanchez J, Goldbaum F, Aluja A, Larralde C (2002) New approaches to improve a peptide vaccine against porcine *Taenia solium* cysticercosis. Arch Med Res 33:371–378

Sher A, Gazzinelli RT, Oswald IP, Clerici M, Kullberg M, Pearce EJ, Berzofsky JA, Mosmann TR, James SL, Morse HC III (1992) Role of T-cell derived cytokines in the downregulation of immune responses in parasitic and retroviral infection. Immunol Rev 127:183–204

Sinisi AA, Pasquali D, Notaro A, Bellastella A (2003) Sexual differentiation. J Endocrinol Invest 26:S23–S28

Sternberg EM (1997) Neural-immune interactions in health and disease. J Clin Invest 100:2641–7

Strohman R (2003) Thermodynamics–old laws in medicine and complex disease. Nat Biotechnol 21(5):477–9

Tella JL, Scheuerlein A, Ricklefs RE (2002) Is cell-mediated immunity related to the evolution of life-history strategies in birds? Proc Royal Soc London Biol Sci 22:1059–1066

Terrazas LI, Bojalil R, Govezensky T, Larralde C (1994) A role for 17-beta-estradiol in immunoendocrine regulation of murine cysticercosis (*Taenia crassiceps*). J Parasitol 80:563–568

Terrazas LI, Bojalil R, Govezensky T, Larralde C (1998) Shift from an early protective Th1-type immune response to a late permissive TH2-type response in murine cysticercosis (*Taenia crassiceps*). J Parasitol 84:74–81

Terrazas LI, Cruz M, Rodriguez-Sosa M, Bojalil R, Garcia-Tamayo F, Larralde C (1999) Th1-type cytokines improve resistance to murine cysticercosis caused by *Taenia crassiceps*. Parasitol Res 85:135–141

Toenjes SA, Spolski RJ, Mooney KA, Kuhn RE (1999) The systemic immune response of BALB/c mice infected with larval *Taenia crassiceps* is a mixed Th1/TH2-type response. Parasitol 118:623–633

Townson S, Tagboto SK (1996) In vitro cultivation and development of *Onchocerca volvulus* and *Onchocerca lienalis* microfilariae. Am J Trop Med Hyg 54:32–7

Travis BL, Osorio Y, Melby PC, Chandrasekar B, Arteaga L, Saravia NG (2002) Gender is a major determinant of the clinical evolution and immune response in hamsters infected with *Leishmania spp*. Infect Immun 70:2288–96

Unnasch TR, Bradley J, Beauchamp J, Tuan R, Kennedy MW (1999) Characterization of a putative nuclear receptor from *Onchocerca volvulus*. Mol Biochem Parasitol 104:259–69

Vacchina P, Valdéz RA, Gómez Y, Revelli S, Romano MC (2008) Steroidogenic capacity of *Trypanosoma cruzi* trypomastigotes. J Steroid Biochem Mol Biol 111:282–6

Valdéz RA, Jiménez P, Cartas AL, Gómez Y, Romano MC (2006) Taenia solium cysticerci synthesize androgens and estrogens in vitro. Parasitol Res 98(5):472–6

Vargas-Villavicencio JA, Larralde C, Morales-Montor J (2006) Gonadectomy and progesterone treatment induce protection in murine cysticercosis. Parasite Immunol 28:667–674

Vargas-Villavicencio JA, Larralde C, De León-Nava MA, Escobedo G, Morales-Montor J (2007) Tamoxifen treatment induces protection in murine cysticercosis. J Parasitol 93:1512–1517

Verthelyi D (2001) Sex hormones as immunomodulators in health and disease. Int Immunopharmacol 1:983–93

Vicogne J, Pin JP, Lardans V, Capron M, Noel C, Dissous C (2003) An unusual receptor tyrosine kinase of *Schistosoma mansoni* contains a *Venus* Flytrap module. Mol Biochem Parasitol 126:51–62

Walker W, Roberts CW, Ferguson DJ, Jebbari H, Alexander J (1997) Innate immunity to Toxoplasma gondii is influenced by gender and is associated with differences in interleukin-12 and gamma interferon production. Infect Immun 65(3):1119–21

Watanabe K, Hamano KS, Noda K, Koga M, Tada I (1999) *Strongyloides ratti*: Additive effect of testosterone implantation and carbon injection on the susceptibility of female mice. Parasitol Res 85:522–526

Webster JP, Brunton CFA, MacDonald DW (1994) Effect of *Toxoplasma gondii* upon neophobic behaviour in wild brown rats, *Rattus norvegicus*. Parasitology 109:37–43

Willis C, Poulin R (2000) Preference of female rats for the odours of non-parasitised males: The smell of good genes? Folia Parasitologica (Praha) 47:6–10

Yates RA, Tuan RS, Shepley KJ, Unnasch TR (1995) Characterization of genes encoding members of the nuclear hormone receptor superfamily from *Onchocerca volvulus*. Mol Biochem Parasitol 70:19–31

Zhang S, Li Z, Cai S, Zeng L (2000) Observation of T lymphocyte subsets in the liver of patients with advanced schistosomiasis and advanced schistosomiasis accompanied with hepatitis B. J Tongji Med Univ 20:137–8

Zuk M (1994) Immunology and the evolution of behavior. In: Real L (ed) Behavioral mechanisms in evolutionary ecology. University of Chicago Press, Chicago, Illinois, pp 354–368

Zuk M, McKean KA (1996) Sex differences in parasite infections: Patterns and processes. Int J Parasitol 26:1009–1023

Chapter 8
Progesterone, Pregnancy, and Innate Immunity

Julia Szekeres-Bartho and Beata Polgar

Abstract Progesterone plays an important role in the feto-maternal immunological relationship. The products of the progesterone-regulated genes Gal-1 and Hox-A10 affect dendritic cell (DC) function and differentiation of decidual NK cells, respectively. Progesterone favors the induction of immature DCs with a tolerogenic phenotype and inhibits the activity of mature DCs. Deficiency interferes with NK cell differentiation, while NK cell migration to the endometrium is supported by progesterone-induced specific endometrial production of chemokines. Progesterone upregulates HLA-G gene expression, which serves as a ligand for NK inhibitory and activating receptors. Many of the immunological effects of progesterone are mediated by a progesterone-induced protein called progesterone-induced blocking factor (PIBF). Progesterone and PIBF act on T cell differentiation to induce a Th2 dominance during pregnancy. Progesterone also promotes the development of LIF- as well as M-CSF-producing T cells. PIBF blocks upregulation of perforin expression in decidual lymphocytes cultured with decidual adherent cells and inhibits NK cell cytotoxicity by blocking granule exocytosis.

8.1 Pregnancy and the Immune System

Pregnancy is a natural model of an optimal immune regulation in a graft–host relationship. Although 50% of fetal antigens are of paternal origin and there is ample evidence that these antigens are recognized, the immune system of the mother tolerates the semiallogeneic fetus. However, while creating a favorable environment for the fetus, the maternal immune system must be prepared to control possible emerging infections. Therefore, a delicate balance is established to satisfy the conflicting interests of mother and fetus.

J. Szekeres-Bartho (✉) and B. Polgar
Department of Medical Microbiology and Immunology, Medical School, Pecs University, H-7643 Pecs, Hungary

S.L. Klein and C.W. Roberts (eds.), *Sex Hormones and Immunity to Infection*,
DOI 10.1007/978-3-642-02155-8_8, © Springer-Verlag Berlin Heidelberg 2010

Early hypotheses to explain the feto-maternal immunological relationship were based on evidence derived from transplantation studies, and thus, focused on the role of T cells. It was proposed that cytotoxic T cells play a crucial role in pregnancy termination and that fetal rejection is prevented by suppression of maternal cytotoxic T cell function (Bertotto et al. 1984; Slapsys and Clark 1982). It soon became evident that the embryonic trophoblast, which forms the interface between mother and fetus, shows a tightly regulated and limited expression pattern of MHC-antigens (Billington and Bell 1983), which creates a unique immunological environment.

With the exception of HLA-C, polymorphic MHC is absent from the human trophoblast. Although decidual macrophages and dendritic cells (DCs) can present fetal antigens to both decidual CD4 and CD8 T cells, trophoblast-presented antigens are unlikely to be recognized in an MHC-restricted fashion. The trophoblast also resists natural killer (NK) cell-mediated lysis, (Zuckerman and Head 1987), because nonconventional MHC antigens serve as ligands for NK cell inhibitory receptors. In the decidua, there are increased numbers of activated $\gamma\delta$ T cell receptor (TCR) positive cells (Liu et al. 1997; Meeusen et al. 1993). $\gamma\delta$ T cells might be candidates for detecting trophoblast-presented antigens, because most of them are capable of recognizing unprocessed foreign antigens without MHC restriction (Heyborne et al. 1994). The main function of innate immunity is to distinguish harmless self from harmful (infectious) non self. Pathogen associated molecular patterns (PAMPs) are recognized by pattern recognition receptors, expressed on monocytes, macrophages, granulocytes, and DCs. Immature DCs reside in the early decidua in humans (Kämmerer et al. 2000; Kämmerer 2005) and mice (Blois et al. 2004; Blois et al. 2005) and serve as sentinels for potential environmental danger signals. Dendritic cells, however, also play a role in maintaining tolerance (Moretta et al. 2005). Immature DCs exhibit a tolerogenic phenotype, characterized by a low expression of costimulatory molecules (CD40, CD80, and CD86), a low production of proinflammatory cytokines, an increased production of IL-10, and a capacity to induce regulatory T (T reg) cells. Upon contact with a PAMP, such as LPS, DCs mature through a process that correlates with upregulation of surface CD40, CD80, CD86 and MHC class II. The crosstalk between DCs and NK cells results in "DC editing," when NK-cells activated by maturing DCs kill those DCs that have failed to undergo complete maturation (Marcenaro et al. 2005). Blois et al. (2005) have demonstrated an association between increased numbers of mature decidual DCs and pregnancy failure in mice. Pregnancy, therefore, initiates a series of events that create a favorable immunological environment for the fetus. These include altered cytokine production and NK cell function.

8.1.1   Pregnancy and NK Cells

NK cells play diverse roles in the reproductive process. $CD56^{bright}CD16^{neg}CD3^{neg}$ granulated decidual NK cells, which constitute a dominant lymphocyte population in the early decidua (King et al. 1998) are not cytotoxic, despite their high perforin content (Rukavina et al. 1995; Crnic et al. 2007), but secrete an array of angiogenic factors and cytokines (Koopman et al. 2003). The dynamics of the appearance of uterine NK cells suggest that one of their functions might be the control of placentation.

Cytotoxic mechanisms exerted by NK cells can potentially damage the trophoblast and induce ablation of placenta. In addition, TNF-α produced by NK cells in response to intrauterine infections can induce prostaglandin synthesis, which can result in uterine contractions and initiate the induction of preterm labor. Decreased expression of HLA-G on the trophoblast may result in an inadequate trophoblast invasion leading to an abnormal interaction with decidual NK cells, which are believed to play a major role in these processes through the production of immunoregulatory cytokines and angiogenic factors (Sargent et al. 2006). Furthermore, recurrent miscarriages are associated with an increased number of endometrial NK cells (Quenby and Farquharson 2006).

In mice, high peripheral NK activity is associated with deleterious effects on fetal development (De Fougerolles and Baines 1987). Transfer of high NK activity spleen cells from poly (I:C)-treated nonpregnant mice to pregnant BALB/c mice induces abortion (Kinsky et al. 1990). In human pregnancy, such a direct relationship between high NK activity and spontaneous pregnancy termination cannot be demonstrated. Normal human pregnancy, however, is characterized by low peripheral NK activity (Aoki et al. 1995), whereas increased NK activity is associated with spontaneous abortions of unknown etiology (Ntrivalas et al. 2001; Ntrivalas et al. 2005; Putowski et al. 2004; Shakhar et al. 2003; Yamada et al. 2003).

8.1.2 Pregnancy and Cytokine Responses

Cytokine secretion by T helper 1 (Th1) and T helper 2 (Th2) cells tends to favor cellular or humoral immune responses, respectively. Successful pregnancy is proposed to be a "Th2 phenomenon," implying that as a result of the bidirectional interaction between mother and fetus the maternal immune response is shifted toward humoral immunity (Wegmann et al. 1993). Consequently, increased immunoglobulin synthesis (Lichtenstein 1942; Dresser 1991) and decreased cell-mediated responses are associated with pregnancy (Tangri et al. 1994). The concept of a Th2 bias also is supported by clinical observations of pregnant women suffering from autoimmune diseases. Approximately 70% of women with rheumatoid arthritis (i.e., a predominantly cell-mediated autoimmune disorder) experience a temporary remission of their symptoms during gestation (Buyon 1998; Ostensen and Villiger 2007; de Man et al. 2008). Conversely, systemic lupus erythematosis (SLE), in which the principal pathology is mediated by excessive autoantibody production, tends to flare up during pregnancy, especially in women with recently active disease before conception (Varner 1991). There also are a number of infectious diseases caused by intracellular pathogens, which appear to be exacerbated by pregnancy. HIV-associated infections (Minkoff et al. 1987), leprosy, malaria,

(Reinhardt 1980; Lagerberg 2008), and toxoplasmosis (Luft and Remington 1982) fall in this category and are discussed more fully in Chap. 9 of this book.

Studies on placental cytokine production reveal that Th1 cytokines are not always harmful to pregnancy development. Th1 cytokines, depending on the time of expression, stage of gestation, and relative concentrations, could positively affect pregnancy (Chaouat 2007). Other cytokines (e.g., leukemia inhibitory factor LIF, M-CSF) produced by T cells are important for the maintenance of pregnancy (Stewart et al. 1992; Clark et al. 1997). IFN-γ is necessary for implantation, but this cytokine also activates cytotoxic T cells and NK cells, which may damage the trophoblast. IFN-γ also inhibits GM-CSF production as well as proliferation of Th2 cells and consequently B cell maturation and Ig synthesis (Piccinni 2007). Low concentrations of IFN-γ slow down intrauterine development in mice, whereas administration of high doses results in abortion (Chaouat et al. 1990).

TNF-α inhibits trophoblast cell proliferation in vitro. Injection of recombinant murine or human TNF-α into pregnant mice results in miscarriage (Chaouat 1994). The murine (CBA/J x DBA/2) mating combination results in 20–50% of the fetuses being resorbed by gestational day 13; thus, this model has been an extensively studied model of immunologically mediated spontaneous fetal loss (Chaouat et al. 1985). CBA/J females mated with DBA/2 males have a high incidence of fetal resorptions; whereas in genetically similar (H-2^kxH-2^d) matings, such as CBA/J females mated with BALB/c males, females have normal low resorption rates. Anti-TNF antibodies or TNF antagonists, normalize the high resorption rates in CBA/J x DBA/2 matings, suggesting that increased TNF-α production might account for the failure of pregnancies in this particular mating combination (Chaouat et al. 1995). In supernatants of trophoblast-activated peripheral lymphocytes of women who recurrently abort, elevated levels of TNF-α and -β have been demonstrated, suggesting that these cytokines might act as mediators in the development of recurrent spontaneous abortion (RSA).

Th2 cytokines appear to play a suppressive role in the feto-maternal relationship. IL-10 inhibits the production of Th1 cytokines and stimulates proliferation of B cells. The levels of LIF, IL-4, IL-10, and M-CSF produced by decidual T cells from women suffering from unexplained spontaneous abortion are lower than those from normal pregnant women, indicating that these cytokines may contribute to the maintenance of pregnancy (Piccinni 2005). IL-10$^{-/-}$ mice are born with significantly lower birth weight than their heterozygous siblings, which is prevented by the administration of rIL10 or anti-TNF treatment (Kuhn et al. 1992). Granulocyte-macrophage colony-stimulating factor is a potentially important mediator of intercellular communication in the female reproductive tract, by targeting the preimplantation embryo and trophoblast cells of the developing placenta. Data by Clark et al. (1997) suggest that GM-CSF acts indirectly to prevent pregnancy loss in abortion-prone DBA/2 x CBA/J mice. Consistent with a role for GM-CSF in successful transplantation, GM-CSF-deficient (GM$^{-/-}$) mice have histological abnormalities in their placentas (Robertson et al. 1996, 1999).

In humans, there is a well-established relationship between peripheral cytokine pattern and the outcome of pregnancy. Phytohaemagglutinin (PHA)-stimulated peripheral lymphocytes of healthy pregnant women produce significantly higher levels of Th2-associated cytokines than those from women with RSA (Raghupathy et al. 2000). Conversely, lymphocytes from RSA patients produce more Th1-associated cytokines than those from healthy pregnant women (Raghupathy et al. 1999). In addition, lymphocytes from healthy pregnant women secrete significantly higher levels of Th2-associated cytokines both in response to autologous placental cells and choriocarcinoma cells than lymphocytes from RSA patients (Krishnan et al. 1996).

8.2 The Effect of Progesterone on the Maternal Immune System

Pregnancy is characterized by overall hormonal changes (see Figure 9.2). Progesterone is critical to the establishment and for the maintenance of pregnancy, as it regulates menstrual bleeding, tissue repair and regeneration, inflammation, and angiogenesis. Compared with the low levels (1–2 nmol l^{-1}) during the follicular phase of the menstrual cycle, progesterone concentrations increase to 15–20, 35–50 and 20–40 nmol l^{-1} in the early- ,mid- ,and late luteal phases, respectively. The major source of progesterone during early pregnancy is the corpus luteum of the ovary but in many species, including humans and rodents, progesterone production is eventually sustained by the placenta. In humans, progesterone production gradually rises during gestation to reach a level of 3 $\mu g/g^{-1}$ of placental tissue (1–10 μM), whereas the serum concentrations of progesterone range from about 100 to500 nM during pregnancy. The need for progesterone in maintaining pregnancy is illustrated by the fact that blocking of progesterone binding sites with RU 486 causes abortion in humans and also in various nonhuman animal species (Baulieu 1987; Ulmann et al. 1987). Besides its endocrine effects, progesterone mediates interactions between the endocrine and immune systems. Ovariectomized female mice treated with progesterone, but not with estrogen, develop uterine abscesses after injection with *Escherichia coli*. The number of leukocytes, as well as immunoglobulin production in the rodent uterus depends on the hormonal status of the female (Tchernitchin et al. 1976; Wira and Sandoe 1980). High local concentrations of progesterone prolong the survival of xenogeneic and allogeneic skin grafts as well as of xenogeneic tumor cells implanted in hamster uteri (Moriyama and Sugawa 1972; Hansen et al. 1986). In vitro treatment of T cells with 10 $\mu g/ml$ of progesterone blocks T cell activation and killing (Mori et al. 1977; Pavia et al. 1979; Wyle and Kent 1977). It is noteworthy that in all of the early human studies, lymphocytes from nonpregnant women were examined to determine progesterone action and as a consequence supra-physiologic concentrations of progesterone were needed to achieve a significant effect. Later studies, however, suggest a dose-dependent decrease of peripheral NK cell activity of lymphocytes from pregnant women at low concentrations (40–500 nM) of progesterone (Szekeres-Bartho et al. 1985a, 1985b). Conversely, experiments using lymphocytes of nonpregnant women fail to demonstrate such effect at similar concentrations (Sulke et al. 1985; Uksila 1985). These data suggested the presence of a population of NK cells with increased progesterone sensitivity among peripheral lymphocytes from pregnant women. Immunocytochemistry has been utilized to demonstrate progesterone receptors (PRs) in peripheral blood mononuclear cells (PBMCs) and peripheral blood $\gamma\delta$ T cells of pregnant women (Szekeres-Bartho et al. 2001).

8.2.1 Progesterone Receptors (PRs)

The biological activity of progesterone is mediated by genomic and nongenomic pathways. The genomic pathway depends on two nuclear progesterone receptor (nPR) isoforms: PR-A and PR-B (Li et al. 2004; Mulac-Jericevic and Conneely 2004), both of which are products of the same gene, but are transcribed under the control of two distinct promoters (Li and O'Malley 2003). The PR-A and PR-B isoforms differ in that PR-B contains an additional N-terminal stretch of approximately 165 amino acids.

Spatial and temporal expression of the PR-A and PR-B vary in reproductive tissues as a consequence of developmental and hormonal status. In mice, null mutation of nPR gene reveals that transcriptional activity of nPR controls the uterine immune environment as well as endometrial receptivity and decidualization (Lydon et al. 1995; Tibbetts et al. 1999). Functionally active nPRs in the thymus are required for thymic involution during pregnancy and for a normal fertility (Mulac-Jericevic et al. 2000). Studies with selective PR-A and PR-B knock out mice demonstrate that progesterone-induced activation of PR-A is both necessary and sufficient for the establishment and the maintenance of pregnancy, but elicits reduced pregnancy-stimulated mammary gland morphogenesis (Conneely et al. 2003). In contrast, progesterone-induced transcriptional activity of the PR-B isoform is insufficient for implantation and the maintenance of pregnancy needed for proliferative effects of progesterone in the mammary gland (Conneely et al. 2003). Consequently, mice lacking PR-A are infertile (Mulac-Jericevic et al. 2000, 2003). These findings imply that the relative expression of the two isoforms is critical to the appropriate reproductive tissue responses to progesterone (Fernandez -Valdevie et al. 2005).

The existence of PRs on lymphocytes is a controversial issue. Kimoto showed the presence of PR mRNA in human lymphocytes. The PR protein has been identified in human multiple myeloma cells, B cells, plasma cells, and macrophages as well as in murine hybridomas (Daniel et al. 1988; Pasanen et al. 1998). Chiu et al. (1996) found increased PR expression on peripheral lymphocytes from pregnant women after immunotherapy for habitual abortion. Recently, Arruvito et al. (2008) showed that peripheral blood NK cells express both classical PR isoforms. Furthermore, RU486 treatment significantly augments NK cell cytolytic activity in vitro and this can be reversed by treatment of NK cells with progesterone (Hansen et al. 1992). Roussev et al. (1993) demonstrated an enrichment of PR positive lymphocytes in the uterus compared with peripheral blood cells from pregnant women.

On the other hand, King et al. (1996) and Stewart et al. (1998) failed to demonstrate PR or ER expression in lymphocytes from endometrial tissue. The latter group, however, found an increased ER expression on lymphocytes collected during endometriosis, suggesting that the expression or the ability to detect steroid receptors might largely depend on reproductive condition. Henderson et al. (2003) demonstrated the absence of PRs in purified decidual NK cells and suggested that progesterone could act on these cells through glucocorticoid receptors (Henderson et al. 2003), putative membrane receptors (Zhu et al. 2003), or calcium-activated K+ channels (Chien et al. 2007). These data suggest that PR expression could be an attribute of peripheral rather than of decidual lymphocytes.

A high proportion of peripheral lymphocytes during pregnancy react with PR-specific monoclonal antibodies, with the rate of PR positive cells increasing throughout gestation (Szekeres-Bartho et al. 1989a, 1989b); whereas resting lymphocytes in nonpregnant women are largely nonreactive to PR antibodies (Szekeres-Bartho et al. 1990). The percentage of PR-expressing lymphocytes from women who recurrently abort their fetuses or among women with clinical symptoms of threatened preterm delivery is significantly lower than is the expression in cells isolated from peripheral blood of pregnant women of corresponding gestational ages (Szekeres-Bartho et al. 1989a, b). These findings suggest a relationship between lymphocyte PR expression and the outcome of pregnancy; thus, the regulation of PRs in lymphocytes might be of biological importance. In contrast to the classical endocrine tissues, in lymphocytes estrogens do not induce PRs (Szekeres-Bartho et al. 1990). This might be due to the absence of estrogen receptors (ERs) in peripheral human lymphocytes. Although ERs have been demonstrated in mouse spleen, rat thymus, guinea pig fetal thymus, human T cells, human mononuclear cells, and macrophages (Defletsen et al. 1977; Screpanti et al. 1982; Cohen et al. 1983; Stimson 1988), ERs are not present in resting human lymphocytes, by an enzyme immunoassay utilizing a rat monoclonal antibody to the human ER (Szekeres-Bartho et al. 1989). On the other hand, exposure of lymphocytes from nonpregnant women to PHA or alloantigenic stimuli increases lymphocyte PR expression in humans (Paldi et al. 1994), suggesting that PR induction is an activation-related phenomenon. Transplantation is an appropriate natural model for addressing whether allogeneic stimulation by itself is the only condition required for inducing PRs or if hormonal and other pregnancy-specific attributes are also involved. The expression of PRs is equally high in transplantation patients and in pregnant women (Szekeres-Bartho et al. 1989a), suggesting that PR expression is not the consequence of pregnancy per se, but that of in vivo allogeneic stimulation by fetal antigens. Further, Chiu et al. (1996) found increased PR expression in lymphocytes from women who recurrently abort their fetuses after immunization with paternal leukocytes. These data indicate that lymphocyte activation is a major factor in the induction of PRs in immune cells. We interpret these data as suggesting that progesterone-dependent immunomodulation is dependent on the mode of fetal antigen presentation and the lymphocytes that recognize the fetal antigen, which ultimately determines the quality and the extent of the immune response.

Following recognition of fetal antigens, activated maternal γδ T cells express nPRs (Szekeres-Bartho et al. 2001) and upon progesterone binding, these cells produce progesterone-induced blocking factor (PIBF) (Szekeres-Bartho et al. 1985b, Szekeres-Bartho et al. 1989b; Polgar et al. 2003). Progesterone-induced blocking factor alters the profile of cytokine secretion by activated lymphocytes (Szekeres-Bartho and Wegmann 1996). Neutralization of endogenous PIBF activity in pregnant mice by specific anti-PIBF antibody terminates pregnancy. Depletion of NK activity with anti-NK cell antibodies counteracts the effects of PIBF on pregnancy (Szekeres-Bartho et al. 1997), suggesting that in mice PIBF contributes to the success of pregnancy and that the major part of its pregnancy-protective effect lies in controlling NK activity. In urine samples of healthy pregnant women, PIBF concentrations continuously increase until the 37th week of gestation, after which concentrations decrease slowly until term. In pregnancies that result in miscarriage or preterm delivery, urinary PIBF levels fail to increase during pregnancy (Polgár et al. 2004). Both human (Kalinka and Szekeres-Bartho 2005) and rodent (Joachim et al. 2003) studies suggest that synthesis of PIBF is an indirect mechanism by which progestagens improve pregnancy outcome.

8.2.2   Progesterone-Regulated Genes

Among the progesterone-regulated genes that have been identified in the periimplantation uterus, Hox-A10, Hox-A11 (both members of homeobox gene family), and galectin-1 (Gal-1) (Choe et al. 1997) are the most relevant for the maternal-fetal immunological interaction. Hox-A10 deficiency in mice leads to severe local immunological disturbances, characterized by a polyclonal proliferation of T cells that occurs in absence of the normal progesterone-mediated immunosuppression in the peri-implantation uterus (Yao et al. 2003). Natural killer cells constitute the predominant leukocyte population present in endometrium at the time of implantation and early pregnancy (Croy et al. 2006). Hox-A10 deficiency in mice alters region-specific gene expression and compromises NK cell differentiation, but not trafficking of NK precursors cells during decidualization (Rahman et al. 2006).

The expression of Gal-1 is downregulated on placental villous tissues from patients with spontaneous miscarriages (Liu et al. 2006). Furthermore, Gal-1-deficient mice show higher rates of fetal loss compared with wild-type (WT) mice in allogeneic matings (Blois et al. 2007). Treatment with recombinant Gal-1 prevents fetal loss and restores tolerance through multiple mechanisms, including the induction of tolerogenic DCs, which in turn promotes the expansion of IL-10-secreting cells in vivo. Progesterone treatment increases the levels of Gal-1 expression in the myometria and deciduas of stress-challenged pregnancies on gestational day 7.5, suggesting the existence of a progesterone-mediated mechanism that regulates Gal-1 expression at the feto-maternal interface (Blois et al. 2007). Reciprocally, Gal-1 treatment prevents the stress-induced decrease in progesterone and PIBF and increased the concentration of these hormones in sera to levels that are significantly higher than those found during normal gestation in mice (Blois et al. 2007).

8.3 Progesterone and Immunity During Pregnancy

8.3.1 Progesterone and DCs During Pregnancy

Data on the effects of progesterone on DC function during pregnancy are scarce. More general effects of progesterone on DCs are discussed in Chaps. 2 and 3 of this book. In humans, the uterus is typically sterile (i.e., pathogen-free), but immune responses can still be generated in response to pathogens (Blaustein and Kurman 2002). Uterine DCs might serve as a "switchboard" between fetal rejection and tolerance by directing either the induction of tolerance to harmless antigens or the initiation of a robust immune response (Kämmerer et al. 2000; Gardner and Moffett 2003; Blois et al. 2005). MHC II positive, costimulated, semimature DCs may contribute to the induction of tolerance, whereas mature DCs can influence the generation of proinflammatory responses (Sallusto and Lanzavecchia 1999; Lutz and Schuler 2002).

Progesterone favors the induction of immature DCs and inhibits the activity of mature DCs (Liang et al. 2006a, 2006b; Hughes and Clark 2007). Immature DCs in the early pregnancy decidua (Kämmerer et al. 2000; Kämmerer 2005; Blois et al. 2004, 2005) are induced by progesterone-regulated Gal-1 (Blois et al. 2007). In murine pregnancies, with high abortion rates, mature antigen presenting cells are predominant (Blois et al. 2005). Progesterone inhibits, in a receptor-mediated fashion, mature DCs and DC-stimulated proliferation of T cells (Butts et al. 2007), and reduces nitric oxide and IL-12 production following stimulation with LPS (Jones et al. 2008). Furthermore, progesterone inhibits TLR9-induced IFN-α production by human and murine DCs (Hughes et al. 2008), suggesting that progesterone might interfere with antiviral responses as discussed in Chap. 4 of this book.

8.3.2 Progesterone and NK Cells During Pregnancy

In rodents and humans, hormonally controlled uterine NK cells play an important role in creating a suitable environment for the establishment of pregnancy (King et al. 1998). Uterine and peripheral NK cells represent two phenotypically and functionally distinct populations. While approximately 90% of human peripheral NK cells express a low density of CD56 (CD56^{dim}) and high levels of the FCgRIII (CD16), the majority of decidual NK cells express a high density of CD56 (CD56^{bright}) and no CD16. Peripheral CD56^{dim} NK cells are granular and known to be cytotoxic, while CD56^{bright} peripheral NK do not contain granules and are noncytotoxic, but display an imunoregulatory role via cytokine production (Cooper et al. 2001). The temporal and spatial distribution of these cells suggests that one of their functions might be the control of placentation. Consistent with this hypothesis, uterine NK cells increase in number during estrus and proestrus in rodents and during the luteal phase in women and are enriched at sites where the fetal trophoblast infiltrates the decidua in early pregnancy (Van den Heuvel et al. 2005).

Decidual NK cells fulfill dual tasks. During normal conditions they contribute to creating a favorable environment for placentation, but at the same time they are equipped with cytotoxic potential to fight intrauterine infections. Among the genes selectively overexpressed in decidual NK cells are those that encode for integrin subunits, NK inhibitory receptors, galectin-1, and progestagen-associated protein 14, suggesting that decidual NK cells might contribute to the generation of an immunosuppressive environment at the maternal fetal interface (Koopman et al. 2003). On the other hand, perforin and granzymes A and B are expressed by decidual NK cells to a similar or higher level than by $CD56^{dim}$ peripheral NK cells (Koopman et al. 2003; Bogovic et al. 2005), suggesting that decidual NK may have cytotoxic potential. Decidual NK cells secrete angiogenic factors (Koopman et al. 2003) and induce vascular growth in the decidua (Quenby et al. 2009). In tgɛ26, NK cell deficient mice (i.e., mice transgenic for high copy numbers (>35) of the human CD3ɛ gene (Wang et al. 1994)), the placenta is abnormally small and birth weights are low (Guimond et al. 1999). Histological abonormalities can be observed in the arteries around which NK cells are localized. By producing the chemokines CXCL8 (IL-8) and CXCL10 (IP-10), decidual NK cells regulate trophoblast invasion (Hanna et al. 2006). The rapid increase of NK cell counts in the early decidua may be caused by proliferation of the resident population, recruitment of $CD56^{bright}$ lymphocytes from the circulation, or both.

Progesterone plays a role in uterine homing of NK cells by promoting NK cell interactions with the endothelium (Van den Heuvel et al. 2005). NK cell migration to the endometrium also is supported by progesterone-induced specific endometrial production of chemokines (Sentman et al. 2004). Deficiency interferes with NK cell differentiation without affecting the migration of NK cell precursors into the decidua (Rahman et al. 2006). Furthermore, gene expression profiling reveals that at the time of implantation, mediates the progesterone-stimulated proliferation of uterine stromal cells (Yao et al. 2003). Carlino et al. (2008) showed that peripheral blood NK cells of pregnant women have a higher migratory ability through progesterone-treated decidual stromal cells than do NK cells from nonpregnant women or male donors. Increased migration correlates with the ability of progesterone to upregulate stromal cell *Cxcl10*, *Cx3cl1* (Fractalkine), and *Ccl2* (MCP-1) mRNA expression. These results illustrate that peripheral NK cell recruitment to the uterus contributes to the accumulation of NK cells during early pregnancy and progesterone plays a crucial role in this event.

NK cells express activating and inhibitory receptors. NK cytotoxic activity results from the balance of inhibitory and activating signals, following the interaction of NK cell activating and inhibitory receptors with ligands expressed on the surface of the target cell (Ljunggren and Karre 1990; Long 1999; Moretta et al.

2000). CD56^{bright} and CD56^{dim} NK cell subsets differ in their pattern of expression of NK cell receptors (El Colsta et al. 2008). Furthermore, decidual NK cells encounter ligands (HLA-G and HLA-E) that are less likely to be available for peripheral NK cells (Trowsdale and Moffett 2008). Three major superfamilies of NK receptors have been described on human NK cells: the killer immunoglobulin (Ig)-like receptor (KIR) superfamily, which recognizes classical MHC class I molecules, the C-type lectin superfamily, which recognizes nonclassical MHC class I or class I-like molecules, and the natural cytotoxicity receptors (Lanier 2005).

In spite of their high perforin content, decidual NK cells show low spontaneous cytotoxic activity (Crnic et al. 2007). When activated by IL-2, however, decidual NK cells kill trophoblast cells (Zuckerman and Head 1987). Exposure to progesterone concentrations (10^{-5} M) believed to be present at the maternal–fetal interface (Stites and Siiteri 1983) inhibits NK cytolytic activity in vitro (Hansen et al. 1992). Progesterone increases IL-15 production by endometrial stromal cells (Okada et al. 2000). Human decidual NK cells express the IL-15 receptor and in response to IL-15 proliferate and augment their cytolytic activity against K562 (Verma et al. 2000). On the other hand, progesterone upregulates HLA-G gene expression through a novel progesterone response element (Yie et al. 2006). Increased expression of HLA-G, the ligand for NK inhibitory and activating receptors, controls NK activity in the presence of progesterone.

Decidual CD56^{+} cells express PIBF (Faust et al. 1999). PIBF found in CD56^{+} cells is likely produced by other decidual cells and internalized by CD56^{+} cells, because decidual NK cells are PR negative (Henderson et al. 2003). The PIBF receptor is expressed on a majority of lymphocytes. PIBF blocks upregulation of perforin expression in decidual lymphocytes cultured with decidual adherent cells and inhibits NK cell cytotoxicity by blocking granule exocytosis without interfering with target conjugation (Laskarin et al. 1999; Faust et al. 1999). Taking into account PIBF positivity of decidual NK cells, we assume that this mechanism might operate locally at the feto-maternal interface and at least partly account for the low lytic activity of decidual NK cells.

Progesterone and PIBF also act on NK cell activity in an indirect way, by induction of cytokine production in the neighboring cells. PIBF enhances IL-10 production and suppresses IL-12 production by human PBMCs (Polgar et al. 2003). Although decidual NK cells do not express the PR, these cells are affected by the PIBF that is secreted by surrounding cells.

8.3.2.1 Progesterone and Cytokine Production During Pregnancy

As noted earlier, a number of studies illustrate that dominance of Th2-associated cytokines favor normal pregnancy, while an excess of Th1-associoated cytokines leads to the termination of pregnancy (Wegmann et al. 1993; Raghupathy 1997). Consequently, women with recurrent miscarriages have a Th1-dominant cytokine profile (Raghupathy et al. 1999, 2000, Rezaei and Dabbagh 2002; Hossein et al. 2004; Wilson et al. 2004). By signaling via a novel form of the IL-4 receptor (Kozma et al. 2006), PIBF induces a Th2-dominant cytokine response, including production of IL-3, IL-4, and IL-10 (Szekeres-Bartho and Wegmann 1996). PIBF induces immediate STAT6 phosphorylation, whereas progesterone treatment requires 24 h to activate STAT6, suggesting that the effect of progesterone is mediated by PIBF. Neutralization of endogenous PIBF activity in pregnant mice by specific anti-PIBF antibody terminates pregnancy, increases splenic NK cell activity, reduce synthesis of IL-10, and increase IFN-γ production in spleen cells (Szekeres-Bartho et al. 1996).

In a prospective clinical trial, Raghupathy et al. (2005) investigated the production of Th1 and Th2 cytokines from peripheral blood lymphocytes isolated from women who recurrently aborted their fetuses and challenged in vitro with progesterone. They showed that progesterone-induced (i.e., Duphaston) PIBF production downregulates peripheral Th1-associated cytokines and stimulates Th2-associated cytokines. Furthermore, Duphaston treatment of women with preterm delivery induces a shift in cytokine bias, by inhibiting proinflammatory cytokine production and increasing antiinflammatory cytokine production (Raghupathy et al. 2007).

In addition to the classical Th1 and Th2 subsets (Mosmann and Coffman 1989; Romagnani 1991) of CD4$^+$ T cells, there are Treg and helper T cell 17 (Th17) cells that express differential transcriptional factors (i.e., FoxP3 and RorγT, respectively) and produce distinct cytokines (i.e., Treg cells produce IL-10 and TGF-β and Th17 cells produce IL-17 and IL-22) (Huber et al. 2004; Liang et al. 2006a, 2006b; Weiner 2001). To date, there is no evidence for the direct effect of progesterone on Treg cell function. In normal (i.e., non abortion-prone), but not in abortion-prone murine mating combinations, increased numbers of Treg cells are observed in the decidua on day 2 of pregnancy, but accumulation of this population is not correlated with progesterone levels (Thuere et al. 2007). Furthermore, in ovariectomized mice neither estrogen nor progesterone alone or in combination affects the frequency of Treg cells, suggesting that the presence of fetal alloantigens, rather than of placental steroids, is responsible for the increase of Treg cells during pregnancy (Zhao et al. 2007).

The hypothesis that Th1/Th2 skewing underlies successful pregnancy stems from examination of cytokine responses in peripheral blood lymphocyte. Studies of placental cytokine production reveal that, at least in mice, Th1 cytokines are not necessarily detrimental to pregnancy development and some cytokines (e.g., LIF, IL-11, and IFN-γ) are necessary for the implantation of the mouse blastocyst (Stewart et al. 1992; Bilinski et al. 1998; Ashkar et al. 2000; Piccinni et al. 2001a, 2001b). At different stages of gestation, practically all cytokines appear in the decidua (Vince and Johnson 1996). Placental *Il4* mRNA expression in mice is five- to ten-fold higher than in peripheral blood (Delassus et al. 1994). Furthermore, defective IL-4, IL-10, LIF, and M-CSF production by decidual T cells are detected in women with unexplained recurrent abortion at the time of miscarriage (Piccini et al. 1998). Hormones present in the decidual microenvironment cells could be responsible, at least in part, for the altered cytokine profile.

Progesterone affects the differentiation of resting human peripheral blood T cells into Th2-like clones. In IFN-γ secreting Th1-like T cell clones, progesterone treatment induces *Il4* mRNA expression and IL-4 production (Piccinni et al. 1995). Progesterone, in concentrations comparable to those present at the feto-maternal interface, induces the production of the Th2 cytokines, IL-4 and IL-5 (Piccinni et al. 1995). Progesterone acts through specific receptors present on a substantial proportion of T blast cells from established T cell clones (Piccinni et al. 1995). Progesterone, via IL-4 induction, also promotes the development of LIF- and M-CSF-producing T cells (Piccinni et al. 1998, 2001). Progesterone-induced LIF is indispensable for implantation (Stewart et al. 1992). A growing body of evidence shows that steroid hormone control of both follicular development and uterine receptivity is mediated by locally acting growth factors and cytokines.

8.4 Conclusion and Future Directions

In humans, progesterone is indispensable for the establishment and maintenance of pregnancy, not only for its endocrine effects, but also because of the role it plays in the feto-maternal relationship that must promote tolerance to the allergenic fetus, while maintaining host defense against pathogens (Fig. 8.1). The bi-directional interaction between the fetus and the maternal immune system consists in presentation of fetal antigens, followed by recognition that results in altered functioning of

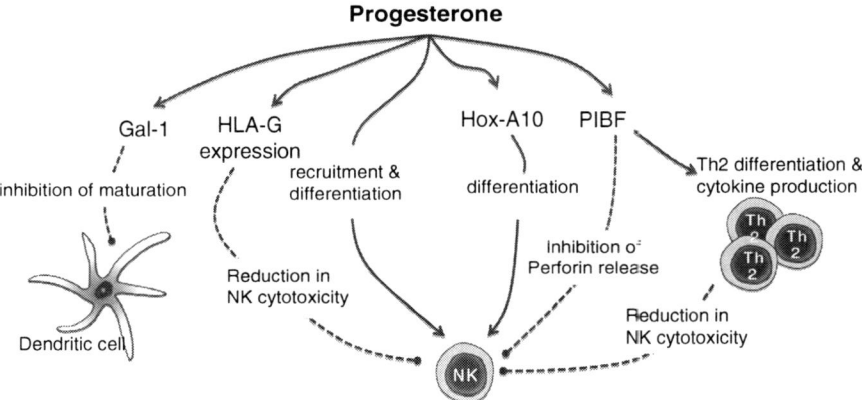

Fig. 8.1 The pregnancy-associated effects of progesterone on the immune system. By inducing Gal-1 synthesis, inhibits dendritic cell maturation. Acts on recruitment, differentiation and cytotoxic activity of decidual NK cells by up-regulating HLA-G and Hox-A10 expression. The progesterone-induced protein; PIBF inhibits degranulation of NK cells and upregulates Th2 type cytokine production

the maternal immune system. Under the influence of progesterone, specialized NK cells accumulate in the endometrium to ensure implantation, decidualization, and proper placentation. Furthermore, progesterone is needed for creating a suitable cytokine environment for the developing fetus. Intrauterine infections represent a threat to fetal well-being and pregnancy outcome. Clinical studies show an association between intrauterine infections and preterm labor (Romero et al. 2003), preeclampsia (von Dadelszen et al. 2003), and IUGR (Goncalves et al. 2002). Immune cells at the maternal-fetal interface and also the trophoblast (Abrahams˙ and Mor 2005) express pattern recognition receptors in order to control pathogens that may compromise the pregnancy. However, the antipathogen immune responses may lead to excessive inflammation or apoptosis at the maternal–fetal interface, which progesterone is clearly unable to overcome. Therefore, although progesterone is crucial in maintaining normal pregnancy, it seems, that in most cases it cannot overcome the deleterious effects of an antipathogen immune response.

References

Abrahams VM, Mor G (2005) Toll-like Receptors and their Role in the Trophoblast Placenta 26:540–547

Aoki K, Kajiura S, Matsumoto Y, Ogasawara M, Okada Y, Tagami Y, Gleicher N (1995) Preconceptional natural-killer-cell activity as a predictor of miscarriage. Lancet 345: 1340–13422

Arruvito L, Giulianelli S, Flores AC, Paladino N, Barboza M, Lanari C, Fainboim L (2008) NK cells expressing a progesterone receptor are susceptible to progesterone-induced apoptosis. J Immunol 180:5746–5753

Ashkar AA, Di Santo JP, Croy BA (2000) Interferon gamma contributes to initiation of uterine vascular modification, decidual integrity, and uterine natural killer cell maturation during normal murine pregnancy. J. Exp. Med. 192:259–270

Baulieu EE (1987) Contraception by the progesterone antagonist RU 486: a novel approach to human fertility control. Res Reprod. 19:3–4

Bertotto A, De Felicis Arcangeli C, Mazzarino I, Vaccaro R, Gerli R, Ferrarese R, Cucchia GC (1984) T-cell subset distribution in chronically aborting women. Lancet 1(8368):108

Bilinski P, Roopenian D, Gossler A (1998) Maternal IL-11Ralpha function is required for normal decidua and fetoplacental development in mice. Genes Dev. 12:2234–2243

Billington WD, Bell SC (1983) Fetal histocompatibility antigens and maternal immune responses. Ciba Found Symp 96:69–88 Review

Blaustein A, Kurman RJ (eds) (2002) Blaustein's Pathology of the Female Genital Tract, 5th edn. Springer, New York, pp 424–431

Blois SM, Alba Soto CD, Tometten M, Klapp BF, Margni RA, Arck PC (2004) Lineage, maturity, and phenotype of uterine murine dendritic cells throughout gestation indicate a protective role in maintaining pregnancy. Biol Reprod; 70:1018–1023

Blois SM, Tometten M, Kandil J, Hagen E, Klapp BF, Margni RA, Arck PC (2005) ICAM-1/LFA-1 cross talk is a proximate mediator capable of disrupting immune integration and tolerance mechanism at the feto-maternal interface in murine pregnancies. J Immunol; 174:1820–1829

Blois S, Ilarregui JM, Tometten M, Garcia M, Orsal AF, Toscano M, Handjiski B, Tirado I, Markert UR, Poirier F, Szekeres-Bartho J, Rabinovich G, Arck P (2007) A pivotal role for galectin-1 in fetal tolerance Nat. Med. 13:1450–1457

Bogovic T, Crncic T, Laskarin G, Juretic K, Strbo N, Dupor J, Srsen S, Randic L, Le Bouteiller P, Tbiasco J, Rukavina D (2005) Perforin and Fas/FasL cytolytic pathways at the maternal–fetal interface. American Journal of Reproductive Immunology 54:241–248

Butts CL, Shukir SA, Duncan KM, Bowers E, Horn C, Belyayskaya E, Tonelli L, Sterbers EM (2007) Progesterone inhibits rat mature dendritic cells in a receptor-mediated fashion Int. Immunol. 19:287–296

Buyon JP (1998) The effects of pregnancy on autoimmune diseases. J Leukoc Biol. 63:281–287

Carlino C, Stabile H, Morrone S, Bulla R, Soriani A, Agostinis C, Bossi F, Mocci C, Sarazani F, Tedesco F, Santoni A, Gismondi A (2008) Recruitment of circulating NK cells through decidual tissues: a possible mechanism controlling NK cell accumulation in the uterus during early pregnancy. Blood. 111:3108–3115

Chaouat G (1994) Synergy of lipopolysaccharide and inflammatory cytokines in murine pregnancy: alloimmunization prevents abortion but does not affect the induction of preterm delivery. Cell Immunol. 157:328–340

Chaouat G (2007) The Th1/Th2 paradigm: still important in pregnancy? Semin Immunopathol. 29:95–113

Chaouat G, Kolb JP, Kiger N, Stanislawski M, Wegmann TG (1985) Immunological consequences of vaccination against abortion in mice. J. Immuol. 134:1594

Chaouat G, Menu E, Kinsky R, Dy M, Minkowsky M, Delage G, Thang MN, Clark DA, Wegmann TG, Szekeres-Bartho J (1990) Lymphokines and non-specific cellular lytic effectors at the feto-maternal interface affect placental size and survival. In: Mettler L, Billington D (eds) Reproductive Immunology 1989. Elsevier, London, pp 283–287

Chaouat G, Assal Meliani A, Martal J, Raghupathy R, Elliott JF, Mosmann T, Wegmann TG (1995) IL-10 prevents naturally occurring fetal loss in the CBA x DBA/2 mating combination, and local defect in IL-10 production in this abortion-prone combination is corrected by in vivo injection of IFN-tau. J Immunol. 154:4261–4268

Chien EJ, Liao CF, Chang CP, Pu HF, Lu LM, Shie MC, Hsieh DJ, Hsu MT (2007) The non-genomic effects on Na(+)/H(+)-exchange 1 by progesterone and 20alpha-hydroxyprogesterone in human T cells. Journal of Cell Physiology 211:544–550

Chiu L, Nishimura M, Ishi Y, Nieda M, Maeshima M, Takedani Y, Tadokoro K, Juji T (1996) Enhancement of the expression of progesterone receptor on progesterone -treated lymphocytes after immunotherapy in unexplained recurrent spontaneous abortion. Am. J. Reprod. Immunol. 35:552–557

Choe YS, Shim C, Choi D, Lee CS, Lee KK, Kim K (1997) Expression of galectin-1 mRNA in the mouse uterus is under the control of ovarian steroids during blastocyst implantation. Mol Reprod Dev 48:261–266

Clark DA, Merali FS, Hoskin DW, Stee-Norwood D, Arck PC, Croitrou K, Murgita RA, Hirte H (1997) Decidua associated suppressor cells in abortion-prone DBA/2-mated CBA/J mice that release bioactive transforming growth factorb2 related immunosuppressive molecules express a bone marrow derived natural suppressor cell marker and gamma delta T-cell receptor. Biol. Reprod. 56:1351–1360

Cohen JHM, Danel L, Cordier G, Saez S, Revillard JP (1983) Sex steroid receptors in peripheral T cells: absence of androgen receptors and restriction of estrogen receptors to OKT8 positive cells. J. Immunol. 131:2767–2771

Conneely OM, Jericevic BM, Lydon JP (2003) Progesterone receptors in mammary gland development and tumorigenesis. J Mammary Gland Biol Neoplasia 8:205–214

Cooper MA, Fehniger TA, Caligiuri MA (2001) The biology of human natural killer-cell subsets. Trends in Immunology 22:633–640

Crnic TB, Laskarin G, Frankovic KJ, Tokmadzic VS, Strobo N, Bedenicki I, Le Bouteiller P, Tabiasco J, Rukavina D (2007) Early pregnancy decidual lymphocytes beside perforin use Fas ligand (FasL) mediated cytotoxicity. J. Reprod. Immunol. 73:108–117

Croy A, van den Heuvel MJ, Borzychowski AM, Tayade C (2006) Uterine natural killer cells: a specialized differentiation regulated by ovarian hormones. Immunol Rev 214:161–185

Daniel L, Vincent C, Rousset F, Klein B, Bataille R, Flacher M, Durie BG, Revillard JP (1988) Estrogen and progesterone receptors in semen human myeloma cell lines and murine hybridomas. J. Steroid Biochem 30:363–367

De Fougerolles R, Baines M (1987) Modulation of natural killer activity influences resorption rates in CBAxDBA/2 matings. J. Reprod. Immunol. 11:147–153

de Man YA, Dolhain RJ, van de Geijn FE, Willemsen SP, Hazes JM (2008) Disease activity of rheumatoid arthritis during pregnancy: results from a nationwide prospective study. Arthritis Rheum 15;59:1241–1248

Defletsen MA, Smith BC, Dickerman HW (1977) A high affinity low-capacity receptor for estradiol in normal and anaemic mouse spleen cytosols. Biochem. Biophys. Res. Commun. 76:1151–1158

Delassus S, Coutinho GC, Saucier C, Darche S, Kourilsky P (1994) Differential cytokine expression in maternal blood and placenta during murine gestation. J Immunol. 152:2411–2420

Dresser DE (1991) The potentiating effect of pregnancy on humoral immune responses of mice. J. Reprod. Immunol. 20:253–266

El Costa H, Casemayou A, Aguerre-Girr M, Rabot M, Berrebi A, Parant O, Clouet-Delannoy M, Lombardelli L, Jabrane-Ferrat N, Rukavina D, Bensussan A, Piccinni MP, Le Bouteiller P, Tabiasco J (2008) Critical and differential roles of NKp46- and NKp30-activating receptors expressed by uterine NK cells in early pregnancy. J Immunol. 181:3009–3017

Faust ZS, Laškarin Szekeres-Bartho J. G, Rukavina D, Szekeres-Bartho J (1999) Progesterone Induced Blocking Factor Inhibits Degranulation of NK Cells. Am J Reprod Immunol 42:71–75

Fernandez-Valdevie R, Mukherjee A, Mulac-Jericevic B, Conneely OM, DeMayo FJ, Amato P, Lydon JP (2005) Revealing progesterone's role in uterine and mammary gland biology: insights from the mouse. Semin Reprod Med; 23:22–37

Gardner L, Moffett A (2003) Dendritic cells in the human decidua. Biol. Reprod; 69:1438–1446

Goncalves LF, Chaiworapongsa T, Romero R (2002) Intrauterine infection and prematurity. Ment Retard Dev Disabil Res Rev 8:3–13

Guimond M-J, Wang B, Croy BA (1999) In vivo demonstrations of lymphocyte-mediated regulation of placental size. Placenta 20:441–450

Hanna J, Goldman-Wohl D, Hamani Y, Avraham I, Greenfield C, Nathanson-Yaron S, Prus D, Cohen-Daniel L, Arnon TI, Manaster I, Gazit R, Yutkin V, Beaharroch D, Porgador A, Keshet E, Yagel S, Decidual MO (2006) NK cells regulate key developmental processes at the human fetal-maternal interface Nat. Med. 12:1065–1074

Hansen PJ, Bazer FW, Segerson EC (1986) Skin graft survival in the uterine lumen of ewes treated with progesterone. Am. J. Reprod. Immunol. Microbiol. 12:48–54

Hansen KA, Opsahl MS, Nieman LK, Baker JR Jr, Klein TA (1992) Natural killer cell activity from pregnant subjects is modulated by RU 486. American Journal of Obstetrics and Gynecology 166:87–90

Henderson TA, Saunders PT, Moffet-King A, Grrome NO, Critchley HO (2003) Steroid receptor expression in uterine natural killer cells. Journal of Clinical Endocrinolology and Metabolism 88:440–449

Heyborne K, Fu Y-X, Nelson A, Farr A, O'Brien R, Born W (1994) Recognition of trophoblasts by γδ T cells. J. Immunol. 153:2918–2926

Hossein H, Mahroo M, Abbas A, Firouzeh A, Nadia H (2004) Cytokine production by peripheral blood mononuclear cells in RM. Cytokine 28:83–86

Huber S, Schramm C, Lehr HA, Mann A, Schmitt S, Becker C, Protschka M, Galle PR, Neurath MF, Blessing M (2004) Cutting edge: TGF-beta signaling is required for the in vivo expansion and immunosuppressive capacity of regulatory CD4+CD25+ T cells. J Immunol; 173:6526–6531

Hughes GC, Clark EA (2007) Regulation of dendritic cells by female sex steroids: relevance to immunity and autoimmunity. Autoimmunity. 40:470–481

Hughes GC, Thomas S, Li C, Kaja MK, Clark EA (2008) Cutting edge: progesterone regulates IFN-alpha production by plasmacytoid dendritic cells. J Immunol. 180:2029–2033

Joachim R, Zenclussen AC, Polgar B, Douglas AJ, Fest S, Knackstedt M, Klapp BF, Arck PC (2003) The progesterone derivative dydrogesterone abrogates murine stress-triggered abortion by inducing a Th2 biased local immune response. Steroids. 68:931–940

Jones LA, Anthony JP, Henriquez FL, Lyons RE, Nickdel MB, Carter KC, Alexander J, Roberts CW (2008) Toll-like receptor-4-mediated macrophage activation is differentially regulated by progesterone via the glucocorticoid and progesterone receptors. Immunology 125(1):59–69

Kalinka J, Szekeres-Bartho J (2005) The impact of dydrogesterone supplementation on hormonal profile and progesterone-induced blocking factor concentrations in women with threatened abortion. Am. J. Reprod. Immunol. 53:166–171

Kämmerer U (2005) Antigen-presenting cells in the decidua. Chem Immunol Allergy 89:96–104

Kämmerer U, Schoppet M, McLellan AD, Kapp M, Huppertz HI, Kampgen E, Dietl J (2000) Human decidua contains potent immunostimulary CD83+ dendritic cells. Am J Pathol 157:159–169

Kimoto Y (1998) A single human cell expresses all messenger ribonucleic acids: the arrow of time in a cell. Mol Gen Genet. 258:233–239

King A, Gardner L, Loke YW (1996) Evaluation of oestrogen and progesterone receptor expression in uterine mucosal lymphocytes. Hum Reprod. 11:1079–1082

King A, Burrows T, Verma S, Hiby S, Loke YW (1998) Human uterine lymphocytes. Hum. Reprod. Update 4:480–485

Kinsky R, Delage G, Rosin N, Thang MN, Hoffmann M, Chaouat G (1990) A murine model for NK mediated resorption. Am. J. Reprod. Immunol. 23:73–75

Koopman LA, Kopcow HD, Rybalov B et al (2003) Human decidual natural killer cells are a unique NK cell subset with immunomodulatory potential. Journal of Experimental Medicine 198:1201–1212

Kozma N, Halasz M, Polgar B, Poehlmann TG, Markert UR, Palkovics T, Keszei M, Kiss K, Szeberenyi J, Par G, Grama L, Szekeres-Bartho J (2006) PIBF activates STAT6 via binding to a novel IL-4 receptor. J. Immunol. 176:819–826

Krishnan L, Guilbert LJ, Wegmann TG, Belosevic M, Mosmann TR (1996) T helper 1 response against Leishmania major in pregnant C57BL/6 mice increases implantation failure and fetal resorptions. Correlation with increased IFN-gamma and TNF and reduced IL-10 production by placental cells. J Immunol 156:653–662

Kuhn R, Rajewsky K, Muller W (1992) IL-4 and IL-10 deficient mice. Eighth International Congress of Immunology (Abstr.) 203

Lagerberg RE (2008) Malaria in pregnancy: a literature review. J Midwifery Womens Health. 53:209–215

Lanier LL (2005) NK cell recognition. Annual Reviews of Immunology 23:225–274

Laskarin G, Strbo N, Sotosek V, Podack RE, Szekeres-Bartho J, Rukavina D (1999) Progesterone directly and indirectly affects perforin expression in cytolytic cells. American Journal of Reproductive Immunology 42:312–320

Li X, O'Malley BW (2003) Unfolding the action of progesterone receptor. J Biol Chem 278:39261–39264

Li X, Lonard DM, O'Malley BW (2004) A contemporary understanding of progesterone receptor function. Mech Ageing Dev 125:669–678

Liang SC, Tan XY, Luxenberg DP, Karim R, Dunussi-Joannopoulos K, Collins M, Fouser LA (2006a) Interleukin (IL)-22 and IL-17 are coexpressed by Th17 cells and cooperatively enhance expression of antimicrobial peptides. J Exp Med 203:2271–2279

Liang J, Sun L, Wang Q, Hou Y (2006b) Progesterone regulates mouse dendritic cells differentiation and maturation. Int Immunopharmacol 6:830–838

Lichtenstein MR (1942) Tuberculin reaction in tuberculosis during pregnancy. Am. Rev. Tuberc. Pulm. Dis. 48:89–93

Liu WJ, Gottshall SL, Hansen PJ (1997) Increased expression of cell surface marker on endometrial g/d T cell receptor intraepithelial lymphocytes induced by the local presence of the sheep conceptus. Am J Reprod Immunol. 37:199–205

Liu AX, Jin F, Zhang WW, Zhou TH, Zhou CY, Yao WM, Qian YL, Huang HF (2006) Proteomic analysis on the alteration of protein expression in the placental villous tissue of early pregnancy loss. Biol Reprod. 75:414–420

Ljunggren HG, Karre K (1990) On search of the 'missing self': MHC molecules and NK cell recognition. Immunology Today 11:237–244

Long EO (1999) Regulation of immune responses through inhibitory receptors. Annual Reviews of Immunology 17:875–904

Luft BJ, Remington JS (1982) Effect of pregnancy on resistance to Listeria monocytogenes and Toxoplasma gondii infections in mice. Infect. Immun. 38:1164–1171

Lutz M, Schuler G (2002) Immature, semi-immature and fully mature dendritic cells: which signals induce tolerance or immunity? Trends in Immunol 23:445–449

Lydon JP, DeMayo FJ, Funk CR, Mani SK, Hughes AR, Montgomery CA Jr, Shyamala G, Conneely OM, O'Malley BW (1995) Mice lacking progesterone receptor exhibit pleiotropic reproductive abnormalities. Genes Dev 9:2266–2278

Marcenaro E, Ferranti B, Moretta A (2005) NK-DC interaction: on the usefulness of auto-aggression. Autoimmun Rev. 4:520–525

Meeusen E, Fox A, Brandon M, Lee CS (1993) Activation of uterine intraepithelial gamma delta T cell receptor positive lymphocytes during pregnancy. Eur. J. Immunol. 23:1112–1117

Minkoff H, Nanda D, Menez R, Fikrig S (1987) Pregnancies resulting in infants with aquired immunodeficiency syndrome or AIDS related complex: follow-up of mothers, children and subsequently born siblings. Obstet. Gynecol. 69:288–291

Moretta L, Biassoni R, Bottino C, Mingari MC, Moretta A (2000) Human NK-cell receptors. Immunology Today. 21:420–422

Moretta A, Marcenaro E, Sivori S, Della Chiesa M, Vitale M, Moretta L (2005) Early liaisons between cells of the innate immune system in inflamed peripheral tissues. Trends Immunol 26:668–675

Mori T, Kobayashi H, Nashimoto H, Suzuki A, Nishimura T (1977) Inhibitory effect of progesterone and 20 -alpha hydoxypregn-4-n-3-one on the phytohaemagglutinin -induced transformation of human lymphocytes. Am. J. Obstet. Gynecol. 127:151–157

Moriyama I, Sugawa T (1972) Progesterone facilitates implantation of xenogenic cultutred cells in hamster uterus. Nature New Biol. 236:150–152

Mosmann TR, Coffman RL (1989) Th1 and Th2 cells: different patterns of lymphokine secretion lead to different functional properties. Annu Rev Immunol 7:145–173

Mulac-Jericevic B, Conneely OM (2004) Reproductive tissue selective actions of progesterone receptors. Reproduction 128:139

Mulac-Jericevic B, Mullinax RA, DeMayo FJ, Lydon JP, Conneely OM (2000) Subgroup of reproductive functions of progesterone mediated by progesterone receptor-B isoform. Science; 289:1751–1754

Mulac-Jericevic B, Lydon JP, DeMayo FJ, Conneely OM (2003) Defective mammary gland morphogenesis in mice lacking the progesterone receptor B isoform. Proc Natl Acad Sci USA 100:9744–9749

Ntrivalas EI, Kwak-Kim JYH, Gilman-Sachs A, Chung-Bang h, NG SC, Beaman KD, Mantou-valos HP, Beer AE (2001) Status of peripheral blood natural killer cells in women with recurrent spontaneous abortions and infertility of unknown aetiology. Hum Reprod 16:855–861

Ntrivalas EI, Bowser CR, Kwak-Kim J, Beaman KD, Gilman-Sachs A (2005) Expression of killer immunoglobulin-like receptors on peripheral blood NK cell subsets of women with recurrent spontaneous abortions or implantation failures. Am J Reprod Immunol 53:215–221

Okada H, Nakajima T, Sanezumi M, Ikuta A, Yasuda K, Kanzaki H (2000) Progesterone enhances interleukin-15 production in human endometrial stromal cells in vitro. Journal of Clinical Endocrinolology and Metabolism 85:4765–4770

Ostensen M, Villiger PM (2007) The remission of rheumatoid arthritis during pregnancy. Semin Immunopathol. 29:185–191

Paldi A, d'Auriol L, Misrahi M, Bakos AM, Chaouat G, Szekeres-Bartho J (1994) Expression of the gene coding for the progesterone receptor in activated human lymphocytes Endocr J 2:317–321

Pasanen S, Ylikomi T, Palojoki E, Syvala H, Pelto-Huikko M, Tuohimaa P (1998) Progesterone receptor in chicken bursa of fabricius and thymus: evidence for expression in B-lymphocytes. Mol Cell Endocrinol 141:119–128

Pavia C, Siiteri PK, Perlman JD, Stites DP (1979) Suppression of murine allogeneic cell interactions by sex hormones. J Reprod Immunol 1(1):33–38

Piccinni MP (2005) T cells in pregnancy. Chem. Immunol. Allergy 89:3–9

Piccinni MP (2007) Role of T-cell cytokines in decidua and in cumulus oophorus during pregnancy. Gynecol Obstet Invest 64:144–148

Piccinni M-P, Giudizi MG, Biagiotti R, Beloni L, Giannarini L, Sampognaro S, Parronchi P, Manetti R, Livi C, Romagnani S, Maggi E (1995) Progesterone favors the development of human T helper cells producing Th2-type cytokines and promotes both IL-4 production and membrane CD30 expression in established Th1 cells clones. J Immunol 155:128–133

Piccinni M-P, Beloni L, Livi C, Maggi E, Scarselli GF, Romagnani S (1998) Defective production of both leukemia inhibitory factor and type 2 T-helper cytokines by decidual T cells in unexplained recurrent abortions. Nat Med 4:1020–1024

Piccinni M-P, Scaletti C, Vultaggio A, Maggi E, Romagnani S (2001a) Defective production of LIF, M-CSF and Th2-type cytokines by T cells at fetomaternal interface is associated with pregnancy loss. J Reprod Immunol 52:35–43

Piccinni MP, Scarletti C, Mavilia C, Lazerri E, Romagnani P, Natali I, Pellegrini S, Livi C, Romagnani S, Maggi E (2001b) Production of IL-4 and leukemia inhibitory factor by T cells of the cumulus oophorus: a favorable microenvironment for pre-implantation embryo development. Eur. J. Immunol. 31:2431–2437

Polgár B, Kispal G, Lachmann M, Paar C, Nagy E, Csere P, Miko E, Szereday L, Varga P, Szekeres-Barth J (2003) Molecular cloning and immunologic characterization of a novel cDNA coding for progesterone-induced blocking factor. Journal of Immunology 171:5956–5963

Polgár B, Nagy E, Mikó É, Varga P, Szekeres-Barthó J (2004) Urinary PIBF (Progesterone Induced Blocking Factor) concentration is related to pregnancy outcome. Biology of Reproduction 71:1699–1705

Putowski L, Darmochwal-Kolarz D, Rolinski J, Oleszczuk J, Jakowicki J (2004) The immunological profile of infertile women after repeated IVF failure (preliminary study). Eur. J. Obstet. Gynecol. Reprod. Biol. Feb 112:192–196

Quenby S, Farquharson R (2006) Uterine natural killer cells, implantation failure and recurrent miscarriage. Reproductive Biomedicine Online 13:24–28

Quenby S, Nik H, Innes B, Lash G, Turner M, Drury J, Bulmer J (2009) Uterine natural killer cells and angiogenesis in recurrent reproductive failure. Hum Reprod. 24:45–54

Raghupathy R (1997) Th-1 type immunity is incompatible with successful pregnancy. Immunology Today 18:478–482

Raghupathy R, Makhseed M, Azizieh F, Hassan N, Al-Azemi M, Al-Shamali E (1999) Maternal Th1- and Th2-type reactivity to placental antigens in normal human pregnancy and unexplained recurrent spontaneous abortions. Cell Immunol 196:122–130

Raghupathy R, Makhseed M, Azizieh F, Omu A, Gupta M, Farhat R (2000) Cytokine production by maternal lymphocytes during normal human pregnancy and in unexplained recurrent spontaneous abortion. Hum Reprod 15:713–718

Raghupathy R, Al Mutawa E, Makhseed M, Azizieh F, Szekeres-Bartho J (2005) Modulation of Cytokine Production by Dydrogesterone in Lymphocytes from Women with Recurrent Abortion. Brit. J. Ob. Gyn. 112:1096–1101

Raghupathy R, Al Mutawa E, Maksheed M, Al-Azemi M, Azizeh F (2007) Redirection of cytokine production by lymphocytes from women with pre-term delivery by dydrogesterone. Am J Reprod Immunol. 58:31–38

Rahman MA, Li M, Li P, Wang H, Dey SK, Das SK (2006) Hoxa-10 deficiency alters region-specific gene expression and perturbs differentiation of natural killer cells during decidualization. Dev Biol. 290:105–117

Reinhardt MC (1980) Effects of parasitic infections in pregnant women. Ciba Found. Symp. 77:149–163

Rezaei A, Dabbagh A (2002) T-helper (1) cytokines increase during early pregnancy in women with a history of recurrent spontaneous abortion. Med Sci Monit 8:CR607–CR610

Robertson SA, Mayrhofer G, Seamark RF (1996) Ovarian steroid hormones regulate granulocyte-macrophage colony-stimulating factor synthesis by uterine epithelial cells in the mouse. Biol Reprod 54:183–196

Robertson SA, Roberts CT, Farr KL, Dunn AR, Seamark RF (1999) Fertility impairment in granulocyte-macrophage colony-stimulating factor-deficient mice. Biol Reprod 60:251–261

Romagnani S (1991) Human Th1 and Th2: doubt no more. Immunol Today; 12:256–257

Romero R, Chaiworapongsa T, Espinoza J (2003) Micronutrients and intrauterine infection, preterm birth and the fetal inflammatory response syndrome. J Nutr 133:1668S–1673S

Roussev RG, Higgins NG, McIntyre JA (1993) Phenotypic characterization of normal human placental mononuclear cells. J Reprod Immunol. 25:15–29

Rukavina D, Rubesa G, Gudelj L, Haller H, Podack ER (1995) Characteristics of perforin expressing lymphocytes within the first trimester of human pregnancy. American Journal of Reproductive Immunology 33:394–404

Sallusto F, Lanzavecchia A (1999) Mobilizing dendritic cells for tolerance priming and chronic inflammation. J Exp Med 189:611–614

Sargent IL, Borzychowski AM, Redman CW (2006) Immunoregulation in normal pregnancy and pre-eclampsia: an overview. Reproductive Biomedicine Online. 13:80–86

Screpanti I, Gulino A, Pasqualini JR (1982) The fetal thymus of guinea pig as an estrogen target organ. Endocrinology 111:1552–1561

Sentman CL, Meadows SK, Wira CR, Eriksson M (2004) Recruitment of uterine NK cells: induction of CXC chemokine ligands 10 and 11 inhuman endometrium by estradiol and progesterone. J Immunol 173:6760–6766

Shakhar K, Ben Eliyahu S, Loewenthal R, Rosenne E, Carp H (2003) Differences in number and activity of peripheral natural killer cells in primary versus secondary recurrent miscarriage. Fertility Sterility 80:368–375

Slapsys RM, Clark DA (1982) Active suppression of host-vs-graft reaction in pregnant mice. IV. Local suppressor cells in decidua and uterine blood. J Reprod Immunol. 4:355–364

Stewart CL, Kaspar P, Brunet LJ, Bhatt H, Gadi I, Kontgen F, Abbondazo SJ (1992) Blastocyst implantation depends on maternal expression of leukaemia inhibitory factor. Nature. 359:76–79

Stewart JA, Bulmer JN, Murdoch AP (1998) Endometrial leucocytes: expression of steroid hormone receptors. J Clin Pathol. 51:121–126

Stimson WH (1988) Oestrogen and human T lymphocytes: Presence of specific receptors in the T suppressor cytotoxic subset. Scand. J. Immunol. 28:345–350

Stites DP, Siiteri PK (1983) Steroids as immunosuppressants in pregnancy. Immunol Rev. 75:117–138

Sulke AN, Jones DB, Wood PJ (1985) Hormonal modulation of human natural killer cell activity in vitro. Journal of Reproductive Immunology 7:105–110

Szekeres-Bartho J, Wegmann TG (1996) A progesterne-dependent immunomodulatory protein alters the Th1/Th2 balance J. Reprod. Immunol. 31:81–95

Szekeres-Bartho J, Hadnagy J, Pacsa AS (1985a) The suppressive effect of progesterone on lymphocyte cytotoxicity: unique progesterone sensitivity of pregnancy lymphocytes. Journal of Reproductive Immunology 7:121–128

Szekeres-Bartho J, Kilar F, Falkay G, Csernus V, Torok A, Pacsa AS (1985b) Progesterone-treated lymphocytes of healthy pregnant women release a factor inhibiting cytotoxicity and prostaglandin synthesis. Am. J. Reprod. Immunol. Microbiol. 9:15–19

Szekeres-Bartho J, Reznikoff-Etievant MF, Varga P, Varga Z, Chaouat G (1989a) Lymphocytic progesterone receptors in human pregnancy. J. Reprod. Immunol. 16:239–247

Szekeres-Bartho J, Weill BJ, Mike G, Houssin D, Chaouat G (1989b) Progesterone receptors in lymphocytes of liver-transplanted and transfused patients Immunol. Letters 22:259–261

Szekeres-Bartho J, Autran B, Debre P, Andreu G, Denver L, Chaouat G (1989c) Immunoregulatory effects of a suppressor factor from healthy pregnant women s lymphocytes after progesterone induction. Cell. Immunol. 122:281–294

Szekeres-Bartho J, Gy S, Debre P, Autran B, Chaouat G (1990) Reactivity of lymphocytes to a progesterone receptor-specific monoclonal antibody. Cell Immunol 125:273–283

Szekeres-Bartho J, Zs F, Varga P, Szereday L, Kelemen K (1996) The immunological pregnancy protective effect of progesterone is manifested via controlling cytokine production. Amer. J. Reprod. Immunol. 35:348–351

Szekeres-Bartho J, Par G, Dombay G, Smart YC, Volgyi Z (1997) The antiabortive effect of progesterone-induced blocking factor in mice is manifested by modulating NK activity. Cellular Immunology 177:194–199

Szekeres-Bartho J, Barakonyi A, Miko E, Polgar B, Palkovics T (2001) The role of T cells in the feto-maternal relationship. Seminars in Immunology 13:229–233

Tangri S, Wegmann TG, Lin H, Raghupathy R (1994) Maternal anti-placental reactivity in natural immunologically-mediated fetal resorptions. J. Immunol. 152:4903–4911

Tchernitchin X, Tchernitchin A, Galand P (1976) Dynamics of eosinophils in the uterus after oestrogen administration. Differentiation 5:151–154

Thuere C, Zenclussen ML, Schumacher A, Langwisch S, Schulte-Wrede U, Teles A, Paeschke S, Volk HD, Zenclussen AC (2007) Kinetics of regulatory T cells during murine pregnancy. Am J Reprod Immunol. 58:514–523

Tibbetts OM Conneely, O'Malley BW (1999) Progesterone via its receptor antagonizes the proinflammatory activity of estrogen in the mouse uterus. Biol Reprod 60:1158

Trowsdale J, Moffett A (2008) NK receptor interactions with MHC class I molecules in pregnancy. Semin Immunol. 20:317–320

Uksila J (1985) Human NK cell activity is not inhibited by pregnancy and cord serum factors and female steroid hormones in vitro. Journal of Reproductive Immunology 7:111–120

Ulmann A, Dubois C, Philibert D (1987) Fertility control with RU 486. Horm Res 28:274–278

Van den Heuvel MJ, Chantakru S, Xumei X, Evans EE, Tekpetey F, Mote PA, Clarke CL, Croy BA (2005) Trafficking of circulating pro-NK cells to the decidualizing uterus: regulatory mechanisms in the mouse and human. Immunol. Invest. 34:273–293

Varner MW (1991) Autoimmune disorders in pregnancy. Semin. Perinatol. 15:238–250

Verma S, Hiby SE, Loke YW, King A (2000) Human decidual natural killer cells express the receptor for and respond to the cytokine interleukin 15. Biology of Reproduction 62:959–968

Vince GS, Johnson PM (1996) Is there a Th2 bias in human pregnancy? J. Reprod. Immunol. 32:101–104

von Dadelszen P, Magee LA, Krajden M, Alasaly K, Popovska V, Devarakonda RM, Money DM, Patrick DM, Brunham RC (2003) Levels of antibodies against cytomegalovirus and *Chlamydophila pneumoniae* are increased in early onset pre-eclampsia. BJOG 110:725–730

Wang B, Biron C, She J, Higgins K, Sunshine MJ, Lacy E, Lonberg N, Terhorst C (1994) A block in both early T lymphocyte and natural killer cell development in transgenic mice with high-copy numbers of the human CD3ε gene. Proc. Natl. Acad. Sci. USA 91:9402–9406

Wegmann TG, Lin H, Guilbert L, Mosmann TR (1993) Bidirectional cytokine interactions in the maternal-fetal relationship: is successful pregnancy a Th2 phenomenon? Immunology Today 14:353–356

Weiner LH (2001) Induction and mechanism of action of transforming growth factor beta secreting Th3 regulatory cells. Immunol Rev. 182:207–214

Wilson B, Moor J, Jenkins C, Miller H, Walker JJ, McLean MA, Norman J, McInnes IB (2004) Abnormal first trimester serum interleukin 18 levels are associated with a poor outcome in women with a history of RM. Am. J. Reprod. Immunol. 51:156–159

Wira CR, Sandoe CP (1980) Hormonal regulation of immunoglobulins: influence of estradiol in immunoglobulins A and G in the rat uterus. Endocrinology 106:1020–1026

Wyle FA, Kent JR (1977) Immunosuppression by sex steroid hormones. I. The effect upon PHA and PPD stimulated lymphocytes. Clin. Exp. Immunol. 27:407–415

Yamada H, Morikawa M, Kato EH, Shimada S, Kobashi G, Minakami H (2003) Pre-conceptional natural killer cell activity and percentage as predictors of biochemical pregnancy and spontaneous abortion with normal chromosome karyotype. American Journal of Reproductive Immunology 50:351–354

Yao MW, Lim H, Schust DJ, Choe SE, Farago A, Ding Y, Michaud S, Church GM, Maas RL (2003) Gene expression profiling reveals progesterone-mediated cell cycle and immunoregulatory roles of Hoxa-10 in the preimplantation uterus. Mol. Endocrinol. 17:610–627

Yie SM, Xiao R, Librach CL (2006) Progesterone regulates HLA-G gene expression through a novel progesterone response element Human Reproduction 21:2538–2544

Zhao JX, Zeng YY, Liu Y (2007) Fetal alloantigen is responsible for the expansion of the CD4(+) CD25(+) regulatory T cell pool during pregnancy. J Reprod Immunol. 75:71–81

Zhu Y, Rice CD, Pang Y et al (2003) Cloning, expression, and characterization of a membrane progestin receptor and evidence it is an intermediary in meiotic maturation of fish oocytes. Proceedings of the National Academy of Sciences USA 100:2231–2236

Zuckerman FA, Head JR (1987) Possible mechanism of non-rejection of the feto-placental allograft: Trophoblast resistance to lysis by cellular immune effectors. Transplant Proc. 1:554–559

Chapter 9
Pregnancy and Susceptibility to Parasites

Fiona L. Henriquez, Fiona M. Menzies, and Craig W. Roberts

Abstract Successful pregnancy is dependent on many dynamic immunological events that occur at the maternal–fetal interface. While the trophoblast initiates implantation through the uterine wall, maternal immune mechanisms limit the extent to which the decidua is invaded by the placenta without disrupting pregnancy. Immunological reactions are not observed at the maternal–fetal interface. Pregnancy-induced modulation of systemic maternal immunity results in amelioration of certain autoimmune diseases and exacerbation of some infectious diseases. Alteration of immunological function during pregnancy in some circumstances might facilitate congenital transmission of certain pathogens. In other circumstances, the immune reactions induced by pathogens might result in disruption of pregnancy. These complex interactions are discussed in the context of a number of diseases caused by infection with taxonomically diverse protozoan parasites, nematodes, and cestodes.

9.1   Introduction

Pregnancy has been known to modulate the immune response since the mid 1930s with the discovery that pregnant rabbits have a lower capacity to produce antibody (Keeler and Castle 1934). Studies during the 1960s elucidated that pregnancy induced inertia of the immune response (Anderson 1965). This has been attributed to the need to not only accept an "allograft," but also to actively nurture this 50% foreign material via an invasive placenta structure, without allowing the placenta to

C.W. Roberts (✉)
Strathclyde Institute of Pharmacy Biomedical Sciences, University of Strathclyde, 27 Taylor St, Glasgow, G4 0NR, Scotland, UK
e-mail: c.w.roberts@strath.ac.uk

S.L. Klein and C.W. Roberts (eds.), *Sex Hormones and Immunity to Infection*,
DOI 10.1007/978-3-642-02155-8_9, © Springer-Verlag Berlin Heidelberg 2010

further encroach into the body. The systemic effects of this immune modulation in pregnant women can be beneficial by ameliorating the severity of certain autoimmune diseases. However, in other circumstances, pregnancy-induced immune modulation has proven detrimental by reducing the ability of the mother to control infectious diseases. This has consequences for the health and longevity of the mother, but also can disrupt pregnancy, result in fetal abnormalities, and facilitate congenital or postpartum transmission of infection to the progeny.

The precise mechanisms of immune modulation during pregnancy have not been fully characterized, but it is widely supported by the literature that alteration in steroid and nonsteroid hormone levels accompanying pregnancy are responsible for many of the observed effects both systemically and at the maternal–fetal interface (Szekeres-Bartho et al. 2005).

9.2 Immunological Interactions Within the Placenta During Normal Pregnancy

The ability of mammals to survive gestation in spite of being comprised of 50% paternally derived alloantigens was recognized by Medawar as an anomaly that challenged the self, nonself theory of immune recognition (Medawar 1982). Consequently, immunologists have strived to understand the mechanisms that are responsible for preventing rejection of the fetus. It is now recognized that the maternal–fetal interface is immunologically active and shares many similarities with inflammation, but has evolved complex mechanisms of regulation. These are sometimes similar to those seen in other anatomical sites, but in other cases unique to the maternal–fetal interface.

9.2.1 Placentation

The process of placentation is an invasion of the uterine decidua by the placental unit containing the chorionic villus, the cytotrophoblast progenitor cells, syncytiotrophoblast, the column cytotrophoblasts, and the invasive trophoblasts. Placentation occurs within 4 days of fertilization when cytotrophoblasts appear on the surface of the blastocyst. These cells are responsible for the production of human chorionic gonadotrophin, which prepares the endometrium for embryo implantation and allows the continuation of progesterone secretion from the granulosa cells of the corpus luteum. The trophoblast also implantation through attachment to the basement membrane. The early placenta at this point commences the invasive process into the uterine wall and into the maternal blood stream. This process not only insures the anchoring of the fetal unit to the uterus but also increases the nutrient and blood supply from the mother to the fetus, including maternal IgG

(Pereira et al. 2005). The placentation process is tightly regulated by the maternal immune system as the trophoblast has a potent invasive capacity, but the fetus must also endeavor to protect itself from maternal immune attack. Therefore, during pregnancy, immune function is dramatically modulated to induce both maternal tolerance of the fetus and control of the trophoblast invasion. The trophoblast induces immunomodulation in the growing placenta through the production of progesterone.

As discussed in Chap. 8 in greater detail, early gestation of the fetus is characterized by a number of decidual leukocytes found within this interface. These cells are composed of uterine NK cells, macrophages, dendritic cells (DCs), and T cells. In the fetal chorionic villus, a number of macrophages (Hofbauer cells) and DCs are present. The macrophages are typically found in the villus from 28 days until term (Castellucci and Kaufmann 1982) and are characterized by antiphagocytic activity and the expression of Fc receptors, CD14 (Saji et al. 1994) and MHC I and II (Castellucci and Kaufmann 1982). In addition, Hofbauer cells produce prostaglandins and thromboxanes (Wetzka et al. 1997) and this, together with the expression of vascular endothelial growth factor (VEGF) (Clark et al. 1998), indicates that Hofbauer cells are involved in the regulation and stabilization of the placenta.

The decidua–placenta interface harbors mature CD83+ DCs, which display a reduced capacity to produce IL-12p70 (Miyazaki et al. 2003) and therefore suppress Th1-type inflammation. In addition, studies in the mouse decidua report that DCs produce IL-10 (Blois et al. 2004a), although DCs are not present at all the stages of pregnancy. There is also evidence that DCs can play an adverse role as an increase in decidual DCs has been found to coincide with spontaneous abortion in humans (Askelund et al. 2004).

During early gestation, a number (20% of all decidual leukocytes) of immature DCs (iDC) expressing CD209 (DCSIGN) are present near the invading trophoblast. They are found in abundance during early pregnancy and proliferate in the decidua (Kammerer et al. 2003; Rieger et al. 2004). These cells also are efficient in antigen uptake (apoptotic particles of the invading extravillous trophoblast) and mature into $CD83^+$ $CD25^+$ DCs with the ability to stimulate naive T cells (Kammerer et al. 2003). Through the action of CTLA4-expressing regulatory T (Treg) cells (Mellor and Munn 2004) in the placenta, DCs express indoleamine 2.3-dioxygenase (IDO), which is involved in the inhibition of T cell proliferation (Munn et al. 2002) and prevention of maternal T cell attack of the fetus (Munn et al. 2002).

Uterine NK (uNK) cells represent 70% of all leukocytes in the maternal–fetal interface (Moffett and Loke 2006). They possess a unique phenotype and they are $CD94^+$ bright exclusively (Ponte et al. 1999). All NK cells express killer-inhibitory receptors (KIR) 2DL4. Both CD94 and KIR2DL4 are crucial in the outcome of successful pregnancy as they bind to HLA-E and HLA-G, respectively. HLA-E and HLA-G are nonclassical MHC I molecules expressed on the fetal unit (King et al. 2000; Rajagopalan et al. 2006). HLA-C is a polymorphic MHC molecule that also is expressed on the surface of trophoblast cells. uNK cells can inhibit maternal immune attack through recognition of this molecule by KIR2D (Parham 2005).

During early gestation, uNK cells come into direct contact with the trophoblast, expressing HLA-E and HLE-G on its surface, thus becoming "non-killing uNK cells." This interaction stimulates uNK to produce antiinflammatory cytokines (e.g., IL-10) (reviewed in Dietl et al. 2006).

It is evident that DC-SIGN$^+$ expressing iDCs and uNK cells are the most dominant leukocyte populations in early pregnancy. The ratio between uNK cells and iDCs, however, appears to be crucial to the outcome of a successful pregnancy. Low ratios of uNK/iDC cultures stimulate DC production of TNF-α and IL-12, thus potentially threatening rejection of the fetus (Piccioli et al. 2002). In contrast, at a high uNK/iDC ratio, NK cells that do not express KIRs can lyse iDCs, but not mature DCs due to the presence of HLA-E (Piccioli et al. 2002; Della Chiesa et al. 2003; Wilson et al. 1999; Carbone et al. 1999). This has a profound effect on immunomodulation in early pregnancy. High numbers of uNK cells can kill autologous DCs, thus protecting the fetus. In contrast, the production of cytokines such as IL-10 and GM-CSF by uNK cells that comes into contact with nonclassical HLA-expressing trophoblast prevent iDC maturation and consequently T cell activation and if uNK cells become activated by the presence of infectious agents, DC maturation will occur (reviewed in Dietl et al. 2006).

The suppression of the Th1 response in the vicinity of the placenta during pregnancy is necessary for the inhibition of the fetal semi-allograft, as production of IFN-γ promotes fetal rejection. In contrast, Th2 cytokines favor the maintenance of a successful pregnancy (Piccinni et al. 1998). The polarization towards a Th2 environment is largely attributed to the presence of sex hormones (Correale et al. 1998; Giron-Gonzalez et al. 2000), such as progesterone and estrogen. Recruitment and activation of T cells in the placenta may be affected by mediators, such as CCL5 (i.e., Regulated on activation normal T cell expressed and secreted, RANTES) and placental IDO (pIDO). In fact, it has been suggested that progesterone-regulated CCL5 may be relevant in early pregnancy and fetal tolerance (Ramhorst et al. 2007) as an increase in serum CCL5 has been observed in successful pregnancies (Ramhorst et al. 2004). On the other hand, pIDO intensifies mesenchymal stem cell immunosuppression of T cells following IFN-γ induction (Jones et al. 2007). Progesterone induces Th2 dominance during pregnancy (Szekeres-Bartho and Wegmann 1996; Szekeres-Bartho et al. 1996; Szekeres-Bartho et al. 1996; Orsal et al. 2006). This is achieved by the production of progesterone-induced blocking factor (PIBF) by T cells and NK cells upon stimulation of progesterone. PIBF mediates IL-4 production and naive T cells through the activation of STAT6 (Kozma et al. 2006) and downregulates STAT4, required for Th1 type responses (Kozma et al. 2006). Furthermore, the expression of NF-κB, necessary for a successful Th1 response is downregulated during pregnancy, thereby favoring Th2 development (McCracken et al. 2004).

Studies in murine models of pregnancy have established that Th2 dominance is not solely responsible for the survival of a semi-allograft graft because IL-4/IL-10 single and double deficient mice achieve a successful pregnancy

(Bonney 2001; Svensson et al. 2001). This focused the attention on CD4$^+$CD25$^+$ Treg cell population, numbers of which are significantly increased during pregnancy in the placenta and return to lower levels postpartum (Somerset et al. 2004). Treg cells are immunosuppressive and play a critical role in the induction of fetal tolerance in both mice and humans (Aluvihare et al. 2004; Zenclussen et al. 2005; Sasaki et al. 2004; Somerset et al. 2004). Their primary site of tolerance action could be in lymph nodes draining the uterus and the placental maternal–fetal interface (Aluvihare and Betz 2006). This is further supported by evidence that PD-L1 (a negative costimulatory molecule) is expressed on Treg cells, decidual cells, and on the synctiotrophoblast (Sandner et al. 2005; Guleria et al. 2005; Guleria and Sayegh 2007). Furthermore, adoptive transfer of Treg cells in murine models from wild type into PD-L1-deficient mice significantly improved fetal survival (Habicht et al. 2007).

In the placenta, the cytotoxic activity of CD8$^+$ T cells is downregulated as a direct consequence of the Th1 suppression (Jones and Robinson 1983; Guntupalli and Steingrub 2005; Constantin et al. 2007). Despite this suppression, there is no impairment of the establishment of memory T cells, and a significant CD8$^+$ T cell infiltration the placenta can occur if a primary infection is established during pregnancy (Constantin et al. 2007). Therefore, other mechanisms are also involved in prevention of fetal rejection by cytotoxic activity. Cytotoxic T cells (CTLs) induce their activity by binding to MHC I class molecules on the target cells surface. The fetal semi-allograft does not express common paternal MHC antigens, but nonclassical MHC I class molecules, defined in humans as HLA-G, HLA-E, and the polymorphic HLA-C (Hutter et al. 1996; King et al. 1996); McMaster et al. 1995; Loke et al. 1997; Prolla et al. 1994; Parham 2005). The expression of HLA-G on the on the maternal–fetal interface and trophoblast has an immunotolerant effect and inhibits antigen-specific, MHC-restricted CTL responses (Rouas-Freiss et al. 1999). In addition to the immunotolerance role of nonclassical HLA molecules in the placenta, CD8$^+$ T cells are responsive to progesterone during pregnancy through a progesterone receptor (PR) that releases PIBF from both CD4$^+$ and CD8$^+$ T cells (Blois et al. 2004b).

During placentation, γδ T cells originate in the decidua (Mincheva-Nilsson et al. 1997), and are upregulated during the duration of pregnancy (Kimura et al. 1995). The importance of the increase in γδ T cells is mainly to induce immunotolerance by the recognition of nonclassical HLA class I molecule (HLA-G and HLA-E) (Barakonyi et al. 2002). This is achieved by the expression and binding of KIRs, such as CD94, to nonclassical HLA molecules displayed on the fetal semi-allograft (Barakonyi et al. 2002).

The downregulation of the Th1-type response, including production of IFN-γ, in the placenta through a Th2-dominant/antiinflammatory response (i.e., elevated production of IL-4, IL-5, and IL-6) and the presence of pregnancy hormones, influences the activity of humoral immunity, including B cell activity and antibody production. The expression of the antiapoptotic molecule Bcl-2 on B cells during pregnancy causes downregulation of B cell lymphopoiesis (Caucheteux et al. 2005). The expression of Bcl-2 is regulated by progesterone which, through the PR, interacts directly with its promoter (Yin et al. 2007). In the placenta, there is an increased presence of asymmetric antibodies, which are characterized by the presence of glycosylation on one branch of the Fab region (Labeta et al. 1986). The antibodies can still bind foreign fetal antigen, but the effector functions, such as the ability to promote phagocytosis and fix complement, are inactivated (Eblen et al. 2000). Therefore, these asymmetric antibodies can act as "blocking antibodies" to mask fetal antigens (Eblen et al. 2000).

9.2.2   Implications of Placentation Being an Immunological Process

The fact that placentation is an immunological process has at least three major implications for infection during pregnancy. First, pathogens or pathogen-derived molecules that alter the immunological balance can disrupt pregnancy. Second, the mechanisms that normally control pathogens can be affected in the vicinity of the placenta. Third, the immunological processes occurring in the placenta can have an effect on systemic immunity. These points, individually or collectively, may account for exacerbation or reactivation of infectious diseases during pregnancy, in some case accompanied with vertical pathogen transmission and/or pathogen-induced abortion (discussed below).

9.3   Impact of Pregnancy on the Systemic Immune System

Many autoimmune disorders, including rheumatoid arthritis and multiple sclerosis, which have been attributed to Th1 cell activation, are alleviated during pregnancy (Al-Shammri et al. 2004). In contrast, Th2-associated immunological disorders, including asthma, are generally increased in pregnant women (Rastogi et al. 2006). Together, these observations suggest that pregnancy exerts a strong influence over certain aspects of systemic immunity. These effects are likely to be due to the effect of pregnancy-associated hormones on systemic immunity. The effects of progesterone on specific immune cells are discussed in Chap. 8 and estrogens and androgens in Chap. 2.

9.4   Interaction of Parasitic Infections and Pregnancy

The severity of a number of infectious diseases is known to be increased if contracted during pregnancy. Pathogens contracted prior to conception may become reactivated or exacerbated during pregnancy. This is a particular concern for chronic debilitating diseases, such as malaria, leishmaniasis, trypanosomiasis, or schistosomiasis, endemic in some of the poorest nations of the world. The reasons for these observations are frequently due to hormonally induced alterations of immunological parameters that would normally limit the extent of infection.

Levels of steroid hormones are dramatically altered during human pregnancy. In nonpregnant women, progesterone levels fluctuate between $.8 \pm 0.34$ and 43 ± 13 nmol l^{-1} during the follicular and luteal phases of the menstrual cycle, respectively, but are raised to 500–610 nmol l^{-1} during the later stages of pregnancy. Testosterone varies between 0.75 ± 0.2 nmol l^{-1} at the follicular phase and 1.28 ± 0.3 nmol l^{-1} at the luteal phase of the menstrual cycle, but is raised to 64.5 ± 33.7 nmol l^{-1} during pregnancy. 17β-estradiol cycles between 0.35 ± 0.18 and 0.64 ± 0.38 nmol l^{-1} during the follicular and luteal phases, respectively, but increases to 3.6 ± 1.5 nmol l^{-1} in the final stages of pregnancy. Cortisol levels exhibit a diurnal cycle in humans with a mean of 190.7 ± 23.4 pg m $^{-1}$ (Hindmarsh et al. 1989). Cortisol levels rise during pregnancy, with mean concentrations 3–6 h prior to the onset of labor of $597 \pm 51 \mu g$ l^{-1} and $468 \pm 29 \mu g$ l^{-1} in primaparous and multiparous women, respectively (Kono et al. 1987). Changes in peptide hormones, such as prolactin also are noted during pregnancy, which increases from a mean of 294 milli-IU l^{-1} in nonpregnant women to a mean of 4092 milli-IU l^{-1} in pregnant women (O'Leary et al. 1991).

Additional physiological consequences of pregnancy-associated hormones as well as the competing demands of pregnancy in an already sick individual also may impact responses to infection during pregnancy. Pregnancy-induced or -associated complications of parasitic infections are a global problem for infectious diseases and some pathogens have evolved to exploit the condition. Notably, toxoplasmosis, which infects around 30% of the world's population, can be transmitted transplacentally and can cause abortion, devastating congenital abnormalities and infant mortality.

9.4.1 *Congenital Transmission and Disruption of Pregnancy by Parasites*

9.4.1.1 Parasite-Mediated Pathology

Parasites typically associated with the disruption of pregnancy and/or congenital transmission include *Trichomonas foetus* in cattle (Fitzgerald 1986; Yule et al. 1989; BonDurant 1997; BonDurant et al. 2003), *Trichomonas vaginalis* in humans (Van Der Pol et al. 2005), *Toxoplasma gondii*, (Menzies et al. 2008), *Plasmodium* spp. (Menendez and Mayor 2007), *Leishmania* spp. (Eltoum et al. 1992), *Schistosomes* (Willingham et al. 1999), and a number of filarial and intestinal nematodes. The pathogenesis of *T. foetus* and *T. vaginalis* infections is predominantly associated with sexual transmission of parasites within the uterus where they are responsible for cytopathology through the adherence to host tissue cells (Silva Filho and de Souza 1988; Benchimol et al. 2008). The pathogenesis of other parasite-mediated complications of pregnancy is more complex, and is discussed below.

9.4.1.2 Immunological Interference

Disruption of pregnancy by many of the parasites that cause systemic infections has been attributed to both parasite-mediated damage to the placenta and immunological interference. It is difficult to determine the relative contribution of parasite-mediated damage versus immune-mediated damage to the placenta, but some parasites are known to infect the placenta, and placental pathology has been noted in some parasitic diseases.

Immune-mediated disruption of the complex immunological interactions occurring at the maternal–fetal interface by pathogens or their products has been well documented. Notably, stimulation of the innate immune system through the toll-like receptor (TLR) 4 with the bacterial product lipopolysaccharide (LPS) can induce abortion in mice. This is a rapid process that is dependent on a TNF-α and nitric oxide (NO) (Gendron and Baines 1988; Gendron et al. 1990a; Gendron et al. 1990b). Similarly, the synthetic mimic of viral double-stranded RNA, poly I:C, that is a ligand for TLR3 that does not signal through MyD88, also can induce abortion through NK-cell-dependent mechanisms (Lin et al. 2006; Zhang et al. 2007). Thus, engagement of both MyD88-dependent and -independent TLRs are sufficient to induce abortion. Whether TLR-induced abortion occurs during infection remains to be determined. However, as more TLR ligands are identified in an increasing number of protozoan and metazoan eukaryotic pathogens, their role in abortion is likely to be further recognized. In addition, alteration of the adaptive immune response may also be sufficient to induce abortion or to prevent implantation. Notably, a strong Th1 and consequently reduced Th2 immunological environment has adverse effects on pregnancy. As a general principle, pregnancy is easier to disrupt in the first trimester than at later stages most probably due the immunological environment conducive to successful pregnancy being less well established. A corollary of this is that in the third trimester, the immunological conditions that are conducive to pregnancy are well established and may interfere with the effective control and facilitate congenital transmission of certain parasites (Fig. 9.1).

9.4.2 Protozoan Infections and Pregnancy

9.4.2.1 T. gondii

T. gondii is typically transmitted through infected meat or the ingestion of oocysts found in cat feces. Following infection, the rapidly dividing tachyzoite stage, that is capable of infecting virtually any cell, predominates until the onset of immunity. Driven by environmental stress associated with the host immune response, tachyzoites undergo transformation to the slowly dividing bradyzoite stages that form quiescent cysts throughout the body (Dubey 1998). Both innate and adaptive aspects of the immune system cooperate to control *T. gondii* tachyzoites and T cells, and an intact immune system necessary to prevent disease reactivation

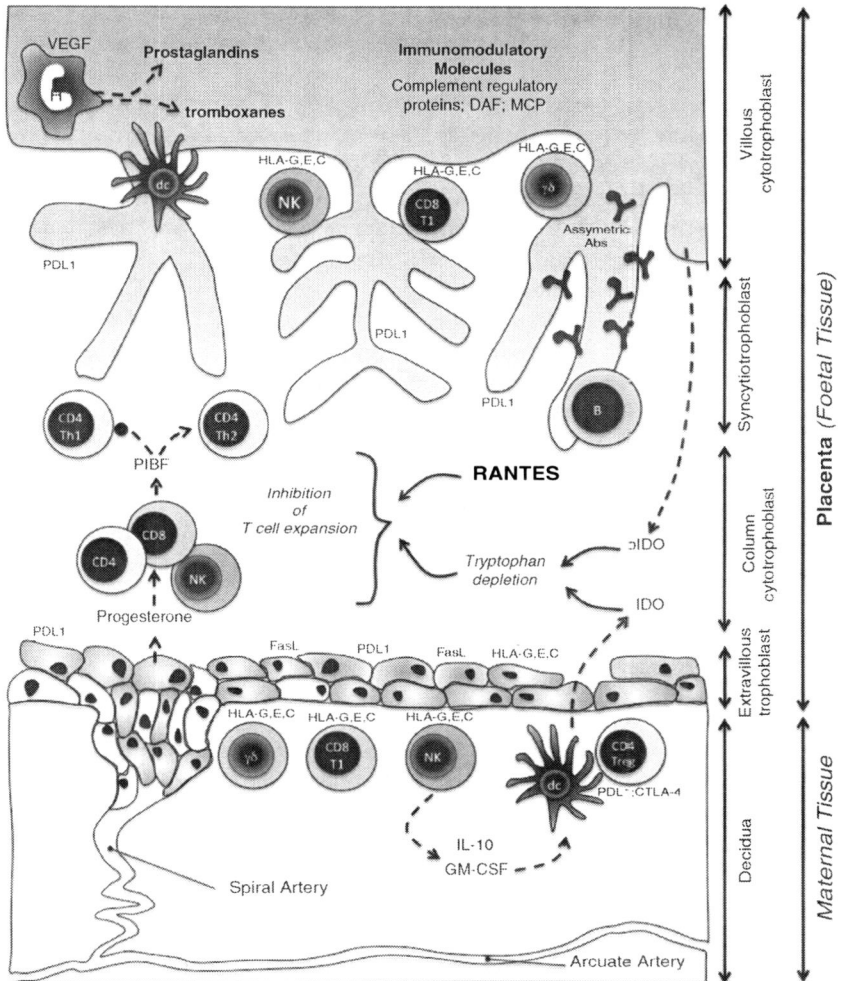

Fig. 9.1 Immunological interactions in the human placenta. The placenta consists of four main areas: the extravillous trophoblast that invades the maternal decidual layer, the column cytotrophoblast, the syncytiotrophoblast, and the villous cytotrophoblast. Progesterone is produced from the extravillous trophoblast in early pregnancy and contributes to the Th2 bias at the maternal–fetal interface. Progesterone binds to receptors on NK cells and T cells (CD4+ and CD8+), which produce progesterone-induced blocking factor (PIBF). PIBF acts on CD4+ T cells to upregulate STAT6, an important signal transducer required for the production of IL-4 and Th2 cell activation. In contrast, PIBF inhibits the expression of the transcription factor NF-kB and STAT4 resulting in inhibition of Th1 cell activation. The placenta has a number of surface-expressed immunomodulatory molecules including FasL and PDL1, which can induce apoptosis and down-regulation of T cell proliferation. Nonclassical HLA molecules (HLA-G, HLA-E, HLA-C) on the

(reviewed in Denkers and Gazzinelli 1998; Denkers et al. 2003). Congenital transmission represents an additional mode of infection and can occur in animals and humans.

Congenital infection can occur either following a primary infection during pregnancy (termed exogenous transplacental transmission) or due to reactivation of a chronic infection as a result of pregnancy (endogenous congenital transmission) (Trees and Williams 2005). Until recently, it was believed that endogenous congenital transmission did not occur in humans, but there are now increasing reports of this occurring (e.g., Awan 1978).

An immune response is initiated to *T. gondii* by a number of pathogen-associated molecular patterns (PAMP) including: (1) GPI-anchors, which are effective ligands for TLR2 and TLR4 (Debierre-Grockiego et al. 2007); (2) HSP70 that can stimulate macrophages and induce maturation of DCs through ligation of the TLR4 (Chen et al. 2002; Aosai et al. 2006); and (3) profilin which interacts with TLR11 in some mammals to induce IL-12 production (Yarovinsky et al. 2005). In addition, *T. gondii* cyclophilin 18 is a chemokine mimic that can bind the CCR5 receptor to induce IL-12 production (Aliberti et al. 2003). These interactions control initial parasite multiplication through NK-cell-produced IFN-γ, macrophage TNF-α production, and the induction of NOS2 with resultant reactive nitrogen intermediates (Aliberti et al. 2003). This immunological environment also effectively drives the development of Th1 cells and CD8 cytolytic T cells (Combe et al. 2005). CD4^{+} T cell production of IL-2 is essential for the development of CD8 cytolytic T cells that are critical for resolution of disease, prevention of encephalitis, and prevention of disease reactivation (Gazzinelli et al. 1992). Th2 cytokines antagonize many of the beneficial effects that the Th1 cytokines play in control of parasites, but are necessary for limiting inflammation and immunopathology (Nickdel et al. 2004; Villard et al. 1995; Roberts et al. 1996; Suzuki et al. 1996; Alexander et al. 1998; Gazzinelli et al. 1996). In immunocompromised and congenitally infected individuals where these processes can be defective, latent infection can reactivate through the periodic rupture of cysts and subsequent proliferation of tachyzoites. Although pathology can occur in many organs, reactivation of infection is most often observed as toxoplasmic encephalitis

←───

Fig. 9.1 (continued) surface of placental cells inhibit the cytotoxic effects of CD8+ T cells and NK cells. $\gamma\delta$ T cells, also present in the placenta bind nonclassical HLA molecules and are likely to suppress the maternal antifetal immune response. Progesterone promotes the production of by the trophoblast in early pregnancy, which suppresses T cell proliferation in the placenta. Placental IDO (pIDO), secreted from the villous cytotrophoblast is important in the downregulation of CD4+ cell proliferation. Dendritic cells also produce IDO through stimulation of negative costimulatory molecules, such as CTLA-4 and PDL1 found on Tregs and the placental surface. The hormonal environment also affects B cells. In the placenta, B cells produce asymmetric antibodies that are not able to carry out their classical effector functions. Hofbauer cells (H) produce prostaglandins and tromboxanes and express VEGF, which stabilizes the placenta. NK cells produce IL-10 and GM-CSF on recognition of nonclassical HLA molecules. Release of soluble mediators is shown by *dashed lines; arrow heads* denote positive effects and *closed circle* denotes negative effects; *solid lines* illustrate effects

(Hunter and Remington 1994) or ocular toxoplasmosis (Garweg et al. 2005b; Jones et al. 2006).

Reactivation during pregnancy has been observed in people with immunological lesions, and in some cases has resulted in exogenous transplacental transmission. For example, reactivation of *T. gondii* infection in a pregnant woman with systemic lupus erythematosus, resulting in congenital transmission of the parasite to the fetus has been described (D'Ercole et al. 1995). Two separate cases of congenital *T. gondii* from HIV-infected women following disease reactivation have been reported (Bachmeyer et al. 2006; Bongain et al. 2002). Most of the studies suggest that reactivation in immunosuppressed individuals, however, is relatively rare (Anonymous 1996; Biedermann et al. 1995).

The extent of reactivation of *T. gondii* in immunocompetent people during pregnancy is not known. Unless it results in clinical disease or congenital transmission, it will be unnoticed; increasing numbers of reports, however, suggest that it is more common than previously acknowledged. An early study described three cases of congenital toxoplasmosis and disease reactivation in chronically infected women (Awan 1978), with a number of similar cases reported since (e.g. Vogel et al. 1996; Silveira et al. 2003; Kodjikian et al. 2004). An additional study examined 18 immunocompetent females with previous episodes of ocular toxoplasmosis over 35 pregnancies and found that reactivation of ocular disease occurred in seven pregnancies (Garweg et al. 2005a).

T. gondii has been found in the placenta of humans and in a variety of nonhuman animals (reviewed, Tenter et al. 2000). Often no pathological lesions are found in human placenta following toxoplasmosis, but necrotizing inflammation or, in rare cases, granulomatous villitis have been observed (Yavuz et al. 2006). The hypothesis that abortion during *T. gondii* infection may be immunologically mediated is gaining momentum. *T. gondii*-derived molecules, including GPI-anchors, HSP70 and profilin have been demonstrated to interact with TLR2, TLR4, and TLR11, respectively (Debierre-Grockiego et al. 2007; Aosai et al. 2006). These interactions may be sufficient to induce abortion in a similar way as demonstrated for LPS and Poly I:C (Gendron and Baines 1988; Gendron et al. 1990a; Gendron et al. 1990b; Lin et al. 2006; Zhang et al. 2007).

The development of a murine model (Roberts and Alexander 1992; Menzies et al. 2008) that mimics many aspects of the human disease has provided a tool to understand the pathogenesis of congenital disease. Similar to in humans, the outcome of murine congenital transmission is determined by the trimester at which infection first occurs. In humans, congenital transmission is most common in those infected in the third trimester, intermediate in the second trimester and low in the first (reviewed, Menzies et al. 2008). Abortion is most likely to occur during infections that occur in the first trimester. A similar pattern is observed in BALB/c mice with infection during the first trimester causing resorption or abortion, whereas infection in the second trimester being more likely to result in congenital transmission (reviewed, Menzies et al. 2008). Infection of BALB/c mice during the third trimester, as observed in humans, is unlikely to result in abortion and, unlike in humans, does not result in congenital transmission (a likely limitation of the relatively short 21-day gestation in mice) (reviewed, Menzies et al. 2008). These observations in both mice and humans support the hypothesis that immunological disruption of pregnancy is relatively simple, as hormonal-induced immuno-modualtion is not fully developed; conversely, abortion in the third trimester is difficult as hormonal immunomodulation is well established. In addition, in the second and third trimesters, the immunological mechanisms are in place, namely Th2 bias with high levels of IL-10-producing Treg cells, therefore, facilitating parasite multiplication and congenital transmission. *T. gondii*-induced abortion is reduced in IFN-γ-deficient mice, despite harboring increased parasite numbers in the uterus and the placentas of the fetuses when compared with mice (Shiono et al. 2007). This supports the hypothesis that a failure to limit *T. gondii* replication is not responsible for abortion. In fact, IFN-γ, a mediator capable of controlling *T. gondii* replication, has adverse effects on pregnancy. It has recently been demonstrated that *T. gondii* can also modulate IFN-γ signaling through induction of suppressor of cytokine signaling (SOCS) proteins (Zimmermann et al. 2006). The role of SOCS during pregnancy remains to be determined; however, it might act to reduce IFN-γ-induced abortion and killing of the parasite and thus promote fetal survival at the cost of congenital transmission.

Pregnancy has the ability to affect maternal immunity to *T. gondii* in the murine model as pregnant mice produce less IFN-γ accompanied with increased mortality compared with nonpregnant females infected with *T. gondii* (Shirahata et al. 1992). This can be reversed by administration of IFN-γ or IL-2 (Shirahata et al. 1993). The skewing towards production of Th2 cytokines during pregnancy favors parasite survival and the transmission of this parasite to the fetus.

The closely related apicomplexan parasite, *Neospora caninum*, a causative agent of abortion in cattle may share similar interactions with the host immune system during pregnancy as described for *T. gondii* (Quinn et al. 2004; Williams and Trees 2006). The immune response to *N. caninum* is similar to that of *T. gondii*, with BALB/c mice serving as a suitable model for studies of congenital infection (Long and Baszler 1996). As with *T. gondii*, the stage of gestation at which infection occurs determines the outcome. Mice and cattle infected before pregnancy do not transmit the parasite to the fetus, whereas infection during early gestation leads to abortion and infection during late gestation leads to transmission of the parasite (Quinn et al. 2004; Williams and Trees 2006). Studies examining the cytokine profiles of pregnant BALB/c mice infected with *N. caninum* show the balance between IFN-γ and IL-4 to be crucial in determining transmission to the fetus. IFN-γ production prevents the parasite crossing the maternal–fetal barrier whereas IL-4 creates an environment conducive to congenital transmission (Long and Baszler 2000; Nishikawa et al. 2003; Quinn et al. 2004; Kano et al. 2005). Subsequent studies in cattle have shown that abortion through infection in early pregnancy is due to IFN-γ production from NK cells, CD4[+], and γδ T cells, leading to destruction of placental tissues (Maley et al. 2006; Klevar et al. 2007).

9.4.2.2 *Trypanosoma cruzi*

Trypanosoma cruzi is the causative agent of Chagas disease, affecting approximately 13 million people across 15 different countries (WHO 2002). This parasite typically causes cardiac problems as well as megaesophagus and megacolon; pathology, however, only appears upon reactivation after a period of clinical latency. Immunity is dependent on both innate and adaptive responses. The innate immune system is stimulated through parasite GPI-anchors and DNA engaging TLR2 and TLR9, respectively (Tarleton 2007; Bafica et al. 2006). Th1 cells and IFN-γ play an important role in controlling parasite multiplication in the early stages. CD8$^+$ T cells play a dichotomous role and have been implicated in protection as well as immunopathology (reviewed in Tarleton 1991; Tarleton 2007).

T. cruzi has been reported to cause preterm births, still birth, and abortion, but the extent of the problem in all likelihood has not been fully recorded. Numerous case reports and some large clinical studies demonstrate that congenital transmission of *T. cruzi* can occur in humans. The reported incidence varies in a number of studies that examined many geographical regions at different times (e.g., Bolivia: 2.64–6% (Billot et al. 2005; Salas et al. 2007) and Argentina: 2.64% (Streiger et al. 1995)). In an Argentinian study, congenital transmission was clustered in families and second-generation transmission from grandmother, through mother to daughter was observed (Sanchez Negrette et al. 2005).

There is little data concerning the effect of *T. cruzi* infection on the severity of maternal infection. Generally, only a relatively small proportion of pregnant mothers transmit *T. cruzi* transplacentally. In one study, mothers who transmitted parasites to the fetus were found to have increased parasitemia and depressed IFN-γ, but similar IL-2, IL-4, and TGF-β production by peripheral blood mononuclear cells stimulated with *T. cruzi* antigens (Hermann et al. 2004). In a separate study, mothers who transmitted *T. cruzi* to their fetus had lower levels of circulating TNF-α than those who did not (Garcia et al. 2008). Although it has been suggested that parasite strain might also be a determinant of the likelihood of congenital transmission, most literature does not support this notion (e.g., Burgos et al. 2007).

T. cruzi is capable of infecting the placenta and of inducing inflammatory changes therein. Studies suggest that *T. cruzi* achieves invasion through binding placental alkaline phospatase, a GPI-anchored protein on the surface of the placenta (Lin et al. 2005). A study of placentas from women with *T. cruzi* infection that resulted in stillbirth or congenital transmission found that the inflammatory infiltrate was mainly CD68$^+$ macrophages, T cells, and a few NK cells (Altemani et al. 2000). In the still birth cases examined, villitis was diffuse and parasites were numerous. In these placentas, trophoblastic necrosis was observed in proximity to macrophages and granulocytes. The placentas from live births had few parasites present (Altemani et al. 2000). Intercellular adhesion molecule has been demonstrated to be upregulated in placentas of infected mothers and may facilitate the inflammatory process (Juliano et al. 2006).

Early studies that examined chronically infected mice, found that infection of the placenta is *T. cruzi* strain dependent, although the ability to cause congenital transmission depended on suppression of phagocyte activity (Andrade 1982; Delgado and Santos-Buch 1978). Chronic *T. cruzi* infection also affected fetal weight and size, but not the reproductive capacity of mice. Maternal parasitemia remained similar to nonpregnant control mice and parasites were absent from fetal blood (Carlier et al. 1987a). In spite of this, pregnancy reduced *T. cruzi*-specific IgG and IgM levels. The greater reduction of IgG2a levels in pregnant mice relative to nonpregnant mice, although originally attributed to a reduction in polyclonal B cell activation, may be interpreted as a reduced Th1 response (Carlier et al. 1987b). Pregnancy increases circulating TNF-α in *T. cruzi*-infected mice (Rivera et al. 1995).

Acute infection with *T. cruzi* has a more obvious effect in mice than in humans and it can induce infertility, placental parasite invasion, and fetal loss. In infected dams, approximately 28% of fetuses were resorbed and the remainder died during gestation or following birth. Notably, parasites were absent from the fetal tissue, but tissues exhibited ischemic necrosis (Mjihdi et al. 2002). A further study demonstrated that *T. cruzi* infection during pregnancy increased circulating TNF-α and IFN-γ and induced placental IFN-γ. Administration of pentoxifylline, an inhibitor of TNF-α and IFN-γ, reduced fetal loss, implicating these mediators in the process (Mjihdi et al. 2004). Reduced fertility during acute infection is due to inhibition of preimplantation embryo development (Id Boufker et al. 2006). Whether the effect of acute infection on pregnancy is dependent on *T. cruzi* strain or the magnitude of the host immune responses as a result of differences in parasite replication requires clarification (Solana et al. 2002). The ability of *T. cruzi* to induce abortion in an IFN-γ and TNF-α dependent manner is consistent with TLR-mediated inflammatory responses, which may be caused by the stimulation of TLR2 and TLR9 by the parasite (Tarleton 2007; Bafica et al. 2006).

In spite of the general inability of *T. cruzi* to cause congenital infection in mice, pups born to infected dams have a reduced ability to control infection later in life. This indicates that maternal infection may affect the immunological response of the progeny to *T. cruzi* infection and has important implications if similar observations are found in humans (Marques de Araujo et al. 1996; Carlier et al. 1992; Carlier et al. 1992). Congenital infection has been reported in acutely infected Wistar rats, which may provide a useful model to study this phenomenon in humans (Moreno et al. 2003).

9.4.2.3 *Leishmania* spp.

Leishmania major is a protozoan parasite that exists in two forms; the intracellular amastigote and the promastigote within the digestive tract of the host, the phlebotamine sandfly. *Leishmania* spp. are the quintessential examples of a pathogen that drives development of Th1 responses (Alexander et al. 1999). Healing leishmaniasis involves the production of IL-12 from infected macrophages, which stimulates NK and CD4$^+$ T cells to produce IFN-γ, and IFN-γ induces TNF-α and NO, which are microbicidal for the parasite. Th2 immunity against *L. major* induces a nonhealing phenotype that is independent of IL-4 (reviewed in Alexander and Bryson 2005). Other species include *L. mexicana* that can cause cutaneous or visceral disease, *L. donovani*, *L. infantum*, and *L. chagasi* which causes visceral leishmaniasis and *L. brasiliensis* that causes mucocutaneous disease.

The extent to which pregnancy affects the progression of disease in females or the ability of infection to affect pregnancy in humans requires further clarification. There is, however, a report of a woman with visceral leishmaniasis (kala-azar) that transmitted parasites to her first child and aborted her second at 5 months of gestation, with *Leishmania* organisms present in the placenta (Eltoum et al. 1992). There have been a number of other cases reported where transplacental transmission of kala-azar has resulted in congenital infection of humans (reviewed, Figueiro-Filho et al. 2004). These cases were largely confirmed as congenital infection due to the relocation of the pregnant woman to a nonendemic geographic location, so the true extent of congenital transmission in endemic areas is not known. As yet there is no large number of studies on pregnant women in endemic areas to determine the rate of congenital transmission or fetal loss due to Leishamaniasis. *L. major*, *L. braziliensis*, and *L. infantum* induce TLR9 signaling and IL-12 production by DCs and NK cell activation (Liese et al. 2008). The role of TLR2 and TLR9 signaling in disruption of pregnancy, as described for TLR3 and TLR4, remains to be determined.

Studies in C57BL/6 mice, which are typically resistant to *Leshmania* infection, have shown that the host–parasite relationship during pregnancy is complex. Krishnan et al. (1996a) demonstrated a mechanism by which pregnancy can alter the immune control of the parasite. In comparison with nonpregnant mice, infected pregnant mice displayed a greater parasite burden associated with decreased IFN-γ and elevated levels of the Th2 cytokines IL-4, IL-5, and IL-10 (Krishnan et al. 1996a). Concurrently, the host immune response against *L. major* adversely affected the pregnancy. Krishnan et al. (1996b) demonstrated low placental levels of IL-4 and IL-10 and high IFN-γ and TNF-α levels as a consequence of *L. major* infection, which can prevent successful pregnancy by causing implantation failure and fetal resorption (Krishnan et al. 1996b). These studies were performed before the characterization of Treg cells, which are now known to produce IL-10 and to be present at the maternal–fetal interface during pregnancy (see above). Future studies should establish whether *L. major* alters Treg activity in the area of the placenta. Finally, the relevance of these studies to human disease as well as the role that hormones play in modulating immune responses to *Leishmania* spp. during pregnancy remains to be clarified.

9.4.2.4 *Plasmodium* spp.

Plasmodium falciparum, P. vivax, P. ovale, and *P. malariae* are the four species that commonly cause malaria in humans. *P. falciparum* accounts for 80% of malarial infections, a disease affecting approximately 515 million people. *P. vivax* infects 130–435 million of the 2.6 billion people living in malaria endemic areas (reviewed in Baird 2007).

P. falciparum is particularly a problem during pregnancy (reviewed, Menendez and Mayor 2007). The incidence of maternal malaria in Africa is estimated to be around 25% (reviewed in Desai et al. 2007). In areas of low transmission, infection in pregnant women results in severe disease, whereas maternal disease is less common in pregnant women who live in areas of high transmission (Menendez 1995). Infected erythrocytes can accumulate in the blood spaces of the placenta and avoid maternal immune responses (Salanti et al. 2004). The accumulation of the parasitized red blood cells (RBCs) in the placenta is achieved through their adhesion to molecules such as chondroitin sulfate A (CSA) and hyalaronic acid (HA) receptors, both of which are expressed on the surface of placental endothelial cells (reviewed, Schofield and Grau 2005). Chemokines mediate the infiltration of a range of cells into the placenta, including macrophages, which release proinflammatory cytokines (Beeson and Duffy 2005). This is detrimental to the development of the fetus and can affect those that survive through childhood to adult. Infection during the first and second trimester most likely results in abortion, while infection in the third results in stillbirth, pre-term birth, and low birth weight (Beeson and Duffy 2005). Infant mortality is increased in offspring of infected mothers and those who survive have increased incidence of developmental and behavioral disorders (reviewed, Desai et al. 2007). *P. vivax* does not cause RBCs to adhere to the placenta, but infection during pregnancy, nonetheless, is responsible for low birth weights (reviewed in Desai et al. 2007). In common with *P. falciparum, P. vivax* infection is more severe in pregnant women compared with nonpregnant women. Data is lacking on the effects of *P. malariae* and *P. ovale* during pregnancy (reviewed, Desai et al. 2007).

The contribution of immune disturbance in *Plasmodium*-induced abortion has been subject to discussion. In humans, polymorphisms in TLR4 and TLR9, which are activated by products of the asexual blood stages, are associated with low birth weight (Mockenhaupt et al. 2006). This would indicate that innate immune mechanisms can have a detrimental effect on pregnancy. The contribution of these TLR-mediated mechanisms to abortion, most prevalent in the first stages of pregnancy, remains to be determined.

Infection induces a strong Th1 response, which has been noted to induce abortion and increase implantation failure (reviewed in Quinn et al. 2002; Fried et al. 1998). In the case of *P. falciparum* infection, parasite sequestration in the placenta has been attributed to placental pathology and fetal loss in the murine model and in humans (van Zon and Eling 1980a; van Zon and Eling 1980b; Sholapurkar et al. 1988; Nair and Nair 1993; Singh et al. 1995; Poovassery and Moore 2006). Nonetheless, Th2-biased immunity has been implicated in abortion in mice, and changes in placental Th2 cytokine production have been associated with poor pregnancy outcomes in humans (Fried et al. 1998; Kabyemela et al. 2008). Production of high levels of TNF-α has been found in the placentas of babies with low birth weight born to infected mothers (Fried et al. 1998; Moormann et al. 1999; Rogerson et al. 2003; Rogerson et al. 2007). Conversely, polymorphisms in the IL-10 promoter that results in increased production are more common in women with pre-term delivery, but this would appear to be linked with increased anemia rather than altered Th1/Th2 balance in the placenta (Ohashi et al. 2002).

Congenital transmission of *Plasmodium* was once thought to be rare event but is now known to be common even in endemic areas. Prevalence of up to 33% has been reported for congenital transmission (reviewed in Menendez 2006). Congenital infection is known to occur in both *P. vivax* and *P. falciparum* infections and in study of 27 cases in Thailand, *P. vivax* was found to be responsible for 81.5% of cases (Wiwanitkit 2006). More studies are required to determine the extent of the problem for all four species of *Plasmodium* that infect humans.

Malaria is not only a risk to the fetus during pregnancy, but is also of increased severity in the expectant mother. A number of mechanisms (many discussed in the context of other infections in this chapter) alone or in combination could account for the observed decrease in maternal immunity. However, two conflicting hypotheses prevail, the "cortisol hypothesis" and the "prolactin hypothesis"(reviewed, Mavoungou 2006). The cortisol hypothesis is based on the observation that cortisol levels are raised during pregnancy and can cause immunosupression. In support of this hypothesis, cortisol levels are highest in primigravid women who are also most susceptible to malaria (reviewed, Bouyou-Akotet et al 2004; Mavoungou 2006 and Pearson 2004). It has been proposed that high levels of cortisol in primagravid women are responsible for reduced NK cell NKp30 surface expression and reduced NK cytotoxicity (reviewed in Mavoungou 2006). Although more recent studies have found a positive correlation with cortisol levels and parasite numbers in pregnant women, these studies have failed to demonstrate a difference in cortisol levels between primagravid and multigravid women (Adam et al. 2007). The prolactin hypothesis is based on the observation that levels of this immunostimulatory peptide hormone are lower in primagravid than in multigravid women. This would suggest that protection is mediated by prolactin rather than susceptibility being mediated by cortisol. In support of this, prolactin has been demonstrated to augment NK cell function through augmentation of NKp30 and NKp46 (Mavoungou et al. 2005). Both these hypotheses are not mutually exclusive, and in reality, it is unlikely that any single one of the plethora of immunological parameters known to be affected by pregnancy is solely responsible for the increased susceptibility of women to malaria during pregnancy.

Plasmodium berghei (ANKA strain) has been used in a murine model of placental malaria. It has a number of similarities with the human disease and adherence of parasitized RBCs is also dependent on CSA and HA. In this model, placental pathology is associated with decreased fetal viability, intrauterine growth retardation, placental pathology, reduced fetal weight, and reduced growth in the progeny. In addition, maternal malaria is more severe in pregnant mice compared with nonpregnant mice (Neres et al. 2008).

9.4.3   Metazoan Infections

9.4.3.1   Schistosomes

The three species of Schistosomes associated with humans, include *Schistosoma mansoni*, *S. japonicum*, and *S. haematobium*, which currently infect in excess of

200 million people, of which approximately 40 million are women of child-bearing age. Infection of nonpregnant women results in nutritional, hematologic, and cognitive deficits and substantial mortality (reviewed, Friedman et al. 2007), and pregnancy is likely to exacerbate both the nutritional and hematological deficits. Murine studies support the hypothesis that schistosomiasis can have an adverse effect on maternal health with moderately infected mice having experienced reduced longevity following pregnancy compared with similarly infected nonparid mothers. This reinforces the need for more human studies.

Although several studies have shown congenital infection to be possible with *S. japonicum* in a variety of hosts (Willingham et al. 1999; Qian et al. 2000; Iburg et al. 2002; Johansen et al. 2002), other members of this parasite family have not yet been shown to cause congenital disease. Infection of pigs with *S. japonicum* in mid to late pregnancy results in congenital transmission, as detected by the presence of eggs in the liver and feces of the piglets (Willingham et al. 1999). The immunological basis for this congenital disease requires more investigation. Schistosomes typically induce Th2 responses in the host, by downregulating Th1 and increasing the production of cytokines, such as IL-4 and IL-5 (Pearce et al. 1991). This facilitates their survival in the host (Farah et al. 2007). Therefore, it could be suggested that the Th2 environment created during pregnancy, most probably as a result of progesterone, will facilitate survival and transmission of the parasite.

9.4.3.2 Filariasis

Wuchereria bancrofti, Brugia malayi, and *B. timori* are the causative agents of lymphatic filariasis and *Onchocerca volvulus* is the causative agent of onchocerciasis in endemic countries, predominately in Africa and the Indian subcontinent. Disease is characterized by the presence of adult worms that infect various tissues, the females of which release numerous microfilaria over a period of years. Congenital transmission is documented and is presumed to be caused by microfilaria transplacental migration into the fetal circulation where they establish infection (Fonticiella et al. 1995; Eberhard et al. 1993; Ufomadu et al. 1990; Anosike and Onwuliri 1993; Brinkmann et al. 1976).

9.4.3.3 Intestinal Helminths

Intestinal helminth infections (caused by *Ascaris lumbricoides, Trichuris trichiura, Necator americanus, Enterobius vermicularis* and *Strongyloides stercoralis*) are common in many countries throughout the world and have adverse effects on pregnancy due to their ability to increase anemia (Rodriguez-Morales et al. 2006). *A. lumbricoides,* the most prevalent parasite of humans, is normally relatively benign, but in pregnant women is associated with biliary ascariasis (Shah et al. 2006). Although host immunity may be the cause of biliary ascariasis during pregnancy, the relaxant effects of progesterone and estrogen on the sphincter of oddi (the muscular valve surrounding the exit of the bile and pancreatic duct into the duodenum) is likely to be responsible for the condition. There are some reported cases where *A. lumricoides* infection has been associated with abortion in humans (Deveci et al. 2001; Shah et al. 2006) and studies in mice have demonstrated that ascaris extracts can induce intrauterine death, poor birth weights, and ill health in surviving progeny (Blaszkowska 2000; Blaszkowska 2003). *Trichinella spiralis* also has been demonstrated to be capable of transplacental transmission (Dubinsky et al. 2001).

The extent to which metazoan parasites can exploit pregnancy is best illustrated in veterinary parasites. The nematode parasite *Toxocara canis*, a common round-worm of dogs, has the ability to reactivate during pregnancy, particularly in the last trimester (Burke and Roberson 1985). Reactivation of *T. canis* larval stages in pregnant dogs is associated with decreased IFN-γ, but increased IL-10 production (Torina et al. 2005). Reactivation often results in neurological and ophthalmolo-gical lesions in pregnant bitches (Hill et al. 1985; (Molk 1983). The larvae of *Ancylostoma caninum*, another nematode parasite of dogs, are capable of arrested development. During pregnancy, TFG-β is increased by an estrogen- and prolactin-dependent manner and is responsible for disease reactivation and completion of the development of the arrested larvae (Arasu 2001). The persistence of these nema-todes in their hosts as relatively quiescent forms until pregnancy provides the perfect opportunity for parasitic infections to become established once again within the host, and in doing so facilitates both congenital and postpartum transmission.

9.5   Conclusions and Future Directions

Pregnancy relies on immunological processes at the maternal–fetal interface for success. However, some of these processes operate systemically. As these process-es are not yet entirely understood in healthy individuals; understanding the effect of infectious agents on pregnancy and the effect of pregnancy on development of infectious diseases presents challenges. Studying the bidirectional interaction of pregnancy with infection offers opportunities to understand how the immune response contributes to healthy pregnancy. Identification of parasite-derived mole-cules that induce immunological processes that interfere with pregnancy may ultimately aid our understanding of pregnancy. To some extent, this has already proven fruitful as pathogen-derived LPS is now known to induce TNF-α and NOS2 expression which can disrupt pregnancy (discussed above). Successful pregnancy is the greatest single evolutionary pressure on humans and, as such, has evolved to generally withstand the maternal infections. The robustness of pregnancy may actually be exploited by certain pathogens that have evolved mechanisms to infect the fetus in utero.

Pregnancy dramatically alters the hypothalamus–pituitary–adrenal (HPA) axis, resulting in increased levels of cortisol (reviewed, Lindsay and Nieman 2005). Similarly, inflammation, a common symptom of many infections, stimulates the HPA-axis, which results in the production of cortisol. The ability of parasitic infections to induce changes in pregnancy-associated hormones is not known and is an obvious future avenue of research. In addition, further studies should examine the interactions of the immune response induced by infection with pregnancy and the HPA-axis. Studies now suggest that exposure to cortisol due to stress during pregnancy can affect the fetal development and fetal HPA-axis (e.g., reviewed in Weinstock 2008). Future studies should assess the potential effects of infection on the development of the fetal immune system and potential long-term behavioral and psychological abnormalities in childhood and/or adulthood.

Pregnancy presents significant challenges to the treatment of diseases, and currently, many antiparasitic drugs employed are far from ideal and are known to pose threats to the fetus. There is a need to balance the likely benefit to the mother and fetus with the possible detrimental effects of treatment on the fetus. In the future, immunotherapies might be possible to eliminate parasites or to prevent abortion. These would, however, also seem to pose a significant risk. Deliberate alteration of the immune response may control parasites while having adverse consequences for pregnancy or, conversely, may stabilize pregnancy, but promote parasite growth and therefore congenital transmission. In this respect, vaccines given prior to pregnancy would be the ideal, albeit challenging solution.

References

Adam I, Nour BY, Almahi WA, Omer ES, Ali NI (2007) Malaria susceptibility and cortisol levels in pregnant women of eastern Sudan. Int J Gynaecol Obstet 98:260–261

Anonymous (1996) Low incidence of congenital toxoplasmosis in children born to women infected with human immunodeficiency virus. European collaborative study and research network on congenital toxoplasmosis. Eur J Obstet Gynecol Reprod Biol 68:93–96

Al-Shammri S, Rawoot P, Azizieh F, AbuQoora A, Hanna M, Saminathan TR, Raghupathy R (2004) Th1/Th2 cytokine patterns and clinical profiles during and after pregnancy in women with multiple sclerosis. J Neurol Sci 222:21–27

Alexander J, Bryson K (2005) T helper (h)1/Th2 and *Leishmania*: paradox rather than paradigm. Immunol Lett 99:17–23

Alexander J, Satoskar AR, Russell DG (1999) *Leishmania* species: models of intracellular parasitism. J Cell Sci 112(18):2993–3002

Alexander J, Jebbari H, Bluethmann H, Brombacher F, Roberts CW (1998) The role of IL-4 in adult acquired and congenital toxoplasmosis. Int J Parasitol 28:113–120

Aliberti J, Valenzuela JG, Carruthers VB, Hieny S, Andersen J, Charest H, Reise Sousa C, Fairlamb A, Ribeiro JM, Sher A (2003) Molecular mimicry of a CCR5 binding-domain in the microbial activation of dendritic cells. Nat Immunol 4:485–490

Altemani AM, Bittencourt AL, Lana AM (2000) Immunohistochemical characterization of the inflammatory infiltrate in placental Chagas' disease: a qualitative and quantitative analysis. Am J Trop Med Hyg 62:319–324

Aluvihare VR, Betz AG (2006) The role of regulatory T cells in alloantigen tolerance. Immunol Rev 212:330–343

Aluvihare VR, Kallikourdis M, Betz AG (2004) Regulatory T cells mediate maternal tolerance to the fetus. Nat Immunol 5:266–271

Anderson JM (1965) Immunological inertia in pregnancy. Nature 206:786–787

Andrade SG (1982) The influence of the strain of *Trypanosoma cruzi* in placental infections in mice. Trans R Soc Trop Med Hyg 76:123–128

Anosike JC, Onwuliri CO (1993) A probable case of vertical transmission of *Onchocerca volvulus* microfilariae. J Helminthol 67:83–84

Aosai F, Rodriguez Pena MS, Mun HS, Fang H, Mitsunaga T, Norose K, Kang HK, Bae YS, Yano A (2006) *Toxoplasma gondii*-derived heat shock protein 70 stimulates maturation of murine bone marrow-derived dendritic cells via Toll-like receptor 4. Cell Stress Chaperones 11:13–22

Arasu P (2001) In vitro reactivation of *Ancylostoma caninum* tissue-arrested third-stage larvae by transforming growth factor-beta. J Parasitol 87:733–738

Askelund K, Liddell HS, Zanderigo AM, Fernando NS, Khong TY, Stone PR, Chamley LW (2004) CD83(+) dendritic cells in the decidua of women with recurrent miscarriage and normal pregnancy. Placenta 25:140–145

Awan KJ (1978) Congenital toxoplasmosis: chances of occurrence in subsequent siblings. Ann Ophthalmol 10:459–465

Bachmeyer C, Mouchnino G, Thulliez P, Blum L (2006) Congenital toxoplasmosis from an HIV-infected woman as a result of reactivation. J Infect 52:e55–e57

Bafica A, Santiago HC, Goldszmid R, Ropert C, Gazzinelli RT, Sher A (2006) Cutting edge: TLR9 and TLR2 signaling together account for MyD88-dependent control of parasitemia in *Trypanosoma cruzi* infection. J Immunol 177:3515–3519

Baird JK (2007) Neglect of Plasmodium vivax malaria. Trends Parasitol 23:533–539

Barakonyi A, Kovacs KT, Miko E, Szereday L, Varga P, Szekeres-Bartho J (2002) Recognition of nonclassical HLA class I antigens by gamma delta T cells during pregnancy. J Immunol 168:2683–2688

Bouyou-Akotet MK, Issifou S, Meye JF, Kombila M, Ngou-Milama E, Luty AJ, Kremsner PG, Mavoungou E (2004) Depressed natural killer cell cytotoxicity against Plasmodium falciparum-infected erythrocytes during first pregnancies. Clin Infect Dis 38:342–347

Beeson JG, Duffy PE (2005) The immunology and pathogenesis of malaria during pregnancy. Curr Top Microbiol Immunol 297:187–227

Benchimol M, de Andrade Rosa I, da Silva Fontes R, Burla Dias AJ (2008) Trichomonas adhere and phagocytose sperm cells: adhesion seems to be a prominent stage during interaction. Parasitol Res 102:597–604

Biedermann K, Flepp M, Fierz W, Joller-Jemelka H, Kleihues P (1995) Pregnancy, immunosuppression and reactivation of latent toxoplasmosis. J Perinat Med 23 191–203

Billot C, Torrico F, Carlier Y (2005) Cost effectiveness study of a control program of congenital Chagas disease in Bolivia. Rev Soc Bras Med Trop 38(Suppl 2):108–113

Blaszkowska J (2000) Disturbances of mouse pregnancy after injection of *Ascaris* homogenate during early organogenesis. Wiad Parazytol 46:369–378

Blaszkowska J (2003) Preliminary evaluation of maternotoxic effect of *Ascaris* extract in mice. Wiad Parazytol 49:187–194

Blois SM, Alba Soto CD, Tometten M, Klapp BF, Margni RA, Arck PC (2004a) Lineage, maturity, and phenotype of uterine murine dendritic cells throughout gestation indicate a protective role in maintaining pregnancy. Biol Reprod 70:1018–1023

Blois SM, Joachim R, Kandil J, Margni R, Tometten M, Klapp BF, Arck PC (2004b) Depletion of CD8+ cells abolishes the pregnancy protective effect of progesterone substitution with dydrogesterone in mice by altering the Th1/Th2 cytokine profile. J Immunol 172:5893–5899

BonDurant RH (1997) Pathogenesis, diagnosis, and management of trichomoniasis in cattle. Vet Clin North Am Food Anim Pract 13:345–361

BonDurant RH, Campero CM, Anderson ML, Van Hoosear KA (2003) Detection of *Tritrichomonas foetus* by polymerase chain reaction in cultured isolates, cervicovaginal mucus, and formalin-fixed tissues from infected heifers and fetuses. J Vet Diagn Invest 15:579–584

Bongain A, Berrebi A, Marine-Barjoan E, Dunais B, Thene M, Pradier C, Gillet JY (2002) Changing trends in pregnancy outcome among HIV-infected women between 1985 and 1997 in two southern French university hospitals. Eur J Obstet Gynecol Reprod Biol 104:124–128

Bonney EA (2001) Maternal tolerance is not critically dependent on interleukin-4. Immunology 103:382–389

Brinkmann UK, Kramer P, Presthus GT, Sawadogo B (1976) Transmission in utero of microfilariae of Onchocerca volvulus. Bull World Health Organ 54:708–709

Burgos JM, Altcheh J, Bisio M, Duffy T, Valadares HM, Seidenstein ME, Piccinali R, Freitas JM, Levin MJ, Macchi L, Macedo AM, Freilij H, Schijman AG (2007) Direct molecular profiling of minicircle signatures and lineages of Trypanosoma cruzi bloodstream populations causing congenital Chagas disease. Int J Parasitol 37:1319–1327

Burke TM, Roberson EL (1985) Prenatal and lactational transmission of Toxocara canis and Ancylostoma caninum: experimental infection of the bitch before pregnancy. Int J Parasitol 15:71–75

Carbone E, Terrazzano G, Ruggiero G, Zanzi D, Ottaiano A, Manzo C, Karre K, Zappacosta S (1999) Recognition of autologous dendritic cells by human NK cells. Eur J Immunol 29:4022–4029

Carlier Y, Rivera MT, Truyens C, Puissant F, Milaire J (1987a) Interactions between chronic murine Trypanosoma cruzi infection and pregnancy: fetal growth retardation. Am J Trop Med Hyg 37:534–540

Carlier Y, Rivera MT, Truyens C, Ontivero M, Flament J, Van Marck E, de Maertelaer V (1992) Chagas' disease: decreased resistance to Trypanosoma cruzi acquired infection in offspring of infected mice. Am J Trop Med Hyg 46:116–122

Carlier Y, Rivera MT, Truyens C, Goldman M, Lambert P, Flament J, Bauwens D, Vray B (1987b) Pregnancy and humoral immune response in mice chronically infected by Trypanosoma cruzi. Infect Immun 55:2496–2501

Castellucci M, Kaufmann P (1982) A three-dimensional study of the normal human placental villous core: II. Stromal architecture. Placenta 3:269–285

Caucheteux SM, Gendron MC, Kanellopoulos-Langevin C (2005) Pregnancy-induced alterations of B cell maturation and survival are differentially affected by Fas and Bcl-2, independently of BcR expression. Int Immunol 17:55–63

Chen M, Aosai F, Norose K, Mun HS, Takeuchi O, Akira S, Yano A (2002) Involvement of MyD88 in host defense and the down-regulation of anti-heat shock protein 70 autoantibody formation by MyD88 in Toxoplasma gondii-infected mice. J Parasitol 88:1017–1019

Clark DE, Smith SK, Licence D, Evans AL, Charnock-Jones DS (1998) Comparison of expression patterns for placenta growth factor, vascular endothelial growth factor (VEGF), VEGF-B and VEGF-C in the human placenta throughout gestation. J Endocrinol 159:459–467

Combe CL, Curiel TJ, Moretto MM, Khan IA (2005) NK cells help to induce CD8(+)-T-cell immunity against Toxoplasma gondii in the absence of CD4(+) T cells. Infect Immun 73:4913–4921

Constantin CM, Masopust D, Gourley T, Grayson J, Strickland OL, Ahmed R, Bonney EA (2007) Normal establishment of virus-specific memory CD8 T cell pool following primary infection during pregnancy. J Immunol 179:4383–4389

Correale J, Arias M, Gilmore W (1998) Steroid hormone regulation of cytokine secretion by proteolipid protein-specific CD4+ T cell clones isolated from multiple sclerosis patients and normal control subjects. J Immunol 161:3365–3374

D'Ercole C, Boubli L, Franck J, Casta M, Harle JR, Chagnon C, Cravello L, Leclaire M, Blanc B (1995) Recurrent congenital toxoplasmosis in a woman with lupus erythematosus. Prenat Diagn 15:1171–1175

Debierre-Grockiego F, Campos MA, Azzouz N, Schmidt J, Bieker U, Resende MG, Mansur DS, Weingart R, Schmidt RR, Golenbock DT, Gazzinelli RT, Schwarz RT (2007) Activation of TLR2 and TLR4 by glycosylphosphatidylinositols derived from Toxoplasma gondii. J Immunol 179:1129–1137

Delgado MA, Santos-Buch CA (1978) Transplacental transmission and fetal parasitosis of Trypanosoma cruzi in outbred white Swiss mice. Am J Trop Med Hyg 27:1108–1115

Della Chiesa M, Vitale M, Carlomagno S, Ferlazzo G, Moretta L, Moretta A (2003) The natural killer cell-mediated killing of autologous dendritic cells is confined to a cell subset expressing CD94/NKG2A, but lacking inhibitory killer Ig-like receptors. Eur J Immunol 33:1657–1666

Denkers EY, Gazzinelli RT (1998) Regulation and function of T-cell-mediated immunity during *Toxoplasma gondii* infection. Clin Microbiol Rev 11:569–588

Denkers EY, Kim L, Butcher BA (2003) In the belly of the beast: subversion of macrophage proinflammatory signalling cascades during *Toxoplasma gondii* infection. Cell Microbiol 5:75–83

Desai M, ter Kuile FO, Nosten F, McGready R, Asamoa K, Brabin B, Newman RD (2007) Epidemiology and burden of malaria in pregnancy. Lancet Infect Dis 7:93–104

Deveci S, Tanyuksel M, Deveci G, Araz E (2001) Spontaneous missed abortion caused by *Ascaris lumbricoides*. Cent Eur J Public Health 9:188–189

Dietl J, Honig A, Kammerer U, Rieger L (2006) Natural killer cells and dendritic cells at the human feto-maternal interface: an effective cooperation? Placenta 27:341–347

Dubey JP (1998) Advances in the life cycle of *Toxoplasma gondii*. Int J Parasitol 28:1019–1024

Dubinsky P, Boor A, Kincekova J, Tomasovicova O, Reiterova K, Bielik P (2001) Congenital trichinellosis? Case report. Parasite 8:S180–S182

Eberhard ML, Hitch WL, McNeeley DF, Lammie PJ (1993) Transplacental transmission of *Wuchereria bancrofti* in Haitian women. J Parasitol 79:62–66

Eblen AC, Gercel-Taylor C, Shields LB, Sanfilippo JS, Nakajima ST, Taylor DD (2000) Alterations in humoral immune responses associated with recurrent pregnancy loss. Fertil Steril 73:305–313

Eltoum IA, Zijlstra EE, Ali MS, Ghalib HW, Satti MM, Eltoum B, el-Hassan AM (1992) Congenital kala-azar and leishmaniasis in the placenta. Am J Trop Med Hyg 46:57–62

Farah IO, Langoi D, Nyaundi J, Hau J (2007) Schistosome-induced pathology is exacerbated and Th2 polarization is enhanced during pregnancy. In Vivo 21:599–602

Figueiro-Filho EA, Duarte G, El-Beitune P, Quintana SM, Maia TL (2004) Visceral leishmaniasis (kala-azar) and pregnancy. Infect Dis Obstet Gynecol 12:31–40

Fitzgerald PR (1986) Bovine trichomoniasis. Vet Clin North Am Food Anim Pract 2:277–282

Fonticiella M, Lopez-Negrete L, Prieto A, Garcia-Hernandez JB, Orense M, Fernandez-Diego J, Gomez JL (1995) Congenital intracranial filariasis: a case report. Pediatr Radiol 25:171–172

Fried M, Muga RO, Misore AO, Duffy PE (1998) Malaria elicits type 1 cytokines in the human placenta: IFN-gamma and TNF-alpha associated with pregnancy outcomes. J Immunol 160:2523–2530

Friedman JF, Mital P, Kanzaria HK, Olds GR, Kurtis JD (2007) Schistosomiasis and pregnancy. Trends Parasitol 23:159–164

Garcia MM, De Rissio AM, Villalonga X, Mengoni E, Cardoni RL (2008) Soluble tumor necrosis factor (TNF) receptors (sTNF-R1 and -R2) in pregnant women chronically infected with *Trypanosoma cruzi* and their children. Am J Trop Med Hyg 78:499–503

Garweg JG, Scherrer J, Wallon M, Kodjikian L, Peyron F (2005a) Reactivation of ocular toxoplasmosis during pregnancy. BJOG 112:241–242

Garweg JG, Ventura AC, Halberstadt M, Silveira C, Muccioli C, Belfort RJ, Jacquier P (2005b) Specific antibody levels in the aqueous humor and serum of two distinct populations of patients with ocular toxoplasmosis. Int J Med Microbiol 295:287–295

Gazzinelli R, Xu Y, Hieny S, Cheever A, Sher A (1992) Simultaneous depletion of CD4+ and CD8+ T lymphocytes is required to reactivate chronic infection with Toxoplasma gondii. J Immunol 149:175–180

Gazzinelli RT, Wysocka M, Hieny S, Scharton-Kersten T, Cheever A, Kuhn R, Muller W, Trinchieri G, Sher A (1996) In the absence of endogenous IL-10, mice acutely infected with *Toxoplasma gondii* succumb to a lethal immune response dependent on CD4+ T cells and accompanied by overproduction of IL-12, IFN-gamma and TNF-alpha. J Immunol 157:798–805

Gendron RL, Baines MG (1988) Infiltrating decidual natural killer cells are associated with spontaneous abortion in mice. Cell Immunol 113:261–267

Gendron RL, Farookhi R, Baines MG (1990a) Resorption of CBA/J x DBA/2 mouse conceptuses in CBA/J uteri correlates with failure of the feto-placental unit to suppress natural killer cell activity. J Reprod Fertil 89:277–284

Gendron RL, Nestel FP, Lapp WS, Baines MG (1990b) Lipopolysaccharide-induced fetal resorption in mice is associated with the intrauterine production of tumour necrosis factor-alpha. J Reprod Fertil 90:395–402

Giron-Gonzalez JA, Moral FJ, Elvira J, Garcia-Gil D, Guerrero F, Gavilan I, Escobar L (2000) Consistent production of a higher TH1:TH2 cytokine ratio by stimulated T cells in men compared with women. Eur J Endocrinol 143:31–36

Guleria I, Sayegh MH (2007) Maternal acceptance of the fetus: true human tolerance. J Immunol 178:3345–3351

Guleria I, Khosroshahi A, Ansari MJ, Habicht A, Azuma M, Yagita H, Noelle RJ, Coyle A, Mellor AL, Khoury SJ, Sayegh MH (2005) A critical role for the programmed death ligand 1 in fetomaternal tolerance. J Exp Med 202:231–237

Guntupalli SR, Steingrub J (2005) Hepatic disease and pregnancy: an overview of diagnosis and management. Crit Care Med 33:S332–S339

Habicht A, Dada S, Jurewicz M, Fife BT, Yagita H, Azuma M, Sayegh MH, Guleria I (2007) A link between PDL1 and T regulatory cells in fetomaternal tolerance. J Immunol 179:5211–5219

Hermann E, Truyens C, Alonso-Vega C, Rodriguez P, Berthe A, Torrico F, Carlier Y (2004) Congenital transmission of *Trypanosoma cruzi* is associated with maternal enhanced parasitemia and decreased production of interferon-gamma in response to parasite antigens. J Infect Dis 189:1274–1281

Hill IR, Denham DA, Scholtz CL (1985) *Toxocara canis* larvae in the brain of a British child. Trans R Soc Trop Med Hyg 79:351–354

Hindmarsh KW, Tan L, Sankaran K, Laxdal VA (1989) Diurnal rhythms of cortisol, ACTH, and beta-endorphin levels in neonates and adults. West J Med 151:153–156

Hunter CA, Remington JS (1994) Immunopathogenesis of toxoplasmic encephalitis. J Infect Dis 170:1057–1067

Hutter H, Hammer A, Blaschitz A, Hartmann M, Ebbesen P, Dohr G, Ziegler A, Uchanska-Ziegler B (1996) Expression of HLA class I molecules in human first trimester and term placenta trophoblast. Cell Tissue Res 286:439–447

Iburg T, Balemba OB, Danizer V, Leifsson PS, Johansen MV (2002) Pathogenesis of congenital infection with *Schistosoma japonicum* in pigs. J Parasitol 88:1021–1024

Id Boufker H, Alexandre H, Carlier Y, Truyens C (2006) Infertility in murine acute *Trypanosoma cruzi* infection is associated with inhibition of pre-implantation embryo development. Am J Pathol 169:1730–1738

Johansen MV, Iburg T, Morad J, Ornbjerg N (2002) Congenital infection with *Schistosoma japonicum* but not with *Schistosoma bovis* in sheep. J Parasitol 88:414–415

Jones BJ, Brooke G, Atkinson K, McTaggart SJ (2007) Immunosuppression by placental indoleamine 2, 3-dioxygenase: a role for mesenchymal stem cells. Placenta 28:1174–1181

Jones CT, Robinson JS (1983) Studies on experimental growth retardation in sheep. Plasma catecholamines in fetuses with small placenta. J Dev Physiol 5:77–87

Jones LA, Alexander J, Roberts CW (2006) Ocular toxoplasmosis: in the storm of the eye. Parasite Immunol 28:635–642

Juliano PB, Blotta MH, Altemani AM (2006) ICAM-1 is overexpressed by villous trophoblasts in placentitis. Placenta 27:750–757

Kabyemela ER, Muehlenbachs A, Fried M, Kurtis JD, Mutabingwa TK, Duffy PE (2008) Maternal peripheral blood level of IL-10 as a marker for inflammatory placental malaria. Malar J 7:26

Kammerer U, Eggert AO, Kapp M, McLellan AD, Geijtenbeek TB, Dietl J, van Kooyk Y, Kampgen E (2003) Unique appearance of proliferating antigen-presenting cells expressing DC-SIGN (CD209) in the decidua of early human pregnancy. Am J Pathol 162:887–896

Kano R, Masukata Y, Omata Y, Kobayashi Y, Maeda R, Saito A (2005) Relationship between type 1/type 2 immune responses and occurrence of vertical transmission in BALB/c mice infected with *Neospora caninum*. Vet Parasitol 129:159–164

Keeler CE, Castle WE (1934) The influence of pregnancy upon the titre of immune (blood group) antibodies in the rabbit. Proc Natl Acad Sci U S A 20:465–470

Kimura M, Hanawa H, Watanabe H, Ogawa M, Abo T (1995) Synchronous expansion of intermediate TCR cells in the liver and uterus during pregnancy. Cell Immunol 162:16–25

King A, Allan DS, Bowen M, Powis SJ, Joseph S, Verma S, Hiby SE, McMichael AJ, Loke YW, Braud VM (2000) HLA-E is expressed on trophoblast and interacts with CD94/NKG2 receptors on decidual NK cells. Eur J Immunol 30:1623–1631

King KE, Kao KJ, Bray PF, Casella JF, Blakemore K, Callan NA, Kennedy SD, Kickler TS (1996) The role of HLA antibodies in neonatal thrombocytopenia: a prospective study. Tissue Antigens 47:206–211

Klevar S, Kulberg S, Boysen P, Storset AK, Moldal T, Bjorkman C, Olsen I (2007) Natural killer cells act as early responders in an experimental infection with *Neospora caninum* in calves. Int J Parasitol 37:329–339

Kodjikian L, Hoigne I, Adam O, Jacquier P, Aebi-Ochsner C, Aebi C, Garweg JG (2004) Vertical transmission of toxoplasmosis from a chronically infected immunocompetent woman. Pediatr Infect Dis J 23:272–274

Kono H, Furuhashi N, Shinkawa O, Takahashi T, Tsujiei M, Yajima A (1987) The maternal serum cortisol levels after onset of labor. Tohoku J Exp Med 152:133–137

Kozma N, Halasz M, Polgar B, Poehlmann TG, Markert UR, Palkovics T, Keszei M, Par G, Kiss K, Szeberenyi J, Grama L, Szekeres-Bartho J (2006) Progesterone-induced blocking factor activates STAT6 via binding to a novel IL-4 receptor. J Immunol 176:819–826

Krishnan L, Guilbert LJ, Wegmann TG, Belosevic M, Mosmann TR (1996a) T helper 1 response against *Leishmania major* in pregnant C57BL/6 mice increases implantation failure and fetal resorptions. Correlation with increased IFN-gamma and TNF and reduced IL-10 production by placental cells. J Immunol 156:653–662

Krishnan L, Guilbert LJ, Russell AS, Wegmann TG, Mosmann TR, Belosevic M (1996b) Pregnancy impairs resistance of C57BL/6 mice to *Leishmania major* infection and causes decreased antigen-specific IFN-gamma response and increased production of T helper 2 cytokines. J Immunol 156:644–652

Labeta MO, Margni RA, Leoni J, Binaghi RA (1986) Structure of asymmetric non-precipitating antibody: presence of a carbohydrate residue in only one Fab region of the molecule. Immunology 57:311–317

Liese J, Schleicher U, Bogdan C (2008) The innate immune response against Leishmania parasites. Immunobiology 213:377–387

Lin S, Sartori MJ, Mezzano L, de Fabro SP (2005) Placental alkaline phosphatase (PLAP) enzyme activity and binding to IgG in Chagas' disease. Placenta 26:789–795

Lin Y, Zeng Y, Zeng S, Wang T (2006) Potential role of toll-like receptor 3 in a murine model of polyinosinic-polycytidylic acid-induced embryo resorption. Fertil Steril 85(Suppl 1): 1125–1129

Lindsay JR, Nieman LK (2005) The hypothalamic-pituitary-adrenal axis in pregnancy: challenges in disease detection and treatment. Endocr Rev 26:775–799

Loke YW, King A, Burrows T, Gardner L, Bowen M, Hiby S, Howlett S, Holmes N, Jacobs D (1997) Evaluation of trophoblast HLA-G antigen with a specific monoclonal antibody. Tissue Antigens 50:135–146

Long MT, Baszler TV (1996) Fetal loss in BALB/C mice infected with *Neospora caninum*. J Parasitol 82:608–611

Long MT, Baszler TV (2000) Neutralization of maternal IL-4 modulates congenital protozoal transmission: comparison of innate versus acquired immune responses. J Immunol 164: 4768–4774

Maley SW, Buxton D, Macaldowie CN, Anderson IE, Wright SE, Bartley PM, Esteban-Redondo I, Hamilton CM, Storset AK, Innes EA (2006) Characterization of the immune response in the placenta of cattle experimentally infected with *Neospora caninum* in early gestation. J Comp Pathol 135:130–141

Marques de Araujo S, Rivera MT, El Bouhdidi A, de Maertelaer V, Carlier Y (1996) Maternal *Trypanosoma cruzi*-specific antibodies and worsening of acute infection in mouse offspring. Am J Trop Med Hyg 54:13–17

Mavoungou E, Bouyou-Akotet MK, Kremsner PG (2005) Effects of prolactin and cortisol on natural killer (NK) cell surface expression and function of human natural cytotoxicity receptors (NKp46, NKp44 and NKp30). Clin Exp Immunol 139:287–296

Mavoungou E (2006) Interactions between natural killer cells, cortisol and prolactin in malaria during pregnancy. Clin Med Res 4:33–41

McCracken SA, Gallery E, Morris JM (2004) Pregnancy-specific down-regulation of NF-kappa B expression in T cells in humans is essential for the maintenance of the cytokine profile required for pregnancy success. J Immunol 172:4583–4591

McMaster MT, Librach CL, Zhou Y, Lim KH, Janatpour MJ, DeMars R, Kovats S, Damsky C, Fisher SJ (1995) Human placental HLA-G expression is restricted to differentiated cytotrophoblasts. J Immunol 154:3771–3778

Medawar PB (1982) Fetal and cancer cells: some intriguing links. Hosp Pract (Off Ed) 17:66, 69, 72 passim

Mellor AL, Munn D (2004) Policing pregnancy: Tregs help keep the peace. Trends Immunol 25:563–565

Menendez C (1995) Malaria during pregnancy: a priority area of malaria research and control. Parasitol Today 11:178–183

Menendez C (2006) Malaria during pregnancy. Curr Mol Med 6:269–273

Menendez C, Mayor A (2007) Congenital malaria: the least known consequence of malaria in pregnancy. Semin Fetal Neonatal Med 12:207–213

Menzies FM, Henriquez FL, Roberts CW (2008) Immunological control of congenital toxoplasmosis in the murine model. Immunol Lett 115:83–89

Mincheva-Nilsson L, Kling M, Hammarstrom S, Nagaeva O, Sundqvist KG, Hammarstrom ML, Baranov V (1997) Gamma delta T cells of human early pregnancy decidua: evidence for local proliferation, phenotypic heterogeneity, and extrathymic differentiation. J Immunol 159:3266–3277

Miyazaki S, Tsuda H, Sakai M, Hori S, Sasaki Y, Futatani T, Miyawaki T, Saito S (2003) Predominance of Th2-promoting dendritic cells in early human pregnancy decidua. J Leukoc Biol 74:514–522

Mjihdi A, Truyens C, Detournay O, Carlier Y (2004) Systemic and placental productions of tumor necrosis factor contribute to induce fetal mortality in mice acutely infected with *Trypanosoma cruzi*. Exp Parasitol 107:58–64

Mjihdi A, Lambot MA, Stewart IJ, Detournay O, Noel JC, Carlier Y, Truyens C (2002) Acute *Trypanosoma cruzi* infection in mouse induces infertility or placental parasite invasion and ischemic necrosis associated with massive fetal loss. Am J Pathol 161:673–680

Mockenhaupt FP, Cramer JP, Hamann L, Stegemann MS, Eckert J, Oh NR, Otchwemah RN, Dietz E, Ehrhardt S, Schroder NW, Bienzle U, Schumann RR (2006) Toll-like receptor (TLR) polymorphisms in African children: Common TLR-4 variants predispose to severe malaria. Proc Natl Acad Sci USA 103:177–182

Moffett A, Loke C (2006) Immunology of placentation in eutherian mammals. Nat Rev Immunol 6:584–594

Molk R (1983) Ocular toxocariasis: a review of the literature. Ann Ophthalmol 15:216–219, 222–217, 230–211

Moormann AM, Sullivan AD, Rochford RA, Chensue SW, Bock PJ, Nyirenda T, Meshnick SR (1999) Malaria and pregnancy: placental cytokine expression and its relationship to intrauterine growth retardation. J Infect Dis 180:1987–1993

Moreno EA, Rivera IM, Moreno SC, Alarcon ME, Lugo-Yarbuh A (2005) Vertical transmission of *Trypanosoma cruzi* in Wistar rats during the acute phase of infection. Invest Clin 44:241–254

Munn DH, Sharma MD, Lee JR, Jhaver KG, Johnson TS, Keskin DB, Marshall B, Chandler P, Antonia SJ, Burgess R, Slingluff CL Jr, Mellor AL (2002) Potential regulatory function of human dendritic cells expressing indoleamine 2, 3-dioxygenase. Science 297:1867–1870

Nair LS, Nair AS (1993) Effects of malaria infection on pregnancy. Indian J Malariol 30:207–214

Neres R, Marinho CR, Goncalves LA, Catarino MB, Penha-Goncalves C (2008) Pregnancy outcome and placenta pathology in *Plasmodium berghei* ANKA infected mice reproduce the pathogenesis of severe malaria in pregnant women. PLoS ONE 3:e1608

Nickdel MB, Lyons RE, Roberts F, Brombacher F, Hunter CA, Alexander J, Roberts CW (2004) Intestinal pathology during acute toxoplasmosis is IL-4 dependent and unrelated to parasite burden. Parasite Immunol 26:75–82

Nishikawa Y, Inoue N, Makala L, Nagasawa H (2003) A role for balance of interferon-gamma and interleukin-4 production in protective immunity against *Neospora caninum* infection. Vet Parasitol 116:175–184

Ohashi J, Naka I, Patarapotikul J, Hananantachai H, Looareesuwan S, Tokunaga K (2002) Lack of association between interleukin-10 gene promoter polymorphism −1082G/A, and severe malaria in Thailand. Southeast Asian J Trop Med Public Health 33(Suppl 3):5–7

O'Leary P, Boyne P, Flett P, Beilby J, James I (1991) Longitudinal assessment of changes in reproductive hormones during normal pregnancy. Clin Chem 37:667–672

Orsal AS, Blois S, Labuz D, Peters EM, Schaefer M, Arck PC (2006) The progesterone derivative dydrogesterone down-regulates neurokinin 1 receptor expression on lymphocytes, induces a Th2 skew and exerts hypoalgesic effects in mice. J Mol Med 84:159–167

Parham P (2005) MHC class I molecules and KIRs in human history, health and survival. Nat Rev Immunol 5:201–214

Pearce EJ, Caspar P, Grzych JM, Lewis FA, Sher A (1991) Downregulation of Th1 cytokine production accompanies induction of Th2 responses by a parasitic helminth, *Schistosoma mansoni*. J Exp Med 173:159–166

Pearson RD (2004) Malaria in pregnancy: the "cortisol" and "prolactin" hypotheses. Clin Infect Dis 39:146–147

Pereira L, Maidji E, McDonagh S, Tabata T (2005) Insights into viral transmission at the uterine-placental interface. Trends Microbiol 13:164–174

Piccinni MP, Beloni L, Livi C, Maggi E, Scarselli G, Romagnani S (1998) Defective production of both leukemia inhibitory factor and type 2 T-helper cytokines by decidual T cells in unexplained recurrent abortions. Nat Med 4:1020–1024

Piccioli D, Sbrana S, Melandri E, Valiante NM (2002) Contact-dependent stimulation and inhibition of dendritic cells by natural killer cells. J Exp Med 195:335–341

Ponte M, Cantoni C, Biassoni R, Tradori-Cappai A, Bentivoglio G, Vitale C, Bertone S, Moretta A, Moretta L, Mingari MC (1999) Inhibitory receptors sensing HLA-G1 molecules in pregnancy: decidua-associated natural killer cells express LIR-1 and CD94/NKG2A and acquire p49, an HLA-G1-specific receptor. Proc Natl Acad Sci USA 96:5674–5679

Poovassery J, Moore JM (2006) Murine malaria infection induces fetal loss associated with accumulation of *Plasmodium chabaudi* AS-infected erythrocytes in the placenta. Infect Immun 74:2839–2848

Prolla TA, Pang Q, Alani E, Kolodner RD, Liskay RM (1994) MLH1, PMS1, and MSH2 interactions during the initiation of DNA mismatch repair in yeast. Science 265:1091–1093

Qian BZ, Bogh HO, Johansen MV, Wang PP (2000) Congenital transmission of *Schistosoma japonicum* in the rabbit. J Helminthol 74:267–270

Quinn HE, Ellis JT, Smith NC (2002) Neospora caninum: a cause of immune-mediated failure of pregnancy? Trends Parasitol 18:391–394

Quinn HE, Miller CM, Ellis JT (2004) The cell-mediated immune response to *Neospora caninum* during pregnancy in the mouse is associated with a bias towards production of interleukin-4. Int J Parasitol 34:723–732

Rajagopalan S, Bryceson YT, Kuppusamy SP, Geraghty DE, van der Meer A, Joosten I, Long EO (2006) Activation of NK cells by an endocytosed receptor for soluble HLA-G. PLoS Biol 4:e9

Ramhorst R, Gutierrez G, Corigliano A, Junovich G, Fainboim L (2007) Implication of RANTES in the modulation of alloimmune response by progesterone during pregnancy. Am J Reprod Immunol 57:147–152

Ramhorst RE, Garcia VE, Corigliano A, Rabinovich GA, Fainboim L (2004) Identification of RANTES as a novel immunomodulator of the maternal allogeneic response. Clin Immunol 110:71–80

Rastogi D, Wang C, Lendor C, Rothman PB, Miller RL (2006) T-helper type 2 polarization among asthmatics during and following pregnancy. Clin Exp Allergy 36:892–898

Rieger L, Honig A, Sutterlin M, Kapp M, Dietl J, Ruck P, Kammerer U (2004) Antigen-presenting cells in human endometrium during the menstrual cycle compared to early pregnancy. J Soc Gynecol Investig 11:488–493

Rivera MT, Marques de Araujo S, Lucas R, Deman J, Truyens C, Defresne MP, de Baetselier P, Carlier Y (1995) High tumor necrosis factor alpha (TNF-alpha) production in *Trypanosoma cruzi*-infected pregnant mice and increased TNF-alpha gene transcription in their offspring. Infect Immun 63:591–595

Roberts CW, Alexander J (1992) Studies on a murine model of congenital toxoplasmosis: vertical disease transmission only occurs in BALB/c mice infected for the first time during pregnancy. Parasitology 104(1):19–23

Roberts CW, Ferguson DJ, Jebbari H, Satoskar A, Bluethmann H, Alexander J (1996) Different roles for interleukin-4 during the course of *Toxoplasma gondii* infection. Infect Immun 64:897–904

Rodriguez-Morales AJ, Barbella RA, Case C, Arria M, Ravelo M, Perez H, Urdaneta O, Gervasio G, Rubio N, Maldonado A, Aguilera Y, Viloria A, Blanco JJ, Colina M, Hernandez E, Araujo E, Cabaniel G, Benitez J, Rifakis P (2006) Intestinal parasitic infections among pregnant women in Venezuela. Infect Dis Obstet Gynecol 2006:23125

Rogerson SJ, Hviid L, Duffy PE, Leke RF, Taylor DW (2007) Malaria in pregnancy: pathogenesis and immunity. Lancet Infect Dis 7:105–117

Rogerson SJ, Brown HC, Pollina E, Abrams ET, Tadesse E, Lema VM, Molyneux ME (2003) Placental tumor necrosis factor alpha but not gamma interferon is associated with placental malaria and low birth weight in Malawian women. Infect Immun 71:267–270

Rouas-Freiss N, Khalil-Daher I, Riteau B, Menier C, Paul P, Dausset J, Carosella ED (1999) The immunotolerance role of HLA-G. Semin Cancer Biol 9:3–12

Saji F, Koyama M, Matsuzaki N (1994) Current topic: human placental Fc receptors. Placenta. 15(5):453–466

Salanti A, Dahlback M, Turner L, Nielsen MA, Barfod L, Magistrado P, Jensen AT, Lavstsen T, Ofori MF, Marsh K, Hviid L, Theander TG (2004) Evidence for the involvement of VAR2CSA in pregnancy-associated malaria. J Exp Med 200:1197–1203

Salas NA, Cot M, Schneider D, Mendoza B, Santalla JA, Postigo J, Chippaux JP, Brutus L (2007) Risk factors and consequences of congenital Chagas disease in Yacuiba, south Bolivia. Trop Med Int Health 12:1498–1505

Sanchez Negrette O, Mora MC, Basombrio MA (2005) High prevalence of congenital *Trypanosoma cruzi* infection and family clustering in Salta, Argentina. Pediatrics 115:e668–e672

Sandner SE, Clarkson MR, Salama AD, Sanchez-Fueyo A, Domenig C, Habicht A, Najafian N, Yagita H, Azuma M, Turka LA, Sayegh MH (2005) Role of the programmed death-1 pathway in regulation of alloimmune responses in vivo. J Immunol 174:3408–3415

Sasaki Y, Sakai M, Miyazaki S, Higuma S, Shiozaki A, Saito S (2004) Decidual and peripheral blood CD4+CD25+ regulatory T cells in early pregnancy subjects and spontaneous abortion cases. Mol Hum Reprod 10:347–353

Schofield L, Grau GE (2005) Immunological processes in malaria pathogenesis. Nat Rev Immunol 5:722–735

Shah OJ, Zargar SA, Robbani I (2006) Biliary ascariasis: a review. World J Surg 30:1500–1506

Shiono Y, Mun HS, He N, Nakazaki Y, Fang H, Furuya M, Aosai F, Yano A (2007) Maternal-fetal transmission of *Toxoplasma gondii* in interferon-gamma deficient pregnant mice. Parasitol Int 56:141–148

Shirahata T, Muroya N, Ohta C, Goto H, Nakane A (1992) Correlation between increased susceptibility to primary *Toxoplasma gondii* infection and depressed production of gamma interferon in pregnant mice. Microbiol Immunol 36:81–91

Shirahata T, Muroya N, Ohta C, Goto H, Nakane A (1993) Enhancement by recombinant human interleukin 2 of host resistance to *Toxoplasma gondii* infection in pregnant mice. Microbiol Immunol 37:583–590

Sholapurkar SL, Gupta AN, Mahajan RC (1988) Clinical course of malaria in pregnancy–a prospective controlled study from India. Trans R Soc Trop Med Hyg 82:376–379

Silva Filho FC, de Souza W (1988) The interaction of *Trichomonas vaginalis* and Tritrichomonas foetus with epithelial cells in vitro. Cell Struct Funct 13:301–310

Silveira C, Ferreira R, Muccioli C, Nussenblatt R, Belfort R Jr (2003) Toxoplasmosis transmitted to a newborn from the mother infected 20 years earlier. Am J Ophthalmol 136:370–371

Singh N, Shukla MM, Srivastava R, Sharma VP (1995) Prevalence of malaria among pregnant and non-pregnant women of district Jabalpur, Madhya Pradesh. Indian J Malariol 32:6–13

Solana ME, Celentano AM, Tekiel V, Jones M, Gonzalez Cappa SM (2002) *Trypanosoma cruzi:* effect of parasite subpopulation on murine pregnancy outcome. J Parasitol 88:102–106

Somerset DA, Zheng Y, Kilby MD, Sansom DM, Drayson MT (2004) Normal human pregnancy is associated with an elevation in the immune suppressive CD25+ CD4+ regulatory T-cell subset. Immunology 112:38–43

Streiger M, Fabbro D, del Barco M, Beltramino R, Bovero N (1995) Congenital Chagas disease in the city of Santa Fe. Diagnosis and treatment. Medicina (B Aires) 55:125–132

Suzuki Y, Yang Q, Yang S, Nguyen N, Lim S, Liesenfeld O, Kojima T, Remington JS (1996) IL-4 is protective against development of toxoplasmic encephalitis. J Immunol 157:2564–2569

Svensson L, Arvola M, Sallstrom MA, Holmdahl R, Mattsson R (2001) The Th2 cytokines IL-4 and IL-10 are not crucial for the completion of allogeneic pregnancy in mice. J Reprod Immunol 51:3–7

Szekeres-Bartho J, Wegmann TG (1996) A progesterone-dependent immunomodulatory protein alters the Th1/Th2 balance. J Reprod Immunol 31:81–95

Szekeres-Bartho J, Faust Z, Varga P, Szereday L, Kelemen K (1996) The immunological pregnancy protective effect of progesterone is manifested via controlling cytokine production. Am J Reprod Immunol 35:348–351

Szekeres-Bartho J, Polgar B, Kozma N, Miko E, Par G, Szereday L, Barakonyi A, Palkovics T, Papp O, Varga P (2005) Progesterone-dependent immunomodulation. Chem Immunol Allergy 89:118–125

Tarleton RL (1991) Regulation of immunity in *Trypanosoma cruzi* infection. Exp Parasitol 73:106–109

Tarleton RL (2007) Immune system recognition of *Trypanosoma cruzi*. Curr Opin Immunol 19:430–434

Tenter AM, Heckeroth AR, Weiss LM (2000) *Toxoplasma gondii*: from animals to humans. Int J Parasitol 30:1217–1258

Torina A, Caracappa S, Barera A, Dieli F, Sireci G, Genchi C, Depiazes P, Salerno A (2005) *Toxocara canis* infection induces antigen-specific IL-10 and IFNgamma production in pregnant dogs and their puppies. Vet Immunol Immunopathol 108:247–251

Trees AJ, Williams DJ (2005) Endogenous and exogenous transplacental infection in Neospora caninum and *Toxoplasma gondii*. Trends Parasitol 21:558–561

Ufomadu GO, Sato Y, Takahashi H (1990) Possible transplacental transmission of *Onchocerca volvulus*. Trop Geogr Med 42:69–71

Van Der Pol B, Williams JA, Orr DP, Batteiger BE, Fortenberry JD (2005) Prevalence, incidence, natural history, and response to treatment of *Trichomonas vaginalis* infection among adolescent women. J Infect Dis 192:2039–2044 van Zon AA, Eling WM (1980a) Depressed malarial immunity in pregnant mice. Infect Immun 28:630–632

van Zon AA, Eling WM (1980b) Pregnancy associated recrudescence in murine malaria (*Plasmodium berghei*). Tropenmed Parasitol 31:402–408

Villard O, Candolfi E, Despringre JL, Derouin F, Marcellin L, Viville S, Kien T (1995) Protective effect of low doses of an anti-IL-4 monoclonal antibody in a murine model of acute toxoplasmosis. Parasite Immunol 17:233–236

Vogel N et al (1996) Congenital toxoplasmosis transmitted from an immunologically competent mother infected before conception. Clin Infect Dis 23:1055–1060

Weinstock M (2008) The long-term behavioural consequences of prenatal stress. Neurosci Biobehav Rev 32:1073–1086

Wetzka B, Clark DE, Charnock-Jones DS, Zahradnik HP, Smith SK (1997) Isolation of macrophages (Hofbauer cells) from human term placenta and their prostaglandin E2 and thromboxane production. Hum Reprod 12:847–852

Williams DJ, Trees AJ (2006) Protecting babies: vaccine strategies to prevent foetopathy in Neospora caninum-infected cattle. Parasite Immunol 28:61–67

Willingham AL 3rd, Johansen MV, Bogh HO, Ito A, Andreassen J, Lindberg R, Christensen NO, Nansen P (1999) Congenital transmission of *Schistosoma japonicum* in pigs. Am J Trop Med Hyg 60:311–312

Wilson JL, Heffler LC, Charo J, Scheynius A, Bejarano MT, Ljunggren HG (1999) Targeting of human dendritic cells by autologous NK cells. J Immunol 163:6365–6370

Wiwanitkit V (2006) Congenital malaria in Thailand, an appraisal of previous cases. Pediatr Int 48:562–565

Yarovinsky F, Zhang D, Andersen JF, Bannenberg GL, Serhan CN, Hayden MS, Hieny S, Sutterwala FS, Flavell RA, Ghosh S, Sher A (2005) TLR11 activation of dendritic cells by a protozoan profilin-like protein. Science 308:1626–1629

Yavuz E, Aydin F, Seyhan A, Topuz S, Karagenc Y, Tuzlali S, Ilhan R, Iplikci A (2006) Granulomatous villitis formed by inflammatory cells with maternal origin: a rare manifestation type of placental toxoplasmosis. Placenta 27:780–782

Yin P, Lin Z, Cheng YH, Marsh EE, Utsunomiya H, Ishikawa H, Xue Q, Reierstad S, Innes J, Thung S, Kim JJ, Xu E, Bulun SE (2007) Progesterone receptor regulates Bcl-2 gene expression through direct binding to its promoter region in uterine leiomyoma cells. J Clin Endocrinol Metab 92:4459–4466

Yule A, Skirrow SZ, Bonduran RH (1989) Bovine trichomoniasis. Parasitol Today 5:373–377

Zenclussen AC, Gerlof K, Zenclussen ML, Sollwedel A, Bertoja AZ, Ritter T, Kotsch K, Leber J, Volk HD (2005) Abnormal T-cell reactivity against paternal antigens in spontaneous abortion: adoptive transfer of pregnancy-induced CD4+CD25+ T regulatory cells prevents fetal rejection in a murine abortion model. Am J Pathol 166:811–822

Zhang J, Wei H, Wu D, Tian Z (2007) Toll-like receptor 3 agonist induces impairment of uterine vascular remodeling and fetal losses in CBA x DBA/2 mice. J Reprod Immunol 74:61–67

Zimmermann S, Murray PJ, Heeg K, Dalpke AH (2006) Induction of suppressor of cytokine signaling-1 by *Toxoplasma gondii* contributes to immune evasion in macrophages by blocking IFN-gamma signaling. J Immunol 176:1840–1847

Chapter 10
Sex Steroids and Risk of Female Genital Tract Infection

Patti Gravitt and Khalil Ghanem

Abstract Sex steroids are central to regulation of the dual roles of the female reproductive tract: support of reproduction and protection against deleterious pathogen invasion. The mechanisms by which sex steroids orchestrate the balance between reproduction and protection against pathogenic microorganisms, however, are complex and poorly understood. Epidemiologic studies have revealed associations between sex steroids and diagnosis, shedding, transmission of sexually transmitted diseases and their clinical sequelae, as well as disruptions to the normal vaginal flora. Experimental approaches, including animal models, have provided some insight into the mechanism of these associations, though often with contradictory results. Inherent limitations in traditional approaches to understanding the physiology of sex steroid regulation of immunity in the female genital tract will require significant interdisciplinary collaboration.

10.1 The Role of Sex Steroids on Female Lower Genital Tract Immunity

The female reproductive tract has evolved to maintain a difficult balance of supporting reproduction and protecting against deleterious pathogen invasion. These roles appear diametrically opposed immunologically. To support conception and fetal development, the female reproductive tract must tolerate allogeneic sperm in both the lower and upper genital tract as well as an immunologically distinct fetus in the upper genital tract. Both of these physiological requirements for successful continuation of the species require transient downregulation of

P. Gravitt (✉)
The Department of Epidemiology, The Johns Hopkins Bloomberg School of Public Health, Baltimore, Maryland 21205
e-mail: pgravitt@jhsph.edu

S.L. Klein and C.W. Roberts (eds.), *Sex Hormones and Immunity to Infection*, DOI 10.1007/978-3-642-02155-8_10, © Springer-Verlag Berlin Heidelberg 2010

immunological response to foreign antigen exposure. Pregnancy is well known to produce a state of transient immunosuppression. Perhaps less well understood, but consistent with a wide body of literature, is a cyclical decrease in immunologic markers in the lower genital tract in the periovulatory period coincident with the estrogen surge (Kutteh et al. 1998).

The immune system of the lower genital tract is tasked with (1) maintaining tolerance to commensal organisms, (2) providing at least transient tolerance to allogeneic sperm, and (3) protecting against pathogenic microorganisms. Understanding the mechanism by which this balance is maintained is critical, because pathogenic genital tract infection can result in mortality and serious morbidity including adverse reproductive outcomes and cancer (Table 10.1). As described below, the prevalence of many sexually transmitted infections (STIs), as well as disruption of normal vaginal flora, is associated with changes in sex steroid concentrations, as a result of the menstrual cycle, use of hormonal contraceptives, or pregnancy. However, there are remarkable inconsistencies in the influence of such hormonal variations reported between studies and these can be attributed at least in part to inherent limitations in the standard experimental approaches. For example, while studies using animal models are informative, translation of results is difficult, because these models (usually in mice) rarely recapitulate completely the full spectrum of the natural history of infections in humans; rather, they model one aspect of infection at a time (e.g., duration of infection or severity of clinical symptoms). Of particular difficulty is the ability to adequately model human sexual transmission. Human epidemiological studies are similarly limited in measuring the full spectrum of risk of transmission, persistence, and reactivation using common study designs. An assessment of hormonal influences on infection is not facile by analysis of groups of women either using hormonal or nonhormonal contraception for a number of reasons. Firstly, complete randomization of women to hormonal contraception is not possible for ethical reasons. Secondly, there are inherent differences between women who use hormonal forms of contraception compared with those who use nonhormonal forms of contraception, such as increased sexual risk-taking behavior (e.g., less frequent condom use) (Woods et al. 2006). These differences result in an inability to exclude noncausal explanations for observed associations between hormonal contraceptive use and STIs, such as residual confounding by sexual behavior. Studies that assess the association of genital tract infections with natural sex steroid variation across the menstrual cycle and during pregnancy are less prone to the confounding effects of sexual behavior. However, variability in measurement error, particularly in the staging of the menstrual cycle, may also contribute to inconsistencies across studies. Most studies use day since last menstrual period in the context of a "normal 28-day cycle" to measure the stages of menstrual cycle, generally using a bivariate categorization of "early" (follicular/menses/proliferative) and "late" (luteal/secretory) stages. There is a high degree of interindividual variability in menstrual cycle length; therefore, in the absence of a hormonal measure of ovulation (e.g., luteinizing hormone (LH)), misclassification using the day of cycle as a proxy for menstrual stage limits the power to detect true associations with sex steroid fluctuation (Harlow and Ephross 1995).

Table 10.1 Infections of the female genital tract, US prevalence estimates, and associated morbidity

Agent	Type	Transmission	Prevalence	Adverse outcomes
Chlamydia trachomatis	Bacteria	Sexual	511.7/100,000[a]	PID, ectopic pregnancy, infertility
Neisseria gonorrhea	Bacteria	Sexual	124.3/100,000[a]	PID, ectopic pregnancy, infertility
Treponema pallidum (syphilis)	Bacteria	Sexual	1.0/100,000[a]	Brain, cardiovascular, organ damage, death
Bacterial vaginosis	Bacteria	Nonsexual	Unreported	PID, premature and/or low birth weight deliveries
HIV	Virus	Sexual and non sexual	9,253 new cases in 2006[b] 80% heterosexual contact	Death
HSV-2	Virus	Sexual	24.0%, NHANES[c]	Genital ulcers, increased risk of HIV infection
HPV	Virus	Sexual	26.8%, NHANES[d]	Genital warts, cervical intraepithelial neoplasia, cervical cancer
Candida	Fungus	Nonsexual, commensal overgrowth	NA	Recurrent vaginitis
Trichomonas vaginalis	Parasite	Sexual	NA	Adverse pregnancy outcomes

[a]Prevalence estimates for Chlamydia, gonorrhea, and syphilis from CDC, Trends in reportable sexually transmitted diseases in the United States, 2006: national surveillance data for Chlamydia, Gonorrhea, and Syphilis, November 2007 (http://www.cdc.gov/std/stats/trends2006.htm, loi: June 11, 2008)
[b]Prevalence estimates for HIV (from http://www.cdc.gov/hiv/topics/surveillance/basic.htm#hivest, loi: June 11, 2008)
[c]Prevalence estimates for HSV-2 from Xu et al. (2006)
[d]Prevalence estimates for HPV from Dunne et al. (2007)

Efforts to improve the measurement of sexual behavior and stage of menstrual cycle are ongoing, and will improve the validity of future epidemiological studies. In parallel, basic scientists are developing systems to understand the diverse molecular immunology of the female genital tract, which has lagged far behind our understanding of the mucosal immunology of the gut and respiratory tracts. Understanding the role of sex steroids in genital tract immunity and risk of pathogenic infection is a clear research priority, as the rates of STIs continue to increase (Workowski and Berman 2006) with women having disproportionately higher incidence rates and associated morbidities than men. Unfortunately, the rising rates of STIs are occurring in parallel with an increase in drug resistance, threatening our ability to adequately control these epidemics. What is needed to overcome the inherent limitations of available data to ensure the appropriate institution of public health measures to minimize risks and sequelae of genital tract infections? One answer would be to define plausible biological phenomena

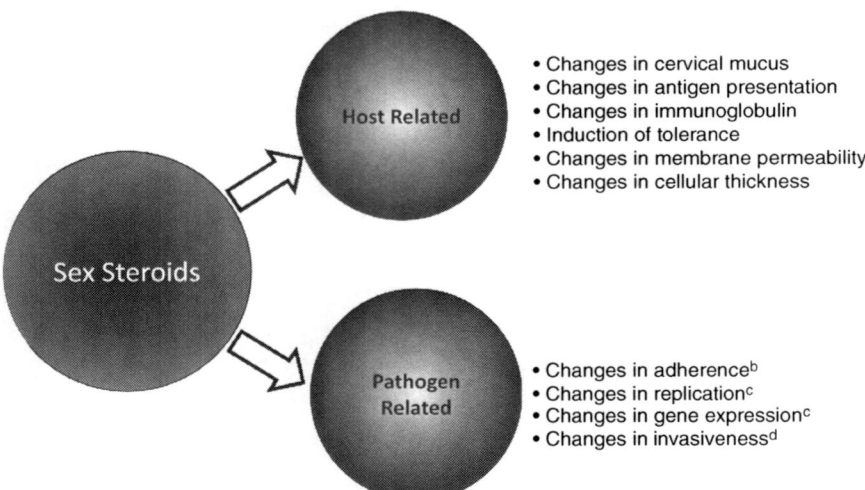

Fig. 10.1 Mechanisms by which sex steroid exposures can increase risk of genital tract infection. Adapted from (**a**) Wira et al. (2005), (**b**) Bose and Goswami (1986), (**c**) Chen et al. (1996), Asin et al. (2008), (**d**) White and Larsen (1997)

that could account for the findings. Figure 10.1 lists some of the biological mechanisms that have been advanced to account for differences in susceptibility and progression of lower genital tract infections in hormone contraceptive users.

Immune responses in the female genital tract are regulated by sex hormones (Wira et al. 2005). In addition to classical immune cells (e.g., monocytes and lymphocytes), the genital tract epithelium is a key component of the genital tract immune system (Wira et al. 2005). Woodworth and Simpson (1993) demonstrated that the normal cervical cells constitutively secrete IL-1α, IL-1β, IL-6, TNF-α, GM-CSF, and CXCL8 (i.e., IL-8) and that decreased expression of cytokines/chemokines by HPV-infected cervical cells may help to explain the persistent nature of HPV infection in vivo. As reviewed in Chap. 2, antigen presentation, cytokine production, immunoglobulin production and transport, and even induction of tolerance are influenced by variations in the levels of sex steroid hormones. The interaction between sex hormones and the immune system is complex and the variation of hormonal effects between species further complicates the true picture as related to humans (Straub 2007). In a rodent model, the induction of tolerance in the lower genital tract is hormonally regulated (Black et al. 2000). In autoimmune diseases, estrogen has a biphasic effect; low levels enhance and high levels inhibit the manifestations of some diseases (e.g., systemic lupus erythematosus), whereas the opposite is observed for other autoimmune syndromes (e.g., rheumatoid arthritis) (Straub 2007). As described in Chap. 2 of this book, low estrogen levels favor the generation of a helper T cell type 1 (Th1) response, whereas high estrogen and testosterone levels favor that of a Th2 response. Such a paradigm is consistent with the opposing effects of pregnancy T cell- and B-cell-mediated immune disease (Straub 2007).

Animal and cell culture models have long suggested that sex hormones modify the risks of some STIs. For example, after depot medroxyprogesterone acetate treatment, there is a 100-fold increase in susceptibility to genital herpes simplex virus (HSV)-2 in mice compared with untreated animals (Kaushic et al. 2003). In addition to altering susceptibility to HSV-2, depot medroxyprogesterone acetate inhibits immune responses to HSV-2 via lowering of IgG and IgA-specific antibodies. A similar increase in susceptibility to human papillomavirus (HPV) pseudovirion infection in mice after progesterone treatment has been reported (Roberts et al. 2007). The time of progesterone exposure also alters the immune responses of mice to HSV; longer exposure to depot medroxyprogesterone acetate reduces innate and adaptive immune responses to HSV-2 (Gillgrass et al. 2003). However, rarely do these animal and tissue culture models examine the physiological scenario of hormone combinations. By failing to capture potentially relevant regulatory mechanisms that are likely instrumental in maintaining the balance between successful reproductive outcomes and protection against pathogenic challenge, the translational significance of these models may be limited.

Using immunological markers in human studies, there is consistency in hormonal effects on lower genital tract immunity, in which immune responses in the cervix are reduced around the time of ovulation. Studies quantitating immunoglobulin and cytokine levels in cervical mucus of pregnant women have shown increased levels of IgA, IgG, and IL-1β (Kutteh and Franklin 2001). Other studies have shown that immunoglobulin and cytokine levels (e.g., IL-10, IL-6, and IL1-β) in nonpregnant women increase approximately 3 days before ovulation corresponding to the time of estrogen surge (Kutteh and Franklin 2001; Kutteh et al. 1998). Concentrations of cytokines in the genital tract versus peripheral blood differ during infection (Al-Harthi et al. 2000). For example, levels of CXCL8 in plasma and IL-6 and IL-1β in the vagina are elevated during the follicular phase when both estrogen and progesterone levels are low. Among adolescents, a negative correlation between estradiol and IgA, IL-2, and IL-10 levels in cervical secretions has been observed (Shrier et al. 2003). In adult women, cervical IL-10 and IL-12 concentrations reach a nadir at ovulation in cycling women and are highest among women using combined oral contraceptives (COCs) (Gravitt et al. 2003). Of note, the correlation between plasma and cervical IL-10 and IL-12 is weak, with higher levels observed in cervical secretions suggesting that local production of genital tract cytokines predominates (Castle et al. 2002). These data emphasize the need for tissue-specific sampling in epidemiologic research.

In the current era, where much of the hope of future STI control lies in vaccine development, the effects of sex hormones on immune responses to vaccination should be critically evaluated (Rupp et al. 2005). Mucosal routes of immunization are attractive because they induce both a local and systemic response. In one study, 21 volunteers received a mucosal vaccine containing cholera toxin B antigen either independently of the stage in the menstrual cycle or on days 10 and 24 in the cycle in different groups of subjects (Johansson et al. 2001). Vaginal and nasal vaccinations both resulted in increased IgA and IgG anticholera toxin B subunit responses in serum in the majority of the volunteers in the various vaccination groups. Only vaginal vaccination given on days 10 and 24 in the cycle induced strong specific antibody responses in the cervix (Johansson et al. 2001).

Virus-like particle (VLP)-based vaccines against genital HPVs have proven highly effective women, (Future II study Group 2007) leading to the approval of Merck & Co.'s Gardasil[R] in 2007. Published data from the Gardasil trials show nearly 100% efficacy following intramuscular immunization, suggesting no modification resulting from sex steroid fluctuations through the menstrual cycle or exogenous contraceptive hormone use (Garland et al. 2007). Earlier studies of HPV VLP intramuscular vaccination, however, report a significant decline in vaccine-induced cervical IgG around the time of ovulation (Nardelli-Haefliger et al. 2003) in normal cycling women, whereas stable cervical IgG levels across the cycle are observed in women using COCs. Plasma IgG levels are correlated with cervical IgG only in the contraceptive, but not in the naturally ovulating, group. These data suggest that cervical immunoglobulins may be a more reliable measure of minimal protective titer for genital tract infections.

10.2 Association of Genital Tract Infections with Menstrual Cycle, Pregnancy, or COC Use in Women

10.2.1 Bacterial Genital Tract Infections

Chlamydia trachomatis. Laboratory diagnosis of *Chlamydia* infection is higher in the luteal phase of the menstrual cycle (measured as weeks 4–5) (Crowley et al. 1997; Rosenthal and Landefeld 1990), correlating with a significant decrease in antichlamydial activity of cervical secretions collected in the later phase of the menstrual cycle (56–68% in weeks 4–5 vs 70–90% in weeks 1–3) (Mahmoud et al. 1994). Taken together, these data suggest that cycling sex hormones modify natural immunity to *Chlamydia* infection.

Does exogenous hormone use modify natural fluctuations in response to bacterial infection? COC users have a significantly higher *Chlamydia* inclusion count (i.e., lower antichlamydial activity) in cervical secretions compared with nonCOC users, independent of the stage of the menstrual cycle (Mahmoud et al. 1994). These data suggest that COC users may be at an increased risk of *Chlamydia* infection; the epidemiologic data to support this hypothesis, however, has been inconsistent. Meta-analyzes reveal a positive association of *C. trachomatis* infection and COC use in the majority of cross-sectional studies reviewed (Mohllajee et al. 2006). In addition, five of six prospective studies report an increased risk of *C. trachomatis* acquisition among COC users (three of which were statistically significant). The single prospective study reporting a nonsignificant protective effect of COC use on *C. trachomatis* infection did not report estimates of risk adjusted for sexual behaviors or age (Rahm et al. 1991). A 5-year follow-up study of women with asymptomatic *C. trachomatis* infection in Colombia reported a significant increased clearance of *C. trachomatis* among COC users, especially in younger women (Molano et al. 2005). Several biological mechanisms have been posited to explain the increased susceptibility to *C. trachomatis* infection resulting from hormonal contraception, including the hormone-induced increase in cervical ectopy, resulting in increased exposure of susceptible columnar cells to *C. trachomatis* infection. When ectopy is directly measured, however, no increase in *C. trachomatis* infection is observed (Jacobson et al. 2000). Examination of the effect of COC use on the duration of *C. trachomatis* infection will be difficult in humans because treatment of *C. trachomatis* infection is initiated after diagnosis, interrupting the natural history of infection. Data on recurrence or reinfection as a function of COC use, however, is valuable and can be obtained. Three prospective studies have estimated the risk of *C. trachomatis* infection associated with injectable depot medroxyprogesterone acetate use. All of these studies which were adjusted for sexual behaviors found a statistically significant increase in the risk of *C. trachomatis* among depot medroxyprogesterone acetate users (Baeten et al. 2001; Lavreys et al. 2004; Morrison et al. 2004).

As with other bacterial and parasitic STIs, *C. trachomatis* infection is associated with an increased risk of adverse pregnancy outcomes. Therefore, while many studies report *C. trachomatis* infection prevalence among pregnant women, few directly compared *C. trachomatis* prevalence in pregnant versus nonpregnant women. A recent study of American Indian and Alaskan Native women from family planning clinics reported that pregnant women had a *C. trachomatis* prevalence equal to symptomatic women, and higher than similar nonpregnant women (Gorgos et al. 2008). *C. trachomatis* prevalence, however, was not the same in pregnant and nonpregnant women attending family planning clinics (Geisler and James 2008) and in a retrospective study at the Baltimore City STD clinics (Johnson et al. 2007). More studies of *Chlamydia* infection in pregnancy are needed.

In vitro models have been used to investigate the role of sex steroids on *C. trachomatis* infection (Sonnex 1998). Tissue culture models reveal an increase in adherence and infectivity of a variety of cell types, including HeLa cervical cancer cells, endometrial glandular cells, and McCoy fibroblasts, by *C. trachomatis* following 17β-estradiol exposure (Bose and Goswami 1986; Maslow et al. 1988; Sugarman and Agbor1986), supporting the association between COC use and increased susceptibility to *C. trachomatis* infection. In guinea pig models, estrogen treatment increases the duration of *C. trachomatis* infection, whereas progesterone has no effect on the duration of *C. trachomatis* infection (Pasley et al. 1985). This is in direct contrast to the epidemiological observation of a shorter duration of *C. trachomatis* infection among COC users in Colombia (Molano et al. 2005). Inconsistencies between the experimental animal models and human observational data suggest deficiencies in one or both areas of research supporting a need for interdisciplinary approaches.

Neisseria gonorrhea. The association between gonorrhea and the menstrual cycle is a longstanding one. Several reports beginning in the 1950s suggested that there is an increased probability of diagnosing *N. gonorrhea* in women shortly after the beginning of menstruation (Putkonen and Ebeling 1950). Several hypotheses were suggested: menstrual blood provided nutritional supplementation to enhance the growth of the organism, favorable characteristics of the cervical mucus, hormonally mediated increased adherence of the organism, or the presence of vaginal flora disruptions (e.g., bacterial vaginosis) enhanced gonococcal infection (Putkonen and Ebeling 1950).

Several prospective studies have been conducted to evaluate the risk of gonococcal cervicitis in hormonal contraceptive users. In a study of 818 women recruited from sexually transmitted diseases clinics in Birmingham, AL, hormonal contraception increased the risk of gonococcal cervicitis by 70% (Louv et al. 1989). Another study of 948 Kenyan commercial sex workers did not reveal an increased risk of gonococcal infection among hormonal contraceptive users (either depot medroxyprogesterone acetate or COC pills) (Baeten et al. 2001). Finally, a prospective study of 819 women in Baltimore, MD, who were followed for 1 year revealed that only depot medroxyprogesterone acetate is significantly associated with a nearly fourfold increased risk of both Chlamydial and Gonococcal cervical infections (Morrison et al. 2004). There are no convincing data that the risk of gonorrhea during pregnancy is altered.

Syphilis. The data on syphilis and sex hormones are very limited. Basic studies are nonexistent because *Treponema pallidum* cannot be grown in the laboratory. There have been no epidemiological associations between menstrual cycle phase and risk of syphilis acquisition. Studies evaluating the effects of hormonal contraception on risk of syphilis acquisition have also been limited. In the prospective study of Kenyan sex workers, hormonal contraceptive use was not associated with the risk of syphilis acquisition (Baeten et al. 2001). There are no convincing data that the risk of syphilis is altered during pregnancy.

Bacterial vaginosis. Bacterial vaginosis is an ecological disease of the vaginal flora characterized by a shift in the vaginal microenvironment from predominantly acid-producing *Lactobacilli* to a predominance of *Gardnerella vaginalis*, anaerobes, and other flora. Although not thought to be sexually transmitted per se, it is a common infection that affects approximately 30% of women. Bacterial vaginosis occurs more frequently during the second week of the menstrual cycle in most women, and a marked increase during the first week is noted in women who have infrequent episodes of bacterial vaginosis (Morrison et al. 2004). In one of the largest studies to evaluate the normal change of the vaginal microflora in response to the menstrual cycle, rate of recovery of heavy *Lactobacillus* growth (normal "healthy" vaginal flora) increased over the menstrual cycle, whereas the concentration of other bacterial species tended to be highest at menses (Eschenbach et al. 2000). This and other studies suggest that the vaginal microflora becomes much less stable during menses. In the prospective Kenyan cohort, women using COC pills and depot medroxyprogesterone acetate were 20–30% less likely to be diagnosed with bacterial vaginosis during follow-up (Baeten et al. 2001). Whether this is due to the impact on menstrual bleeding or whether this is a direct hormonal effect on bacterial flora is unclear. Bacterial vaginosis has been associated with serious sequelae during pregnancy. There are, however, no conclusive data to suggest increased risk of bacterial vaginosis acquisition during pregnancy.

Pelvic inflammatory disease. Pelvic inflammatory disease (PID) is characterized by chronic inflammation of the fallopian tubes (salpingitis) and/or uterus (endometritis), and is linked etiologically to bacterial genital tract infections, particularly *C. trachomatis* and *N. gonorrhea*. The association between PID and sex steroids is controversial. The data are difficult to interpret, not only because of confounding variables in behavior of the groups studied, but also because the methods employed to diagnose PID vary. The gold standard for diagnosing PID is laparoscopy, but it is seldom used in clinical practice. Most clinicians diagnose PID based on clinical signs and symptoms that do not necessarily correlate with laparoscopic findings. Furthermore, the criteria for diagnosing PID have become less stringent recently, resulting in increased sensitivity, but decreased specificity (Workowski and Berman 2006).

Salpingitis occurs more frequently within 1 week after the onset of menses (Sweet et al. 1986). This finding correlates with the increased risk of gonococcal infection during that time (see above). Early studies showed a protective effect of COC on incidence of PID (Rubin et al. 1982; Senanayake and Kramer 1980). Most of these studies are hospital based, suggesting a selection bias in favor of more severe and symptomatic disease. In addition, most do not differentiate between the etiological agents of PID. Additional studies show that COC use protects against symptomatic PID among women infected with *C. trachomatis*, but not among those coinfected with *N. gonorrhea* (Wolner-Hanssen et al. 1990). A prospective study in Kenyan sex workers showed a decreased risk for PID in women using depot medroxyprogesterone acetate (Baeten et al. 2001). Another study revealed an increased incidence of silent endometritis and salpingitis in oral contraceptive users (Wolner-Hanssen 1986). The suggestion that hormonal contraceptive use may not alter susceptibility to PID, but may alter symptoms associated with PID has been advocated. The largest prospective study of 563 patients enrolled in a treatment trial for PID does not report a relationship between any hormonal contraceptive use and PID; diagnosis, however, was based on clinical criteria (Ness et al. 2001). There are no convincing data that the risk of PID during pregnancy is altered.

10.2.2 Viral Genital Tract Infections

Human immunodeficiency virus. There have been several findings to suggest differences in the course of human immunodeficiency virus (HIV) infection between the sexes. For example, women have lower HIV-RNA titers at the time of HIV seroconversion (Farzadegan et al. 1998; Sterling et al. 2001), lower HIV-RNA set points, and may have altered susceptibility to HIV-mediated central nervous system sequelae (Wilson et al. 2006) compared with men. A recent study demonstrated a direct effect of both 17β-estradiol and progesterone on HIV-1 replication in peripheral blood cells; midproliferative phase conditions increase and midsecretory phase conditions decrease HIV-1 replication (Asin et al. 2008). In addition, decreased levels of HIV-1 integration in the midsecretory phase and increased levels of viral transcription in the midproliferative phase are reported.

The heterosexual transmission of HIV is the most common route of spread of the virus worldwide. The amount of virus shed in cervicovaginal secretions affects the risk of HIV transmission to the discordant partner. Thus, characterizing risk factors associated with increased cervicovaginal HIV viral shedding would allow for targeted prevention strategies aimed at decreasing HIV transmission. Attempts to correlate the HIV-RNA levels in secretions of the lower genital tract with variations in sex hormones during the course of the menstrual cycle have yielded mixed results (Cu-Uvin 2005). The lack of standardization of methods makes comparisons difficult. Sampling techniques, sampling frequency, techniques used to detect HIV-1 in the lower genital tract compartment, and population characteristics differed significantly across studies. A prospective study of 55 HIV-1-infected women (74% on highly active antiretroviral therapy) sampled (using Sno-Strips and cervicovaginal lavage fluid) weekly over 8 weeks, reported that genital HIV-RNA is highest during menses and lowest immediately after (Reichelderfer et al. 2000). There also is no effect of menstrual cycle on HIV-RNA in the blood compartment. Another study of 17 women sampled daily for the duration of a menstrual cycle found that cervical HIV-RNA is significantly correlated with the number of days from the midcycle surge in LH (Benki et al. 2004). The lowest levels of cervical HIV-1 RNA are present at the LH surge, and this nadir is followed by an increase in virus levels that reached a maximum before the start of menses. In contrast, there is no significant association between the number of days from the LH surge and the level of HIV-1 RNA in vaginal secretions. Finally, examination of 17 HIV-infected women who underwent daily sampling over the course of a menstrual cycle revealed no correlation between viral shedding and hormonal fluctuations (Mostad 1998).

The effects of hormonal contraception on HIV viral shedding have also yielded mixed results (McClelland 2005). The presence of lower genital tract inflammation in both men and women has been shown to increase HIV-RNA in that compartment (Wright et al. 2001). One study suggested that the use of hormonal contraceptives is an independent predictor of cellular inflammation in the lower genital tract, providing a mechanism by which HIV-RNA shedding may occur (Ghanem et al. 2005). Another study demonstrated a threefold higher odds of HIV cervicovaginal shedding in women on depot-medroxyprogesterone acetate and fourfold and 12-fold higher odds with low-dose and high-dose COCs, respectively (McClelland et al. 2002; Wang et al. 2004). The second study in the same patient population, which was a longitudinal study in 213 women that compared viral shedding pre- and post-hormonal contraceptive initiation (total of 2 visits) reported increased viral shedding after the initiation of hormonal contraceptives (McClelland et al. 2002; Wang et al. 2004).

The role of hormonal contraceptives as a risk factor for HIV acquisition has been suggested by several studies, but controversy still exists. Both cross-sectional and longitudinal studies are plagued by behavioral confounding rendering the interpretation of findings difficult. In the largest prospective study of 6,109 HIV-uninfected women, aged 18–35 years, recruited from family planning clinics in Uganda and Zimbabwe, neither the use of COC (HR, 0.99; 95% CI, 0.69–1.42) nor depot-medroxyprogesterone acetate (HR, 1.25; 95% CI, 0.89–1.78) was associated with risk of HIV acquisition overall (Morrison et al. 2007a). In a subgroup analysis of HSV-2-seronegative participants, however, both COC (HR, 2.85; 95% CI, 1.39–5.82) and depot-medroxyprogesterone acetate (HR, 3.97; 95% CI, 1.98–8.00) users had an increased risk of HIV acquisition compared with the nonhormonal user group. This latter finding suggests a possible interaction between HSV-2 and HIV. Over 100 million women worldwide use some form of hormonal contraception. Although the role of sex hormones in mediating HIV acquisition/transmission and disease progression is still unclear, a better understanding of the effects of hormonal contraceptives on lower genital tract viral shedding and inflammation would allow for the incorporation of hormonal markers into guidelines for HIV prevention.

During pregnancy (reviewed by Watts 2005), plasma HIV-RNA levels are not increased, but a slight rebound is observed in the immediate postpartum period. The three largest cross-sectional studies conducted during pregnancy suggest a 2.6–4.6 increased odds of cervicovaginal HIV-RNA shedding as compared with nonpregnant women (Clemetson et al. 1993; Henin et al. 1993; Kreiss et al. 1994). As for risk of HIV acquisition during pregnancy, the largest prospective study of 4,439 HIV-uninfected women from family planning sites in Uganda and Zimbabwe did not reveal an increased risk of HIV-acquisition during pregnancy (Morrison et al. 2007b).

In summary, there are conflicting data on the effects of sex hormones and susceptibility to HIV infection. If such an association exists, it is at best small. A minority of pregnant women in resource-limited settings has access to antiretroviral therapy to decrease vertical HIV-1 transmission; therefore, it has been argued that the use of hormonal contraception in such settings has led to a significant reduction in the number of vertical transmissions and, thus, the number of HIV-1 infections, in general. There are no data, to date, to justify altering hormonal contraceptive recommendations in women, even those at highest risk for HIV.

HSV-2. Both HSV-1 and HSV-2 are associated with recurrent anogenital infections in humans. Based on the latest National health and nutrition examination survey (NHANES) data, 24% of the United States population is infected with HSV-2 (Xu et al. 2006), while among adolescents in some areas, epidemiologic data suggest that HSV-1 is becoming a more frequent cause of anogenital infection than HSV-2 (Kortekangas-Savolainen and Vuorinen 2007). Most studies evaluating the effects of sex hormones on HSV susceptibility and progression have focused on HSV-2.

In a cross-sectional study of 273 women who were seropositive for HSV-2, women using either depot-medroxyprogesterone acetate and COCs are more likely to experience HSV-2 shedding (Mostad et al. 1997). A longitudinal study of women initiating hormonal contraception, however, did not reveal increased HSV-2 shedding (Wang et al. 2004). The data on hormonal contraception and HSV-2 acquisition are much more limited. In a study of 302 Kenyan commercial sex workers, hormonal contraception increased the risk of genital ulcer disease acquisition, but HSV-2 was likely responsible for only 50% of the genital ulcers (Baeten et al. 2007). In a prospective study of 948 Kenyan commercial sex workers, neither depot-medroxy-progesterone acetate nor COC use increased incident genital ulcerative disease acquisition (Baeten et al. 2001). Shedding of genital HSV-DNA is nearly eightfold higher during pregnancy (Mostad et al. 2000). There are no prospective studies demonstrating increased risk of HSV acquisition during pregnancy.

Significant data on the effect of sex hormones on susceptibility and course of HSV-2 infection have been obtained from mouse models. In ovariectomized mice administered estradiol, progesterone, a combination of both hormones, or saline, administration of saline or progesterone increases susceptible to HSV-2 infection, with extensive infection of the vaginal epithelium and induction of inflammatory chemokine and chemokine receptors (Gillgrass et al. 2005a). 17β-estradiol treatment, however, protects mice from HSV-2 infection. A slower progression of genital pathology is noted in the ovariectomized mice receiving a combination of hormones. Immunization with an attenuated HSV-2 vaccine protects ovariecto-mized females injected with progesterone from subsequent rechallenge, but not females treated with 17β-estradiol (Gillgrass et al. 2005b). These findings suggest that protection against challenge is dependent upon the ability of the attenuated strain to cause infection, which, in turn, is dependent upon the hormonal environ-ment. Lymphoid aggregates consisting of dendritic cells and T-cells form in the vagina of the progesterone-treated, but not estradiol-treated, mice. Finally, the duration of exposure to hormones significantly affects responses to HSV-2; mice that were exposed to progesterone for 5 days and immunized with an attenuated HSV-2 vaccine are more protected from rechallenge with a virulent strain of HSV-2 than are those exposed to progesterone for 15 days prior to immunization (Gillgrass et al. 2005b). These data suggest that hormone type as well as duration of exposure influence susceptibility to infection. Treatment of human female primary epithelial and stromal cocultures with 17β-estradiol increases HSV-2 infection in endometrial cells, whereas treatment with progesterone reduces viral shedding (MacDonald et al. 2007).

Perhaps, the most intriguing association between sex hormones and HSV-2 are the data from the glycoprotein-D-subunit vaccine which, overall, does not protect against acquisition of HSV infection (Stanberry et al. 2002). However, in subgroup analyses, women who were initially seronegative for both HSV-1 and HSV-2 were protected against incident herpes infection after receiving the vaccine but no similar association was noted in men. Whether this finding can be attributed to the effect of sex hormones on the immune response to the vaccine is unclear but interesting.

HPV. HPV detection increases in the follicular phase of the cycle (van Ham et al. 2002) or in the luteal phase (Schneider et al. 1992) of the menstrual cycle in some studies, but not others (Brabin et al. 2005; Fairley et al. 1994; Harper et al. 2003; Wheeler et al. 1996). Inconsistencies across studies may be attributed to variability in assignment of stage of menstrual cycle and method of HPV detection. In a large analysis of 3,739 HPV-positive samples collected in the ASC-US LSIL triage study (ALTS), increased HPV viral load was observed in samples collected during the periovulatory period (Sherman et al. 2006). Vaccine-induced HPV16-specific cervical IgG also varies accordingly to the stage of the menstrual cycle in ovulating women, reaching a nadir immediately prior to the LH surge at ovulation (Nardelli-Haefliger et al. 2003); thus, immune suppression around the time of ovulation may result in a transient increase in viral load, as was observed in the ALTS study.

The association of COC use with prevalent detection of HPV is inconsistent and varies considerably by age, study design, and HPV detection method (Green et al. 2003; Mohllajee et al. 2006). Increased prevalence of any and multiple HPV infections is reported among younger (20–30 years), but not older (40–50 years), women in Denmark (Nielsen et al. 2008). A pooled analysis of prevalence surveys of over 20,000 women with mean age 32–47 years found no association with COC use or duration of use and prevalence of HPV (Vaccarella et al. 2006). Sex steroids are hypothesized to increase the risk of HPV acquisition via increased cervical ectopy (Critchlow et al. 1995; Jacobson et al. 2000) or through modulating the host immune response to the HPV infection, particularly during extreme hormonal disruptions, such as those observed in pregnancy and the menopausal transition. Studies have addressed the risk of HPV acquisition by COC use with both positive (Winer et al. 2003) and negative (Moscicki et al. 2001) associations reported. While most studies report no association between COC use and HPV DNA persistence (Hildesheim et al., 1994; Ho et al. 1998; Moscicki et al. 1998; Richardson et al. 2005), these studies are limited in power to detect meaningful differences in the duration of infection and nonstandardized definitions of HPV persistence.

HPV prevalence increases during pregnancy (Banura et al. 2008; Fife et al. 1999; Hernandez-Giron et al. 2005; Morrison et al. 1996; Ziegler et al. 2003) with a "catch-up" clearance in the early postpartum period (Fife et al. 1999; Nobbenhuis et al. 2002); the increase in HPV prevalence during pregnancy, however, is not uniformly observed (de Roda Husman et al. 1995; Kemp et al. 1992). An increase in HPV acquisition during pregnancy is unlikely to explain the increased HPV prevalence, and it is hypothesized that the observations reflect increases in HPV viral load to levels above the limit of detection resulting from hormone-mediated immunosuppression. Interestingly, though high parity significantly increases the risk of cervical cancer (International Collaboration of Epidemiological Studies of Cervical Cancer 2006), HPV prevalence is inversely correlated with number of live births (Vaccarella et al. 2006).

Sex steroid exposures (e.g., long duration COC use and multiparity) are associated with an increased risk of invasive cervical cancer (ICC) caused by infection with high-risk (HR)-HPV (Appleby et al. 2007). COC use can mediate ICC by increasing the risk of acquisition and/or persistence of HR-HPV or by increasing the risk of neoplastic transformation of HR-HPV infected epithelium. Tissue culture models show an increase in HPV oncogene (E6/E7) expression in HeLa cells following sex steroid exposure (Chen et al. 1996). 17β-estradiol exposure in K14E6/E7 double transgenic mice, which express HPV16 E6/E7 from a keratin promoter thereby targeting expression to epithelial cells, increases the persistence and progression of epithelial tumors (Brake and Lambert 2005), suggesting a biologically plausible effect of COC on cervical cancer progression. The effects of sex steroids on susceptibility or host response to infection, however, cannot be determined from these in vitro models.

Due to the species specificity of papillomavirus infectivity, animal models of genital tract HPV infection are not available, limiting the ability to conduct controlled experiments designed to assess the effects of sex steroids on host immune response to infections. Rabbit oral papillomavirus (Wilgenburg et al. 2005) and cottontail rabbit papillomaviruses (Selvakumar et al. 1997) are good general models to study the immune response to mucosal and cutaneous papillomavirus infections, respectively. However, to our knowledge, the influence of sex steroids in these models has not been evaluated.

10.2.3   Fungal Genital Tract Infections

Candidiasis. Vulvovaginal candidiasis (VVC) is rare among prepubertal and post-menopausal women, suggesting a role for sex steroids in susceptibility to VVC. Consistent with this hypothesis, candidiasis occurs more commonly during the luteal phase of the menstrual cycle, which is associated with both high estrogen and progesterone (Nomelini et al. 2007). Similar results are found in studies of recurrent VVC (RVVC), with increased diagnosis in the luteal phase of the cycle and a significantly lower urine pregnanediol glucuronide concentration in patients with RVVC (Spacek et al. 2007). These data suggest that a hormonal imbalance, possibly favoring estrogen, may be involved in the etiology of VVC. High 17β-estradiol levels increase glycogen content in vaginal epithelial cells, providing a hospitable environment for *Candida* growth in vitro (Larsen and Galask 1984). In addition, candidal invasiveness is increased by progesterone and estrogen treatment which induces switching to the hyphal form (White and Larsen 1997). Estrogen also enhances persistent VVC, whereas progesterone alone does not, in mouse models (Fidel et al. 2000).

Some studies find an increase in VVC among women using COC (Cetin et al. 2007; Cotch et al. 1998; Geiger and Foxman 1996; Grigoriou et al. 2006; Parazzini et al. 2000; Spinillo et al. 1995), while others reported no association between COC use and VVC (Bauters et al. 2002; Beigi et al. 2004; Peddie et al. 1984). An increase in prevalence of *Candida* colonization, however, is associated with current pregnancy and postmenopausal HRT use (Bauters et al. 2002). Although this study does not report an increased prevalence among current COC users in general, women using higher dose of estrogen formulations have a higher *Candida* prevalence compared with women using lower dose formulations. In the same study, higher colonization was reported in the preovulatory, or follicular, phase of the menstrual cycle, which is inconsistent with that reported by others (Nomelini et al. 2007; Spacek et al. 2007).

VVC is a common side effect of pregnancy (Grigoriou et al. 2006; Xu and Sobel 2004). Studies which prospectively tested the *Candida* species throughout pregnancy find that pregnancy-associated symptomatic VVC is associated with prepregnancy commensal *Candida* strains (Daniels et al. 2001; Maffei et al. 1997). Decreased lymphocyte transformation in response to candida antigen in pregnancy is reported in both current carriers and noncarriers, particularly in the third trimester (Brunham et al. 1983). These data suggest that hormonal changes during pregnancy are associated with a decrease in cell-mediated immune function necessary for control of VVC.

10.2.4   Parasitic Genital Tract Infections

Trichomoniasis. Trichomonas vaginalis infection is rarely symptomatic in men, but can have serious adverse outcomes in women including preterm birth and an increased risk of HIV acquisition. *T. vaginalis* detection is reduced at menstruation (Demes et al. 1988a; Fox 1988), potentially resulting from an increase in parasite killing from menstrual blood complement (Demes et al. 1988b). Some studies have found that oral contraceptive use is associated with a decreased risk of *T. vaginalis* infection whereas others have found no association (Mohllajee et al. 2006). In a prospective study of female sex workers, COC use had no effect on *T. vaginalis* acquisition, whereas depot medroxyprogesterone acetate use was significantly protective (Baeten et al. 2001).

 T. vaginalis infection is associated with increased total IgG, IgA, and neutrophils among ovulatory women, but not among women taking COC (Chipperfield and Evans 1975). Women taking COC have a generalized increase in total IgG and IgA during *T. vaginalis* infection (Chipperfield and Evans 1975) that is similar to the heightened antiHPV16 IgG and IgA levels reported in COC users (Nardelli-Haefliger et al. 2003). *T. vaginalis* infection among pregnant women was associated with an increase in neutrophil counts, concentrations of IL1-β and CXCL8, and production of defensins (Cauci and Culhane 2007; Simhan et al. 2007). Conversely, normal pregnancies are associated with a decrease in IL-1R antagonist (Donders et al. 2003), suggesting that increased IL-1β from *T. vaginalis* infection may interact with hormonal dysregulation of cervical immunity during pregnancy to increase inflammatory microenvironment, possibly contributing to adverse pregnancy outcomes.

10.3   Summary and Outstanding Research Questions

The evidence for sex steroids having a central role in female genital tract immunology is clear. The mechanisms by which sex steroids orchestrate the balance between reproduction and protection against pathogenic microorganisms, however, are complex and poorly understood. It is striking that the most significant morbidity associated with bacterial and parasitic infection is adverse pregnancy outcome, suggesting a loss in the homeostatic balance of immunity in the genital tract.

The precise role that sex steroids play in inducing or maintaining this imbalance is as yet undefined. The homeostatic balance of the genital tract involves significant cross-talk among the epithelium, the underlying stromal cells, and immune cells. As a result, simple cell-specific models with individual steroid exposures, while informative, will be limited in their ability to provide insight into the complex interplay that likely explains the pathology of genital tract infections. Better in vitro models of the interactions between combined sex steroids and the immune cells of the genital tract (including epithelium, stroma, monocytes, and others) are needed. In addition, studies directly comparing the response to genital tract infection and sex steroid exposures in rodents and humans are necessary to better understand the validity of the data from animal models. The balance and regulation (or dysregulation) of the immune response to genital tract infection are likely more critical than the magnitude of response, limiting the inferences gained from cross-sectional studies and prospective studies with infrequent sampling. The development of more precise measures of sex steroid exposures and markers of local immunity are needed to improve the information obtained in epidemiologic research. The genital tract is relatively easily accessible allowing for more intensive sample collection over time. Application of improved biomarkers to samples collected in small, well-defined, and intensively sampled cohorts may prove more valuable than larger studies with less frequent sampling. Obtaining a comprehensive understanding of female genital tract immune regulation by sex steroids will require significant interdisciplinary collaboration.

References

Al-Harthi L, Wright DJ, Anderson D, Cohen M, Matity Ahu D, Cohn J, Cu-Unvin S, Burns D, Reichelderfer P, Lewis S, Beckner S, Kovacs A, Landay A (2000) The impact of the ovulatory cycle on cytokine production: evaluation of systemic, cervicovaginal, and salivary compartments. J Interferon Cytokine Res 20(8):719–724

Appleby P, Beral V, Berrington de Gonzalez A, Colin D, Franceschi S, Goodhill A, Green J, Peto J, Plummer M, Sweetland S (2007) Cervical cancer and hormonal contraceptives: collaborative reanalysis of individual data for 16, 573 women with cervical cancer and 35, 509 women without cervical cancer from 24 epidemiological studies. Lancet 370(9599):1609–1621

Asin SN, Heimberg AM, Eszterhas SK, Rollenhagen C, Howell AL (2008) Estradiol and progesterone regulate HIV type 1 replication in peripheral blood cells. AIDS Res Hum Retroviruses 24(5):701–716

Baeten JM, Benki S, Chohan V, Lavreys L, McClelland RS, Mandaliya K, Ndinya-Achola JO, Jaoko W, Overbaugh J (2007) Hormonal contraceptive use, herpes simplex virus infection, and risk of HIV-1 acquisition among Kenyan women. AIDS 21(13):1771–1777

Baeten JM, Nyange PM, Richardson BA, Lavreys L, Chohan B, Martin HL Jr, Mandaliya K, Ndinya-Achola JO, Bwayo JJ, Kreiss JK (2001) Hormonal contraception and risk of sexually transmitted disease acquisition: results from a prospective study. Am J Obstet Gynecol 185 (2):380–385

Banura C, Franceschi S, Doorn LJ, Arslan A, Wabwire-Mangen F, Mbidde EK, Quint W, Weiderpass E (2008) Infection with human papillomavirus and HIV among young women in Kampala, Uganda. J Infect Dis 197(4):555–562

Bauters TG, Dhont MA, Temmerman MI, Nelis HJ (2002) Prevalence of vulvovaginal candidiasis and susceptibility to fluconazole in women. Am J Obstet Gynecol 187(3):569–574

Beigi RH, Meyn LA, Moore DM, Krohn MA, Hillier SL (2004) Vaginal yeast colonization in nonpregnant women: a longitudinal study. Obstet Gynecol 104(5 Pt 1):926–930

Benki S, Mostad SB, Richardson BA, Mandaliya K, Kreiss JK, Overbaugh J (2004) Cyclic shedding of HIV-1 RNA in cervical secretions during the menstrual cycle. J Infect Dis 189 (12):2192–2201

Black CA, Rohan LC, Cost M, Watkins SC, Draviam R, Alber S, Edwards RP (2000) Vaginal mucosa serves as an inductive site for tolerance. J Immunol 165(9):5077–5083

Bose SK, Goswami PC (1986) Enhancement of adherence and growth of Chlamydia trachomatis by estrogen treatment of HeLa cells. Infect Immun 53(3):646–650

Brabin L, Fairbrother E, Mandal D, Roberts SA, Higgins SP, Chandick S, Wood P, Barnard G, Kitchener HC (2005) Biological and hormonal markers of chlamydia, human papillomavirus, and bacterial vaginosis among adolescents attending genitourinary medicine clinics. Sex Transm Infect 81(2):128–132

Brake T, Lambert PF (2005) Estrogen contributes to the onset, persistence, and malignant progression of cervical cancer in a human papillomavirus-transgenic mouse model. Proc Natl Acad Sci USA 102(7):2490–2495

Brunham RC, Martin DH, Hubbard TW, Kuo CC, Critchlow CW, Cles LD, Eschenbach DA, Holmes KK (1983) Depression of the lymphocyte transformation response to microbial antigens and to phytohemagglutinin during pregnancy. J Clin Invest 72(5):1629–1638

Castle PE, Hildesheim A, Bowman FP, Strickler HD, Walker JL, Pustilnik T, Edwards RP, Crowley-Nowick PA (2002) Cervical concentrations of interleukin-10 and interleukin-12 do not correlate with plasma levels. J Clin Immunol 22(1):23–27

Cauci S, Culhane JF (2007) Modulation of vaginal immune response among pregnant women with bacterial vaginosis by Trichomonas vaginalis, Chlamydia trachomatis, Neisseria gonorrhoeae, and yeast. Am J Obstet Gynecol 196(2):133 e1-7

Cetin M, Ocak S, Gungoren A, Hakverdi AU (2007) Distribution of Candida species in women with vulvovaginal symptoms and their association with different ages and contraceptive methods. Scand J Infect Dis 39(6–7):584–588

Chen YH, Huang LH, Chen TM (1996) Differential effects of progestins and estrogens on long control regions of human papillomavirus types 16 and 18. Biochem Biophys Res Commun 224 (3):651–659

Chipperfield EJ, Evans BA (1975) Effect of local infection and oral contraception on immunoglobulin levels in cervical mucus. Infect Immun 11(2):215–221

Clemetson DB, Moss GB, Willerford DM, Hensel M, Emonyi W, Holmes KK, Plummer F, Ndinya-Achola J, Roberts PL, Hillier S et al (1993) Detection of HIV DNA in cervical and vaginal secretions. Prevalence and correlates among women in Nairobi, Kenya. JAMA 269(22):2860–2864

Cotch MF, Hillier SL, Gibbs RS, Eschenbach DA (1998) Epidemiology and outcomes associated with moderate to heavy Candida colonization during pregnancy. Vaginal Infections and Prematurity Study Group. Am J Obstet Gynecol 178(2):374–380

Critchlow CW, Wolner-Hanssen P, Eschenbach DA, Kiviat NB, Koutsky LA, Stevens CE, Holmes KK (1995) Determinants of cervical ectopia and of cervicitis: age, oral contraception, specific cervical infection, smoking, and douching. Am J Obstet Gynecol 173(2):534–543

Crowley T, Horner P, Hughes A, Berry J, Paul I, Caul O (1997) Hormonal factors and the laboratory detection of Chlamydia trachomatis in women: implications for screening? Int J STD AIDS 8(1):25–31

Cu-Uvin S (2005) Effect of the menstrual cycle on virological parameters. J Acquir Immune Defic Syndr 38(Suppl 1):S33–S34

Daniels W, Glover DD, Essmann M, Larsen B (2001) Candidiasis during pregnancy may result from isogenic commensal strains. Infect Dis Obstet Gynecol 9(2):65–73

de Roda Husman AM, Walboomers JM, Hopman E, Bleker OP, Helmerhorst TM, Rozendaal L, Voorhorst FJ, Meijer CJ (1995) HPV prevalence in cytomorphologically normal cervical scrapes of pregnant women as determined by PCR: the age-related pattern. J Med Virol 46(2):97–102

Demes P, Gombosova A, Valent M, Fabusova H, Janoska A (1988a) Fewer Trichomonas vaginalis organisms in vaginas of infected women during menstruation. Genitourin Med 64(1):22–24

Demes P, Gombosova A, Valent M, Janoska A, Fabusova H, Petrenko M (1988b) Differential susceptibility of fresh Trichomonas vaginalis isolates to complement in menstrual blood and cervical mucus. Genitourin Med 64(3):176–179

Donders GG, Vereecken A, Bosmans E, Spitz B (2003) Vaginal cytokines in normal pregnancy. Am J Obstet Gynecol 189(5):1433–1438

Dunne EF, Unger ER, Sternberg M, McQuillan G, Swan DC, Patel SS, Markowitz LE (2007) Prevalence of HPV infection among females in the United States. JAMA 297:813

Eschenbach DA, Thwin SS, Patton DL, Hooton TM, Stapleton AE, Agnew K, Winter C, Meier A, Stamm WE (2000) Influence of the normal menstrual cycle on vaginal tissue, discharge, and microflora. Clin Infect Dis 30(6):901–907

Fairley CK, Robinson PM, Chen S, Tabrizi SN, Garland SM (1994) The detection of HPV DNA, the size of tampon specimens and the menstrual cycle. Genitourin Med 70(3):171–174

Farzadegan H, Hoover DR, Astemborski J, Lyles CM, Margolick JB, Markham RB, Quinn TC, Vlahov D (1998) Sex differences in HIV-1 viral load and progression to AIDS. Lancet 352 (9139):1510–1514

Fidel PL Jr, Cutright J, Steele C (2000) Effects of reproductive hormones on experimental vaginal candidiasis. Infect Immun 68(2):651–657

Fife KH, Katz BP, Brizendine EJ, Brown DR (1999) Cervical human papillomavirus deoxyribonucleic acid persists throughout pregnancy and decreases in the postpartum period. Am J Obstet Gynecol 180(5):1110–1114

Fox H (1988) Fewer Trichomonas vaginalis organisms in vaginas of infected women during menstruation. Genitourin Med 64(4):280

FUTURE II Study Group (2007) Quadrivalent vaccine against human papillomavirus to prevent high-grade cervical lesions. N Engl J Med 356(19):1915–1927

Garland SM, Hernandez-Avila M, Wheeler CM, Perez G, Harper DM, Leodolter S, Tang GW, Ferris DG, Steben M, Bryan J, Taddeo FJ, Railkar R, Esser MT, Sings HL, Nelson M, Boslego J, Sattler C, Barr E, Koutsky LA (2007) Quadrivalent vaccine against human papillomavirus to prevent anogenital diseases. N Engl J Med 356(19):1928–1943

Geiger AM, Foxman B (1996) Risk factors for vulvovaginal candidiasis: a case-control study among university students. Epidemiology 7(2):182–187

Geisler WM, James AB (2008) Chlamydial and gonococcal infections in women seeking pregnancy testing at family-planning clinics. Am J Obstet Gynecol 198(5):502 e1-4

Ghanem KG, Shah N, Klein RS, Mayer KH, Sobel JD, Warren DL, Jamieson DJ, Duerr AC, Rompalo AM (2005) Influence of sex hormones, HIV status, and concomitant sexually transmitted infection on cervicovaginal inflammation. J Infect Dis 191(3):358–366

Gillgrass AE, Ashkar AA, Rosenthal KL, Kaushic C (2003) Prolonged exposure to progesterone prevents induction of protective mucosal responses following intravaginal immunization with attenuated herpes simplex virus type 2. J Virol 77(18):9845–9851

Gillgrass AE, Fernandez SA, Rosenthal KL, Kaushic C (2005a) Estradiol regulates susceptibility following primary exposure to genital herpes simplex virus type 2, while progesterone induces inflammation. J Virol 79(5):3107–3116

Gillgrass AE, Tang VA, Towarnicki KM, Rosenthal KL, Kaushic C (2005b) Protection against genital herpes infection in mice immunized under different hormonal conditions correlates with induction of vagina-associated lymphoid tissue. J Virol 79(5):3117–3126

Gorgos L, Fine D, Marrazzo J (2008) Chlamydia positivity in American Indian/Alaska native women screened in family planning clinics, 1997–2004. Sex Transm Dis 35(8):753–757

Gravitt P, Hildesheim A, Herrero R, Schiffman M, Sherman M, Bratti M, Rodriguez A, Morera L, Cardenas F, Bowman F, Shah K, Crowley-Nowick P (2003) Correlates of IL-10 and IL-12 concentrations in cervical secretions. J Clin Immunol 23(3):175–183

Green J, Berrington de Gonzalez A, Smith JS, Franceschi S, Appleby P, Plummer M, Beral V (2003) Human papillomavirus infection and use of oral contraceptives. Br J Cancer 88(11):1713–1720

Grigoriou O, Baka S, Makrakis E, Hassiakos D, Kapparos G, Kouskouni E (2006) Prevalence of clinical vaginal candidiasis in a university hospital and possible risk factors. Eur J Obstet Gynecol Reprod Biol 126(1):121–125

Harlow SD, Ephross SA (1995) Epidemiology of menstruation and its relevance to women's health. Epidemiol Rev 17(2):265–286

Harper DM, Longacre MR, Noll WW, Belloni DR, Cole BF (2003) Factors affecting the detection rate of human papillomavirus. Ann Fam Med 1(4):221–227

Henin Y, Mandelbrot L, Henrion R, Pradinaud R, Coulaud JP, Montagnier L (1993) Virus excretion in the cervicovaginal secretions of pregnant and nonpregnant HIV-infected women. J Acquir Immune Defic Syndr 6(1):72–75

Hernandez-Giron C, Smith JS, Lorincz A, Lazcano E, Hernandez-Avila M, Salmeron J (2005) High-risk human papillomavirus detection and related risk factors among pregnant and non-pregnant women in Mexico. Sex Transm Dis 32(10):613–618

Hildesheim A, Schiffman MH, Gravitt PE, Glass AG, Greer CE, Zhang T, Scott DR, Rush BB, Lawler P, Sherman ME et al (1994) Persistence of type-specific human papillomavirus infection among cytologically normal women. J Infect Dis 169(2):235–240

Ho GY, Bierman R, Beardsley L, Chang CJ, Burk RD (1998) Natural history of cervicovaginal papillomavirus infection in young women. N Engl J Med 338(7):423–428

International Collaboration of Epidemiological Studies of Cervical Cancer (2006) Cervical carcinoma and reproductive factors: collaborative reanalysis of individual data on 16, 563 women with cervical carcinoma and 33, 542 women without cervical carcinoma from 25 epidemiological studies. Int J Cancer 119(5):1108–1124

Jacobson DL, Peralta L, Farmer M, Graham NM, Gaydos C, Zenilman J (2000) Relationship of hormonal contraception and cervical ectopy as measured by computerized planimetry to chlamydial infection in adolescents. Sex Transm Dis 27(6):313–319

Johansson EL, Wassen L, Holmgren J, Jertborn M, Rudin A (2001) Nasal and vaginal vaccinations have differential effects on antibody responses in vaginal and cervical secretions in humans. Infect Immun 69(12):7481–7486

Johnson HL, Erbelding EJ, Zenilman JM, Ghanem KG (2007) Sexually transmitted diseases and risk behaviors among pregnant women attending inner city public sexually transmitted diseases clinics in Baltimore, MD, 1996–2002. Sex Transm Dis 34(12):991–994

Kaushic C, Ashkar AA, Reid LA, Rosenthal KL (2003) Progesterone increases susceptibility and decreases immune responses to genital herpes infection. J Virol 77(8):4558–4565

Kemp EA, Hakenewerth AM, Laurent SL, Gravitt PE, Stoerker J (1992) Human papillomavirus prevalence in pregnancy. Obstet Gynecol 79(5 (Pt 1)):649–656

Kortekangas-Savolainen O, Vuorinen T (2007) Trends in herpes simplex virus type 1 and 2 infections among patients diagnosed with genital herpes in a Finnish sexually transmitted disease clinic, 1994–2002. Sex Transm Dis 34(1):37–40

Kreiss J, Willerford DM, Hensel M, Emonyi W, Plummer F, Ndinya-Achola J, Roberts PL, Hoskyn J, Hillier S, Kiviat N (1994) Association between cervical inflammation and cervical shedding of human immunodeficiency virus DNA. J Infect Dis 170(6):1597–1601

Kutteh WH, Franklin RD (2001) Quantification of immunoglobulins and cytokines in human cervical mucus during each trimester of pregnancy. Am J Obstet Gynecol 184(5):865–872 discussion 872-4

Kutteh WH, Moldoveanu Z, Mestecky J (1998) Mucosal immunity in the female reproductive tract: correlation of immunoglobulins, cytokines, and reproductive hormones in human cervical mucus around the time of ovulation. AIDS Res Hum Retroviruses 14(Suppl 1):S51–S55

Larsen B, Galask RP (1984) Influence of estrogen and normal flora on vaginal candidiasis in the rat. J Reprod Med 29(12):863–868

Lavreys L, Chohan V, Overbaugh J, Hassan W, McClelland RS, Kreiss J, Mandaliya K, Ndinya-Achola J, Baeten JM (2004) Hormonal contraception and risk of cervical infections among HIV-1-seropositive Kenyan women. AIDS 18(16):2179–2184

Louv WC, Austin H, Perlman J, Alexander WJ (1989) Oral contraceptive use and the risk of chlamydial and gonococcal infections. Am J Obstet Gynecol 160(2):396–402

MacDonald EM, Savoy A, Gillgrass A, Fernandez S, Smieja M, Rosenthal KL, Ashkar AA, Kaushic C (2007) Susceptibility of human female primary genital epithelial cells to herpes simplex virus, type-2 and the effect of TLR3 ligand and sex hormones on infection. Biol Reprod 77(6):1049–1059

Maffei CM, Paula CR, Mazzocato TS, Franceschini S (1997) Phenotype and genotype of Candida albicans strains isolated from pregnant women with recurrent vaginitis. Mycopathologia 137(2):87–94

Mahmoud EA, Hamad EE, Olsson SE, Mardh PA (1994) Antichlamydial activity of cervical secretion in different phases of the menstrual cycle and influence of hormonal contraceptives. Contraception 49(3):265–274

Maslow AS, Davis CH, Choong J, Wyrick PB (1988) Estrogen enhances attachment of Chlamydia trachomatis to human endometrial epithelial cells in vitro. Am J Obstet Gynecol 159(4): 1006–1014

McClelland RS (2005) Effect of exogenous hormones. J Acquir Immune Defic Syndr 38(Suppl 1): S38–S39

McClelland RS, Wang CC, Richardson BA, Corey L, Ashley RL, Mandaliya K, Ndinya-Achola J, Bwayo JJ, Kreiss JK (2002) A prospective study of hormonal contraceptive use and cervical shedding of herpes simplex virus in human immunodeficiency virus type 1-seropositive women. J Infect Dis 185(12):1822–1825

Mohllajee AP, Curtis KM, Martins SL, Peterson HB (2006) Hormonal contraceptive use and risk of sexually transmitted infections: a systematic review. Contraception 73(2):154–165

Molano M, Meijer CJ, Weiderpass E, Arslan A, Posso H, Franceschi S, Ronderos M, Munoz N, van den Brule AJ (2005) The natural course of Chlamydia trachomatis infection in asymptomatic Colombian women: a 5-year follow-up study. J Infect Dis 191(6):907–916

Morrison CS, Bright P, Wong EL, Kwok C, Yacobson I, Gaydos CA, Tucker HT, Blumenthal PD (2004) Hormonal contraceptive use, cervical ectopy, and the acquisition of cervical infections. Sex Transm Dis 31(9):561–567

Morrison CS, Richardson BA, Mmiro F, Chipato T, Celentano DD, Luoto J, Mugerwa R, Padian N, Rugpao S, Brown JM, Cornelisse P, Salata RA (2007a) Hormonal contraception and the risk of HIV acquisition. Aids 21(1):85–95

Morrison CS, Wang J, Van Der Pol B, Padian N, Salata RA, Richardson BA (2007b) Pregnancy and the risk of HIV-1 acquisition among women in Uganda and Zimbabwe. AIDS 21(8): 1027–1034

Morrison EA, Gammon MD, Goldberg GL, Vermund SH, Burk RD (1996) Pregnancy and cervical infection with human papillomaviruses. Int J Gynaecol Obstet 54(2):125–130

Moscicki AB, Hills N, Shiboski S, Powell K, Jay N, Hanson E, Miller S, Clayton L, Farhat S, Broering J, Darragh T, Palefsky J (2001) Risks for incident human papillomavirus infection and low-grade squamous intraepithelial lesion development in young females. JAMA 285 (23):2995–3002

Moscicki AB, Shiboski S, Broering J, Powell K, Clayton L, Jay N, Darragh TM, Brescia R, Kanowitz S, Miller SB, Stone J, Hanson E, Palefsky J (1998) The natural history of human papillomavirus infection as measured by repeated DNA testing in adolescent and young women. J Pediatr 132(2):277–284

Mostad SB (1998) Prevalence and correlates of HIV type 1 shedding in the female genital tract. AIDS Res Hum Retroviruses 14(Suppl 1):S11–S15

Mostad SB, Kreiss JK, Ryncarz AJ, Mandaliya K, Chohan B, Ndinya-Achola J, Bwayo JJ, Corey L (2000) Cervical shedding of herpes simplex virus in human immunodeficiency virus-infected women: effects of hormonal contraception, pregnancy, and vitamin A deficiency. J Infect Dis 181(1):58–63

Mostad SB, Overbaugh J, DeVange DM, Welch MJ, Chohan B, Mandaliya K, Nyange P, Martin HL Jr, Ndinya-Achola J, Bwayo JJ, Kreiss JK (1997) Hormonal contraception, vitamin A deficiency, and other risk factors for shedding of HIV-1 infected cells from the cervix and vagina. Lancet 350(9082):922–927

Nardelli-Haefliger D, Wirthner D, Schiller JT, Lowy DR, Hildesheim A, Ponci F, De Grandi P (2003) Specific antibody levels at the cervix during the menstrual cycle of women vaccinated with human papillomavirus 16 virus-like particles. J Natl Cancer Inst 95(15):1128–1137

Ness RB, Soper DE, Holley RL, Peipert J, Randall H, Sweet RL, Sondheimer SJ, Hendrix SL, Amortegui A, Trucco G, Bass DC, Kelsey SF (2001) Hormonal and barrier contraception and risk of upper genital tract disease in the PID Evaluation and Clinical Health (PEACH) study. Am J Obstet Gynecol 185(1):121–127

Nielsen A, Kjaer SK, Munk C, Iftner T (2008) Type-specific HPV infection and multiple HPV types: prevalence and risk factor profile in nearly 12, 000 younger and older Danish women. Sex Transm Dis 35(3):276–282

Nobbenhuis MA, Helmerhorst TJ, van den Brule AJ, Rozendaal L, Bezemer PD, Voorhorst FJ, Meijer CJ (2002) High-risk human papillomavirus clearance in pregnant women: trends for lower clearance during pregnancy with a catch-up postpartum. Br J Cancer 87(1):75–80

Nomelini RS, Pansani PL, Murta EF (2007) Frequency of cervical intraepithelial neoplasia and infectious agents for vaginitis in menstrual cycle phase. Eur J Gynaecol Oncol 28(5):389–393

Parazzini F, Di Cintio E, Chiantera V, Guaschino S (2000) Determinants of different Candida species infections of the genital tract in women. Sporachrom Study Geoup. Eur J Obstet Gynecol Reprod Biol 93(2):141–145

Pasley JN, Rank RG, Hough AJ Jr, Cohen C, Barron AL (1985) Absence of progesterone effects on chlamydial genital infection in female guinea pigs. Sex Transm Dis 12(3):155–158

Peddie BA, Bishop V, Bailey RR, McGill H (1984) Relationship between contraceptive method and vaginal flora. Aust N Z J Obstet Gynaecol 24(3):217–218

Putkonen T, Ebeling K (1950) Gonococci and the menstrual cycle. J Vener Dis Inf 31(10):263–267

Rahm VA, Odlind V, Pettersson R (1991) Chlamydia trachomatis in sexually active teenage girls. Factors related to genital chlamydial infection: a prospective study. Genitourin Med 67 (4):317–321

Reichelderfer PS, Coombs RW, Wright DJ, Cohn J, Burns DN, Cu-Uvin S, Baron PA, Coheng MH, Landay AL, Beckner SK, Lewis SR, Kovacs AA (2000) Effect of menstrual cycle on HIV-1 levels in the peripheral blood and genital tract. WHS 001 Study Team. AIDS 14 (14):2101–2107

Richardson H, Abrahamowicz M, Tellier PP, Kelsall G, du Berger R, Ferenczy A, Coutlee F, Franco EL (2005) Modifiable risk factors associated with clearance of type-specific cervical human papillomavirus infections in a cohort of university students. Cancer Epidemiol Biomarkers Prev 14(5):1149–1156

Roberts JN, Buck CB, Thompson CD, Kines R, Bernardo M, Choyke PL, Lowy DR, Schiller JT (2007) Genital transmission of HPV in a mouse model is potentiated by nonoxynol-9 and inhibited by carrageenan. Nat Med 13(7):857–861

Rosenthal GE, Landefeld CS (1990) The relation of chlamydial infection of the cervix to time elapsed from the onset of menses. J Clin Epidemiol 43(1):15–20

Rubin GL, Ory HW, Layde PM (1982) Oral contraceptives and pelvic inflammatory disease. Am J Obstet Gynecol 144(6):630–635

Rupp RE, Stanberry LR, Rosenthal SL (2005) Vaccines for sexually transmitted infections. Pediatr Ann 34(10):818–820 822-4

Schneider A, Kirchhoff T, Meinhardt G, Gissmann L (1992) Repeated evaluation of human papillomavirus 16 status in cervical swabs of young women with a history of normal Papanicolaou smears. Obstet Gynecol 79(5 (Pt 1)):683–688

Selvakumar R, Schmitt A, Iftner T, Ahmed R, Wettstein FO (1997) Regression of papillomas induced by cottontail rabbit papillomavirus is associated with infiltration of CD8+ cells and persistence of viral DNA after regression. J Virol 71(7):5540–5548

Senanayake P, Kramer DG (1980) Contraception and the etiology of pelvic inflammatory disease: new perspectives. Am J Obstet Gynecol 138(7 Pt 2):852–860

Sherman ME, Carreon JD, Schiffman M (2006) Performance of cytology and human papilloma-virus testing in relation to the menstrual cycle. Br J Cancer 94(11):1690–1696

Shrier LA, Bowman FP, Lin M, Crowley-Nowick PA (2003) Mucosal immunity of the adolescent female genital tract. J Adolesc Health 32(3):183–186

Simhan HN, Anderson BL, Krohn MA, Heine RP, Martinez de Tejada B, Landers DV, Hillier SL (2007) Host immune consequences of asymptomatic Trichomonas vaginalis infection in pregnancy. Am J Obstet Gynecol 196((1):59 e1-5

Sonnex C (1998) Influence of ovarian hormones on urogenital infection. Sex Transm Infect 74(1):11–19

Spacek J, Buchta V, Jilek P, Forstl M (2007) Clinical aspects and luteal phase assessment in patients with recurrent vulvovaginal candidiasis. Eur J Obstet Gynecol Reprod Biol 131(2):198–202

Spinillo A, Capuzzo E, Nicola S, Baltaro F, Ferrari A, Monaco A (1995) The impact of oral contraception on vulvovaginal candidiasis. Contraception 51(5):293–297

Stanberry LR, Spruance SL, Cunningham AL, Bernstein DI, Mindel A, Sacks S, Tyring S, Aoki FY, Slaoui M, Denis M, Vandepapeliere P, Dubin G (2002) Glycoprotein-D-adjuvant vaccine to prevent genital herpes. N Engl J Med 347(21):1652–1661

Sterling TR, Vlahov D, Astemborski J, Hoover DR, Margolick JB, Quinn TC (2001) Initial plasma HIV-1 RNA levels and progression to AIDS in women and men. N Engl J Med 344(10): 720–725

Straub RH (2007) The complex role of estrogens in inflammation. Endocr Rev 28(5):521–574

Sugarman B, Agbor P (1986) Estrogens and Chlamydia trachomatis. Proc Soc Exp Biol Med 183(1):125–131

Sweet RL, Blankfort-Doyle M, Robbie MO, Schacter J (1986) The occurrence of chlamydial and gonococcal salpingitis during the menstrual cycle. JAMA 255(15):2062–2064

Vaccarella S, Herrero R, Dai M, Snijders PJ, Meijer CJ, Thomas JO, Hoang Anh PT, Ferreccio C, Matos E, Posso H, de Sanjose S, Shin HR, Sukvirach S, Lazcano-Ponce E, Ronco G, Rajkumar R, Qiao YL, Munoz N, Franceschi S (2006) Reproductive factors, oral contraceptive use, and human papillomavirus infection: pooled analysis of the IARC HPV prevalence surveys. Cancer Epidemiol Biomarkers Prev 15(11):2148–2153

van Ham MA, Melchers WJ, Hanselaar AG, Bekkers RL, Boonstra H, Massuger LF (2002) Fluctuations in prevalence of cervical human papillomavirus in women frequently sampled during a single menstrual cycle. Br J Cancer 87(4):373–376

Wang CC, McClelland RS, Overbaugh J, Reilly M, Panteleeff DD, Mandaliya K, Chohan B, Lavreys L, Ndinya-Achola J, Kreiss JK (2004) The effect of hormonal contraception on genital tract shedding of HIV-1. AIDS 18(2):205–209

Watts DH (2005) Effect of pregnancy. J Acquir Immune Defic Syndr 38(Suppl 1):S36–S38

Wheeler CM, Greer CE, Becker TM, Hunt WC, Anderson SM, Manos MM (1996) Short-term fluctuations in the detection of cervical human papillomavirus DNA. Obstet Gynecol 88(2):261–268

White S, Larsen B (1997) Candida albicans morphogenesis is influenced by estrogen. Cell Mol Life Sci 53(9):744–749

Wilgenburg BJ, Budgeon LR, Lang CM, Griffith JW, Christensen ND (2005) Characterization of immune responses during regression of rabbit oral papillomavirus infections. Comp Med 55(5):431–439

Wilson ME, Dimayuga FO, Reed JL, Curry TE, Anderson CF, Nath A, Bruce-Keller AJ (2006) Immune modulation by estrogens: role in CNS HIV-1 infection. Endocrine 29(2):289–297

Winer RL, Lee SK, Hughes JP, Adam DE, Kiviat NB, Koutsky LA (2003) Genital human papillomavirus infection: incidence and risk factors in a cohort of female university students. Am J Epidemiol 157(3):218–226

Wira CR, Grant-Tschudy KS, Crane-Godreau MA (2005) Epithelial cells in the female reproductive tract: a central role as sentinels of immune protection. Am J Reprod Immunol 53(2):65–76

Wolner-Hanssen P (1986) Oral contraceptive use modifies the manifestations of pelvic inflammatory disease. Br J Obstet Gynaecol 93(6):619–624

Wolner-Hanssen P, Eschenbach DA, Paavonen J, Kiviat N, Stevens CE, Critchlow C, DeRouen T, Holmes KK (1990) Decreased risk of symptomatic chlamydial pelvic inflammatory disease associated with oral contraceptive use. JAMA 263(1):54–59

Woods JL, Shew ML, Tu W, Ofner S, Ott MA, Fortenberry JD (2006) Patterns of oral contraceptive pill-taking and condom use among adolescent contraceptive pill users. J Adolesc Health 39(3):381–387

Woodworth CD, Simpson S (1993) Comparative lymphokine secretion by cultured normal human cervical keratinocytes, papillomavirus-immortalized, and carcinoma cell lines. Am J Pathol 142(5):1544–1555

Workowski KA, Berman SM (2006) Sexually transmitted diseases treatment guidelines, 2006. MMWR Recomm Rep 55(RR-11):1–94

Wright TC Jr, Subbarao S, Ellerbrock TV, Lennox JL, Evans-Strickfaden T, Smith DG, Hart CE (2001) Human immunodeficiency virus 1 expression in the female genital tract in association with cervical inflammation and ulceration. Am J Obstet Gynecol 184(3):279–285

Xu F, Sternberg MR, Kottiri BJ, McQuillan GM, Lee FK, Nahmias AJ, Berman SM, Markowitz LE (2006) Trends in herpes simplex virus type 1 and type 2 seroprevalence in the United States. JAMA 296(8):964–973

Xu J, Sobel JD (2004) Candida Vulvovaginitis in Pregnancy. Curr Infect Dis Rep 6(6):445–449

Ziegler A, Kastner C, Chang-Claude J (2003) Analysis of pregnancy and other factors on detection of human papilloma virus (HPV) infection using weighted estimating equations for follow-up data. Stat Med 22(13):2217–2233

Chapter 11
Sex, Pregnancy and Measles

Allison C. Brown and William J. Moss

Abstract The World Health Organization rescinded the recommendation that young children receive high-titer measles vaccine in 1992 following evidence that, in certain regions of the world, girls receiving this vaccine had increased mortality compared with those receiving standard-titer vaccines. It has been suggested that the high-titer vaccine might induce immunomodulation similar to that observed following wild-type measles virus infection. Additionally, maternal infection during pregnancy with malaria or HIV has been shown to reduce passive transfer of antimeasles antibodies to the fetus and lower the age at which infants become susceptible to measles. These data demonstrate that vaccine efficacy and adverse effects can vary between the sexes and this should be studied before new vaccines are widely used.

11.1 Introduction

Systematic investigation of human sex differences in morbidity and mortality began in the seventeenth century when John Graunt published *Natural and Political Observations Made Upon the Bills of Mortality* (Graunt 1662), in which he reported excess male births and deaths in London using parish records. In 1786, Joseph Clarke presented evidence that mortality in males exceeded that of females at all ages but that differences were most marked in early childhood (Clarke 1786). In 1929, Harry Bawkin published a detailed analysis of sex differences in mortality during infancy and early childhood in the United States, describing excess all-cause mortality in boys during the first 3 years of life (Bawkin 1929). Excess mortality

W.J. Moss (✉)

Department of Epidemiology and W. Harry Feinstone Department of Molecular Microbiology and Immunology, Johns Hopkins University Bloomberg School of Public Health, 615 North Wolfe Street, Baltimore, MD 21205, USA

e-mail: wmoss@jhsph.edu

S.L. Klein and C.W. Roberts (eds.), *Sex Hormones and Immunity to Infection*,   281
DOI 10.1007/978-3-642-02155-8_11, © Springer-Verlag Berlin Heidelberg 2010

among males has been attributed, in part, to increased susceptibility to and severity of infectious diseases, particularly in early childhood. Reviewing the published literature and case records at Johns Hopkins Hospital from 1930 to 1963, Washburn et al. reported an increased incidence of bacterial meningitis among boys younger than 15 years and an increased incidence of bacterial sepsis among boys in early infancy compared with girls (Washburn et al. 1965). The authors postulated genetic differences in loci on the X chromosome to account for these observations. Distinguishing genetic from other biological and behavioral risk factors to account for human sex differences in morbidity and mortality has been challenging, and more recent research continues to struggle with these issues (Lopez and Ruzicka 1983). Anatomical reasons can explain sex differences in the incidence of urinary tract infections, but the underlying mechanisms are less well-defined for other infectious diseases. A recent review of sex differences in the incidence and severity of respiratory tract infections found that males had a higher incidence than females of lower respiratory tract infection and more severe disease in all age groups, but the authors were unable to distinguish anatomical, lifestyle, behavioral, and socioeconomic risk factors (Falagas et al. 2007). Sex differences in infectious disease incidence may be due to factors unrelated to host susceptibility. For example, sex differences in the incidence of tuberculosis have been attributed to underreporting among women (Holmes et al. 1998) or to characteristics of the underlying transmission dynamics (Martinez et al. 2000). Much recent work has focused on sex differences in the natural history and response to therapy among HIV-infected persons, including observations of lower HIV-1 viral loads in women compared with men at similar stages of disease (Farzadegan et al. 1998; Gandhi et al. 2002). Furthermore, female fetuses have been demonstrated to be at greater risk of congenital HIV infection than male fetuses (Thorne et al. 2004). The mechanisms responsible for these differences have not been determined.

Measles provides an intriguing example of potential sex differences in response to infection. Firstly, there are reports of increased mortality in girls compared with boys following infection with wild-type measles virus (the terms "girls" and "boys" will be used to refer to prepubertal children). Secondly, girls in receipt of high-titer measles vaccines had an increased incidence of delayed mortality compared with boys vaccinated with high-titer measles vaccines. These two trends run counter to the more common observations of increased severity of infectious diseases in boys compared with girls.

11.2 Measles and Measles Vaccines

11.2.1 Measles

Measles is a highly contagious disease caused by infection with measles virus (Family: *Paramyxoviridae*) and characterized by a prodromal illness of fever, cough, coryza (runny nose), and conjunctivitis followed by the appearance of a generalized maculopapular rash. Measles is predominantly a disease of young children who become susceptible at 6–9 months of age when passively-acquired maternal antibodies wane. Most children recover and develop life-long immunity to reinfection (Duke and Mgone 2003). However, measles remains a leading cause of vaccine-preventable mortality worldwide and has caused millions of deaths since its emergence thousands of years ago (World Health Organization 2007). Measles mortality varies depending upon the average age of infection, nutritional status of the population, measles vaccine coverage, and access to health care. In developed countries, such as the United States, less than 1 in 1,000 children with measles die, but in endemic areas in sub-Saharan Africa and Asia measles case fatality may be 5–10% (Perry and Halsey 2004).

11.2.2   Vaccinations for Measles

Widespread coverage with standard-titer, attenuated measles vaccines has resulted in a marked reduction in global measles mortality and the elimination of measles virus transmission in large geographical areas, including the Americas. Standard-titer measles vaccines, which are attenuated live viral vaccines that replicate within the host to induce protective immunity, are administered at 9 months of age in many countries according to the schedule recommended by the Expanded Program on Immunization. In some countries, measles vaccines are combined with other live attenuated virus vaccines, such as those for mumps, rubella (MMR), and varicella (MMR-V).

The protective efficacy of these vaccines is reduced when administered early in infancy because of the inhibitory effects of maternally acquired antibodies (Albrecht et al. 1977) and immunological immaturity (Gans et al. 1998). Several studies in Africa suggested that high vaccine coverage rates with a single dose of standard-titer measles vaccine at 9 months of age was insufficient to control measles virus transmission because of the high incidence of measles in infants younger than 9 months (Dabis et al. 1988; Taylor et al. 1988). Efforts to immunize young infants were based on high-titer measles vaccines to overcome the inhibitory effect of maternal antibodies. High-titer measles vaccines have ≥ 4.7 $\log_{10}$ infectious units/dose, 10–100 times the standard dose of vaccine virus. Trials of these vaccines in the 1980s demonstrated that they were immunogenic, efficacious, and no more reactogenic than standard-titer vaccines when administered at 4–6 months of age. In 1989, high-titer measles vaccines were recommended for use by the World Health Organization in countries with a high incidence of measles in children younger than 9 months of age (Expanded Programme on Immunization 1990). However, recommendations on the use of these vaccines were subsequently withdrawn after observations of increased mortality in girls who received these vaccines compared with those who received standard-titer measles vaccines.

11.3 Immunity to Measles Virus Infection

11.3.1 Protective Immunity

Host immune responses to measles virus are responsible for viral clearance, clinical recovery, and the establishment of long-term immunity. The early nonspecific (innate) immune responses that occur during the prodromal phase of the illness include activation of natural killer (NK) cells and production of IFN-α and β (Griffin et al. 1990). The protective efficacy of antibodies to measles virus is illustrated by the immunity conferred to infants from passively acquired maternal antibodies and the protection of exposed, susceptible individuals following administration of anti-measles virus immune globulin (Black and Yannet 1960). IgG antibodies to the hemagglutinin and fusion proteins contribute to virus neutralization and are sufficient to provide protection against infection (de Swart et al. 2005). Evidence for the importance of cellular immunity to measles virus is demonstrated by the ability of children with agammaglobulinemia to fully recover from measles, whereas children with severe defects in T lymphocyte function often develop severe or fatal disease (Good and Zak 1956). CD4$^+$ T lymphocytes are activated in response to measles virus infection and secrete cytokines capable of modulating the humoral and cellular immune responses. Plasma cytokine profiles show increased levels of IFN-γ during the acute phase of illness, followed by a shift to high levels of interleukin (IL)-4 and IL-10 during convalescence when disease has resolved (Moss et al. 2002b). Both humoral and cellular immune responses are induced by standard-titer measles vaccines, but are of lower magnitude and shorter duration compared to those following wild-type measles virus infection (Ward et al. 1995)

11.3.2 Immune Suppression Following Measles Virus Infection

Importantly, much of the morbidity and mortality due to measles is attributed to a state of immune suppression that lasts for weeks to months beyond resolution of the acute illness. This state of immune suppression enhances susceptibility to secondary bacterial and viral infections causing pneumonia and diarrhea (Beckford et al. 1985; Greenberg et al. 1991). These secondary infections contribute to a substantial proportion of deaths due to measles (Duke and Mgone 2003). Immunological evidence for this state of immune suppression includes suppressed delayed-type hypersensitivity (DTH) responses to recall antigens, such as tuberculin (Tamashiro et al. 1987), and impaired cellular and humoral responses to new antigens (Coovadia et al. 1978). However, sex differences in the magnitude of immune suppression following wild-type measles virus infection have not been described.

The immunological mechanisms responsible for the immune suppression following measles are not fully defined, but abnormalities of both the innate and adaptive immune responses have been described. Transient lymphopenia with a reduction in CD4$^+$ and CD8$^+$ T-lymphocytes occurs in children with measles (Ryon et al. 2005). Functional abnormalities have also been detected, including decreased lymphocyte proliferative responses (Hirsch et al. 1984). Dendritic cells mature poorly, lose the ability to stimulate proliferative responses in lymphocytes, and undergo apoptosis when infected with measles virus in vitro (Servet-Delprat et al. 2000). The dominant Th2 response in children recovering from measles can inhibit Th1 responses (Griffin et al. 1985; Griffin and Ward 1993). The production of IL-12, important for the generation of a Th1-type immune response, decreases following binding of the virus to the CD46 receptor (Karp et al. 1996), and is low for several weeks in children with measles (Atabani et al. 2001). This diminished ability to produce IL-12 could further result in a limited Th1 immune response to other pathogens. Furthermore, engagement of CD46 and CD3 on monocytes in vitro induces production of high levels of IL-10 and transforming growth factor-β (Kemper et al. 2003), an immunomodulatory and immunosuppressive cytokine profile characteristic of regulatory T cells that may be important in measles pathogenesis. The role of these cytokines in the immune suppression following measles is supported by in vivo evidence of elevated levels of IL-10 in the plasma of children after measles virus infection (Moss et al. 2002b). Although vaccination with standard-titer measles vaccines induces similar immunological changes as those observed in natural infections, clinically significant immune suppression has not been reported following receipt of standard-titer measles vaccines.

11.4 Sex Differences in Measles-Induced Mortality

Evidence for sex differences in measles mortality was first reported by Garenne (1994). This study was initiated following the observation that mortality was increased among girls in rural Senegal following receipt of high-titer measles vaccine (Garenne et al. 1991). The analysis was based on reported age and sex for measles deaths from 1950 to 1989 collected by the World Health Organization from 78 countries with national vital registration systems. Importantly, because these data originated from vital registration, many of the most populous countries where the majority of global measles deaths occurred were excluded from analysis, including China, India, Pakistan, Indonesia, and Brazil, as well as most of sub-Saharan Africa. These exclusions obviously limit the generalizability of the findings. Female sex ratios for measles death rates were estimated for 5-year age groups and by region (with 100 indicating equivalent death rates for females and males and values greater than 100 excess female mortality) and compared with expected death rates using model life tables.

Female sex ratios for measles death rates for all countries and years studied were greater than 100 from birth to 50 years of age (Fig. 11.1). Excess female mortality due to measles was 4.2% in girls 0–4 years of age, 10.9% in girls 5–14 years, and 42.6% in women 15–44 years of age (Garenne 1994). Excess measles mortality in

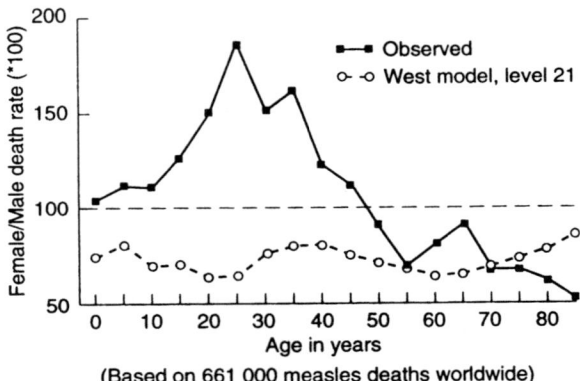

Fig. 11.1 Female sex ratios of measles deaths. From Garenne (1994)
Source: World Health Statistics, 1950–1989

Table 11.1 Female sex ratio of measles deaths rates by region

Region	Number of countries	Number of measles deaths	Female sex ratio of death rates			Expected
			0–4	5–14	15–44	
Europe	28	62,662	101.7	114.7	106.3	76.6
North America[a]	4	10,696	102.6	117.4	110.8	68.6
Far East	3	50,029	104.1	125.6	103.2	72.0
Latin America	22	413,168	102.8	108.3	164.9	81.2
Middle East	8	55,620	121.3	114.3	94.3	79.9
South East Asia	4	66,334	96.6	109.4	74.9	80.9
Islands	9	339	77.7	129.2	–	78.6
Total	78	658,848	104.2	110.9	142.6	75.1

[a]includes Australia and New Zealand
Adapted from Garenne (1994).

females peaked at 88% in women 25–29 years of age, and was even more marked when compared with expected death rates using model life tables because of the expected lower overall mortality of women and girls. Compared with these expected death rates, excess female mortality due to measles was 39% for girls younger than 4 years of age, 43% for girls 5–14 years of age, and 88% for women 14–44 years of age. No differences between observed and expected sex ratios for measles mortality were observed for women older than 45 years. Higher female sex ratios were observed in all regions studied (Table 11.1), providing further consistency to the observation.

Although this study omitted data from sub-Saharan Africa because of a lack of vital registration data, subsequent studies from the region suggest possible increased mortality due to measles virus infection in girls, although the data are not conclusive. In a study of 356 children hospitalized with clinically diagnosed measles in Lusaka, Zambia, and in whom antibodies to HIV-1 were measured, the highest measles case fatality ratio was in HIV-1-seropositive girls vaccinated with standard-titer measles vaccine. However, few girls were included in this subgroup and the finding did not achieve statistical significance ($P = 0.18$). In a subsequent observational study of 546 Zambian children hospitalized with laboratory-confirmed measles and known HIV-1 infection status (determined by detection of HIV-1 RNA), the risk of in-hospital mortality was higher for girls than for boys (OR 3.2, 95% confidence intervals (CI): 1.2, 8.3), with the greatest difference in unvaccinated children (Moss et al. 2002a). However, in a follow-up study of 1,227 hospitalized children with laboratory-confirmed measles (including the 546 previously reported children), the risk of in-hospital mortality due to measles was not increased in girls (OR 1.2, 95% CI: 0.7, 2.0) (Moss et al. 2008). The reasons for these differences between studies are not clear.

Not all studies support the observation of increased measles mortality in girls and women compared with boys and men. Some older historical data, in fact, described increased measles mortality in males compared with females (Aaby et al. 1992b; Babbott and Gordon 1954), including Bawkin's data based on United States mortality statistics from 1915 to 1924 in which the female sex ratio for measles death rates among white infants was 85 (Bawkin 1929). More recent surveillance data from the United States between 1987 and 2000 did not identify sex differences in measles mortality (Perry and Halsey 2004). Increased mortality in girls after measles virus infection was reported in Bangladesh, but was attributed to preferential care of young boys (Bhuiya et al. 1987). Thus, there is highly suggestive but inconclusive evidence for increased measles mortality among girls compared with boys. Evidence for such excess mortality in girls would be strengthened if underlying immunological differences in the responses to measles virus were identified.

11.5 Sex Differences in Immune Responses to Wild-Type Measles Virus

Several studies have reported differences between the immune responses of boys and girls to wild-type measles virus infection. Although most studies have not found sex differences in the quantity of antibody produced after measles, qualitative differences have been described. Measles virus-specific antibody-dependent cellular cytotoxicity (ADCC) was found to differ by age and sex in a study of 114 Fillipino children with measles (Forthal et al. 1995). ADCC results from antigen-specific antibodies that link cytotoxic effector cells expressing Fc receptors and infected target cells. ADCC antibodies appear later postinfection than neutralizing antibodies, but are present at higher levels during convalescence. Girls older than 24 months of age had a higher prevalence of ADCC antibodies than younger girls (67% vs 17%, OR = 9.6, 95% CI: 3, 31), but no significant differences were observed between older and younger boys (55% vs 42%, OR = 1.6, 95% CI: 0.6, 5). Older girls also had higher mean ADCC antibody levels than younger girls (585 vs 27, $P = 0.001$), but no differences in levels were observed between older and younger boys. As a consequence, boys younger than 24 months had higher ADCC antibody titers than girls of similar age. These findings were corroborated by a study of Zambian children with measles, in whom percent cytotoxicity in an ADCC assay was higher in boys than in girls (22.6% vs 12.4%; $P=0.03$) (Permar et al. 2001). Thus, younger girls had a diminished ADCC antibody response to wild-type measles virus infection, perhaps, contributing to sex-related differences in clinical outcomes.

A study of Zambian children with measles reported subtle differences in lymphocyte subsets during acute measles and convalescence (Ryon et al. 2002). Although lymphopenia typically occurs during acute measles, it was more pronounced in girls than in boys as measured by total lymphocyte counts at hospitalization (3326 vs 3868, $P=0.03$). Furthermore, $CD8^+$ T lymphocytes tended to increase more rapidly in boys compared with girls during recovery. This resulted in lower $CD4^+/CD8^+$ T lymphocyte ratios among boys than in girls at hospital discharge, a pattern maintained at 1-month follow-up ($CD4^+/CD8^+$ T-lymphocyte ratio at follow-up: 1.7 for girls and 1.3 for boys, $P=0.03$). An earlier study of measles in Senegal also noted the tendency for girls to have higher percentages of $CD4^+$ T lymphocytes and higher $CD4^+/CD8^+$ T lymphocyte ratios than boys during convalescence (Lisse et al. 1998). As further evidence of greater stimulation and proliferation of $CD8^+$ T lymphocytes in response to measles virus infection among boys, mean soluble CD8 receptor levels were higher at hospitalization in boys (1,056 $U ml^{-1}$) than in girls (761 $U ml^{-1}$; $P=0.02$).

Another potentially important sex difference in the immune responses to measles virus involves regulatory T cells. Regulatory T cells are CD4+ T lymphocytes capable of suppressing immune responses to maintain homeostasis following antigen challenge and to regulate tolerance to selfantigens. Two broad groups of regulatory T cells have been defined: innate regulatory T cells arising from the thymus and acquired regulatory T cells that differentiate in the periphery following antigenic stimulation (Belkaid 2007). Regulatory T cells are best characterized by the presence of the transcription factor *FoxP3* that is uniquely expressed in both natural and acquired $CD4^+CD25^+$ regulatory T cells (Hori et al. 2003), and is encoded on the X chromosome (Bennett et al. 2001). Unpublished observations of Zambian children with measles suggest increased expression of *Foxp3* mRNA during acute illness and that this increased *Foxp3* expression persisted for at least an additional 5–6 weeks in girls but not in boys (M Ota, personal communication). Differential induction of regulatory T cells could underlie the sex differences in mortality following wild-type measles virus infection and high-titer measles vaccine by inducing a prolonged state of immune suppression in girls.

Although immunological studies suggest impaired immune responses to measles virus in girls, a study that used RT-PCR to detect measles virus RNA in children that had been hospitalized for confirmed measles, found that measles virus shedding approximately 1 month later was more common in boys (73% of 26 boys) than in girls (48% of 21 girls; $P=0.07$) (Permar et al. 2001). However, when these studies were extended to approximately 3 months after acute measles, no differences were observed between the proportion of boys and girls with prolonged shedding of measles virus RNA (35% of 26 boys versus 39% of 23 girls; $P = 0.78$) (Riddell et al. 2007).

Reanalyses of data on measles in Guinea-Bissau, Kenya, The Gambia, Denmark, and Senegal all suggest that cross-sex transmission (boy-to-girl and girl-to-boy) of measles virus results in greater disease severity and higher mortality (Aaby et al. 1986; Aaby et al. 1992a; Aaby and Leeuwenburg 1991; Aaby and Lamb 1991; Aaby 1991; Aaby 1992; Pison et al. 1992). In Guinea-Bissau, the case–fatality ratio was highest for girls infected by boys, and in Senegal, the risk of infection through cross-sex transmission was greater for girls than for boys (OR: 1.3, 95% CI: 1.1, 1.5). However, these differences were not consistent across studies. Although it is plausible that the sexes differ in the duration and intensity of measles virus shedding and consequently contagiousness, a biological mechanism to account for potential differences in cross-sex transmission remains to be determined.

11.6 Sex Differences in Immune Responses to Standard-Titer Measles Vaccine

As with wild-type measles virus infection, standard-titer measles vaccines are associated with immunosuppression; however, this immune suppression is less pronounced than after wild-type infection and resolves within weeks after vaccination (Okada et al. 2001). Manifestations include decreased lymphoproliferative responses to mitogens and antigens, altered patterns of cytokine production, and suppression of DTH skin test responses. Tuberculin skin test reactivity may be abrogated for 4–6 weeks after immunization, but unlike wild type measles virus infection, measles vaccine does not exacerbate tuberculosis.

Many studies of the immunogenicity, efficacy, and effectiveness of measles vaccines have been conducted, but few have reported sex differences in the response to standard-titer measles vaccines (Cook 2008). Interest in potential sex differences in response to measles vaccines was stimulated by observations of increased mortality in girls following receipt of high-titer measles vaccine. Sex differences in antibody responses to measles vaccine were observed in a study of 223 men and 66 women following receipt of standard-titer Schwarz measles vaccine (Green et al. 1994). After controlling for differences in prevaccination antibody levels, women had 50% higher geometric mean antibody titers than men at 2 and 4 weeks after vaccination ($P < 0.001$). However, girls were less likely to seroconvert than boys (OR 0.34, 95% CI: 0.15, 0.76.) in a randomized, placebo-controlled trial to assess the effect of vitamin A supplementation on seroconversion after measles vaccination at 6 months of age in Indonesia (Semba et al. 1995). The main finding of this study, that vitamin A lowered the likelihood of seroconversion after measles vaccination, was not corroborated in subsequent studies (Benn et al. 1997;Cherian et al. 2003).

Analogous to the studies of ADCC antibodies following wild-type measles virus infection, measles virus-specific ADCC responses were lower among six girls who received medium-titer Edmondston-Zagreb vaccine at 4 months of age compared with seven boys who received the same vaccination schedule ($P = 0.04$) (Atabani et al. 2000). However, neutralizing antibody levels did not differ between boys and girls. In a study assessing differential reactogenicity of measles–mumps–rubella vaccine among boys and girls, the relative risk of fever and rash during the month after vaccination was 2.4 in girls based on parental reports of 755 infants (Shohat et al. 2000). Geometric mean antibody levels were not different between boys and girls in a subgroup of 237 infants. More firmly established is the increased risk of immune thrombocytopenic purpura in boys following receipt of measles–mumps–rubella vaccine. Using the United States Vaccine Safety Datalink, 1,036,689 children were reported to have received 1,107,814 measles–mumps–rubella vaccinations, of whom 259 children developed immune thrombocytopenic purpura (France et al. 2008). The incident rate ratio for boys 12–15 months of age was 14.6 compared with 3.2 for girls in the same age group. In summary, of the hundreds of studies of immune responses to standard-titer measles vaccines, only a few have reported significant sex differences.

11.7   Sex Differences in Mortality Following High-Titer Measles Vaccine

The most surprising and intriguing observations related to sex differences and measles were those of excess mortality in girls for several years following receipt of high-titer measles vaccines. These observations resulted in a major shift in measles vaccination policy, but remain unexplained and open to different interpretations and hypotheses.

In 1990, investigators first reported increased delayed mortality in recipients of high-titer measles vaccines in Guinea Bissau (Aaby et al. 1993b). Increased mortality subsequently was reported among high-titer vaccine recipients in Senegal (Garenne et al. 1991) (Fig. 11.2) and Haiti (Holt et al. 1993). Most striking was that the delayed mortality appeared to be restricted to girls who received the high-titer vaccines. These observations led the World Health Organization, in 1992, to rescind the recommendation for the use of high-titer measles vaccines and these vaccines have not been used since. Because none of the trials were designed to measure long-term mortality as an outcome, assessing the validity of the observations has not been straightforward. An initial review of the data in 1991 concluded that the information was not sufficiently persuasive to recommend a change in vaccination policy (Expanded Programme on Immunization 1992). However, the data became more convincing with the observations from Haiti and further analysis. In a meta-analysis of trials of high-titer measles vaccines conducted in West Africa (Guinea-Bissau, The Gambia, and Senegal), the adjusted mortality ratio comparing high-titer vaccine recipients with standard-titer vaccine recipients was 1.86 for girls (95% CI: 1.28, 2.7) and 0.91 for boys (95% CI: 0.61, 1.35) (Knudsen et al. 1996). The excess mortality was not statistically significant in any of the individual trials when evaluated from time of the first vaccination at 4–5 months of age, and was only statistically significant in one trial when evaluated after receipt of the control

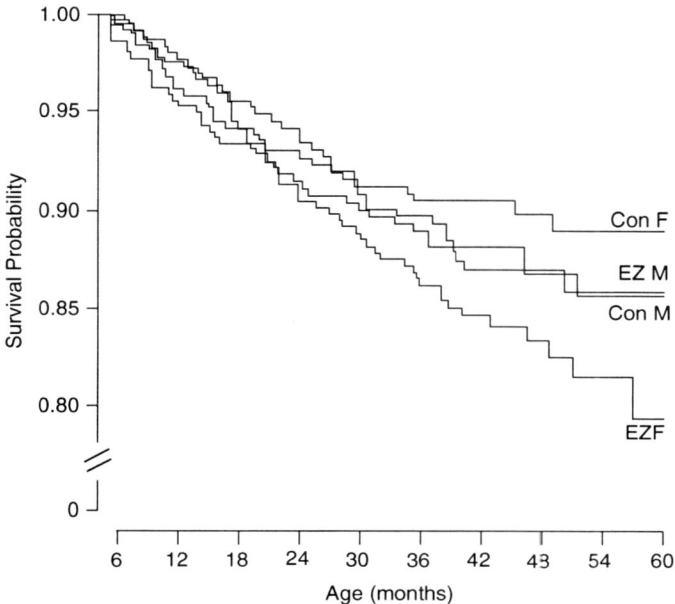

Fig. 11.2 Kaplan-Meier survival curves for recipients of high-titer Edmonston-Zagreb measles vaccine (EZ) and standard-titer Schwarz measles vaccine (Con) in Senegal. From Aaby et al. (2003b) F = female, M = male

vaccine at 9–10 months of age. No excess mortality in girls occurred in the first year of life (crude mortality ratio (MR) 1.0, 95% CI: 0.66, 1.52), increased slightly from 1 to 2 years of age (MR 1.06, 95% CI: 0.71, 1.58), and was most marked from 2 to 3 years of age (MR 2.12, 95% CI: 1.28, 3.51) (Knudsen et al. 1996).

In Haiti, where both medium-titer and high-titer measles vaccines were administered, there was a trend for girls who received high-titer measles vaccine to have greater mortality than those who received medium-titer vaccine (risk ratio 1.71, 95% CI: 0.91, 3.24), whereas no differences were observed for boys. Interestingly, no increased mortality was observed in studies of high-titer measles vaccines in the Philippines, Peru, and Mexico, countries with higher socioeconomic status and lower overall child mortality than many of the other countries studied (Halsey 1993).

Several features of these observations are striking. First, excess mortality was observed only in girls. There was no precedent for such a finding among studies of childhood vaccines. Second, excess mortality was delayed several years after receipt of the vaccine. Third, the causes of deaths were not due to measles, but other common causes of child mortality. Although high-titer measles vaccines were associated with increased mortality in girls compared with those immunized with standard-titer vaccines, mortality rates were lower than in unimmunized girls (Halsey 1993).

The first hypothesis put forward to explain these observations, and one that continues to have adherents, was that high-titer measles vaccines induced a prolonged state of immune suppression similar to that described after wild-type measles virus infection, and thus resulted in increased susceptibility to secondary infections. Children who developed a rash after high-titer vaccination were at particularly high risk of mortality, suggesting that a large inoculum of vaccine virus may mimic the immunosuppressive effects of wild-type measles virus (Seng et al. 1999).

Several studies attempted to identify immunological abnormalities in girls who received high-titer vaccines, but these were conducted long after vaccination. Although excess mortality was not observed in Peru, children who received high-titer measles vaccine were found to have lower percentages of CD4+ T-lymphocytes and decreased lymphoproliferative responses to phytohemagglutinin 2–3 years after vaccination (Leon et al. 1993). A similar decrease in CD4+/CD8+ T-lymphocyte ratios was observed in girls in Guinea-Bissau 3–5 years after vaccination with high-titer measles vaccine (Lisse et al. 1994). However, these findings were not confirmed in Senegal (Samb et al. 1995) and the increased risk of mortality did not extend to more than 36 months after vaccination in any study.

An alternative explanation for the observed increase in mortality following high-titer measles vaccine is that standard-titer, but not high-titer, measles vaccines conferred nonspecific beneficial effects on childhood mortality that were more pronounced in girls than in boys (Aaby et al. 1995; Kristensen et al. 2000; Shann 2000). Evidence presented for a nonspecific beneficial effect of standard-titer measles vaccine included: (1) the protective efficacy against death after measles vaccination (30–86%) exceeded the expected efficacy attributable to a reduction in measles deaths (10%) (Aaby et al. 1995); (2) in the absence of a history of measles, children vaccinated against measles have a lower mortality rate than children who do not receive measles vaccine (Aaby et al. 1995); and (3) measles vaccination at 6 months of age results in a lower overall mortality than vaccination at 9 months of age despite a reduction in seroconversion rates, presumably due to a reduction in nonmeasles deaths (Aaby et al. 1993a). The immunological basis of such an effect has been speculated to be nonspecific immune activation or Th1 skewing of the immune response (Aaby et al. 1995). However, the hypothesis that standard-titer measles vaccination results in a nonspecific reduction in childhood mortality (that is more pronounced in girls) remains controversial and unproven (Fine 2000).

Aaby and colleagues subsequently offered a detailed critique of the hypotheses that high-titer measles vaccines either induce a prolonged state of immune suppression or that standard-titer measles vaccines exert a nonspecific beneficial effect on child mortality (Aaby et al. 2003b), claiming that neither of these hypotheses fully explain the observations. More recently, they have suggested that the sex differences in mortality after high-titer measles vaccine were, in fact, due to nonspecific harmful effects of inactivated diphtheria–tetanus–pertussis and poliovirus vaccines that were administered at 9–10 months of age in many of the high-titer vaccine trials (Aaby et al. 2003a; Aaby et al. 2006b; Aaby et al. 2006a; Aaby et al. 2007a; Aaby et al. 2007b; Veirum et al. 2005). Although this hypothesis, if proven to be correct, exonerates the high-titer vaccine, it too remains conjectural and lacks mechanistic proof that can account for sex differences.

11.8 Impact of Maternal Infection During Pregnancy on Measles Immunity

11.8.1 Maternal Infections and the Transfer of Antimeasles Antibodies

Young infants in the first months of life are protected against measles by maternally acquired IgG antibodies. An active transport mechanism in the placenta is responsible for the transfer of IgG antibodies from the maternal circulation to the fetus starting at about 28 weeks of gestation and continuing until birth. All IgG subclasses cross the placenta, but IgG1 is preferentially transported via placental Fc receptors (Simister 2003). Antigen-specific IgG levels in the newborn are often equivalent to, and sometimes exceed, maternal antibody levels, and high-avidity antibodies may be preferentially transported across the placenta (Avanzini et al. 1998).

Several studies have reported that children born to HIV-1 infected women have lower levels of passively acquired antibodies to measles virus at birth (de Moraes-Pinto et al. 1993; de Moraes-Pinto et al. 1996; de Moraes-Pinto et al. 1998; Scott et al. 2005a). Lower levels of passively acquired antibodies should result in increased susceptibility to measles at a younger age. A 3.8-fold increase in the risk of acquiring measles before 9 months of age was reported among Kenyan infants born to HIV-seropositive women compared with infants born to HIV-seronegative women (Embree et al. 1992). In Zambia, neutralizing antibodies to measles virus were measured in 652 plasma samples collected from 448 infants, of whom 13.6% were HIV-1 infected, 53.4% were HIV-seropositive, but uninfected, and 33% were HIV-seronegative (Scott et al. 2007). The best fitting model suggested that HIV-1-infected infants have lower levels of passively acquired antibodies to measles virus at birth than HIV-seronegative children, but their antibody levels decay more slowly. The mechanisms by which maternal HIV-1 infection impairs transplacental transfer of antibodies are not known, but may be due to impaired placental integrity, immune complex formation, or competition for Fc receptors by antigen-nonspecific antibodies. There is little evidence that HIV-1-infected pregnant women have lower levels of antibodies to measles virus than uninfected women, and the sex of the fetus has not been demonstrated to be a factor capable of influencing maternal antibody.

Evidence that maternal infections other than HIV-1 influence the passive transfer of antimeasles antibodies is less well established (Caceres et al. 2000). In Malawi, maternal malaria was associated with lower infant levels of antimeasles antibodies (Moraes-Pinto et al. 1998), and in The Gambia, maternal malaria was associated with a 30% reduction in cord antimeasles antibody levels (Owens et al. 2006). However, in a study of pregnant women in Kilifi, Kenya where malaria is endemic, cord blood from HIV-1-infected women had 35% lower levels of measles antibodies than cord blood from HIV-1-uninfected women, but no association was found between measles antibody levels and placental malaria infection (Scott et al. 2005b).

11.8.2   Maternal Infections and Cellular Immune Responses in the Infant

No published studies have investigated the impact of maternal infections on cellular immune responses to measles virus in infants. This is due, in part, to the fact that measurements of cellular immune responses are much more complex than measurement of antibody responses, which is particularly challenging given the small blood volumes obtainable from young infants. However, it is likely that maternal infections, and the associated immune responses, have a profound impact on cellular immune responses during infancy, and that the range of maternal infections that can impact offspring cellular immune responses extend beyond HIV-1 and malaria.

Maternal infections can modulate immune responses in their offspring through multiple mechanisms (Fig. 11.3). In mammals with hemochorial placentas, including primates and rodents, the fetal chorion is in direct contact with maternal blood allowing the transfer of IgG antibodies (Vercruysse and Gabriel 2005) and exposure to antigens, lymphocytes, cytokines, and other immunomodulatory proteins. If the immunomodulatory effect of maternal infection persists for years after birth, as suggested by one study of schistosomiasis (Malhotra et al. 1999), then the transplacental transfer of antigen, antibodies, or cytokines must induce profound and persistent changes in the developing immune system.

11.8.3   Maternal Schistosomiasis and Offspring Immune Responses

Schistosomiasis provides one example of a highly prevalent maternal infection that could potentially have an important public health impact through modulation of cellular immune responses in infants to both homologous (schistosomal) and heterologous (nonschistosomal) antigens. Maternal schistosomiasis has been shown to modulate immune responses to schistosomal antigens in offspring (Carlier and Truyens 1995) and maternal IgG antibodies to schistosomal antigens cross the placenta and can be identified in the blood of newborns (Carlier et al. 1980). Cord blood mononuclear cells from newborns of women infected with *Schistosoma mansoni* exhibit strong proliferative responses to antibodies against soluble egg antigen (SEA), suggesting that idiotypic sensitization occurs in utero (Eloi-Santos et al. 1989). In a murine model of *S. mansoni* infection, neonatal idiotypic exposure induced B-lymphocyte (SEA-specific IgG production) and T-lymphocyte

Mother   **Fetus**

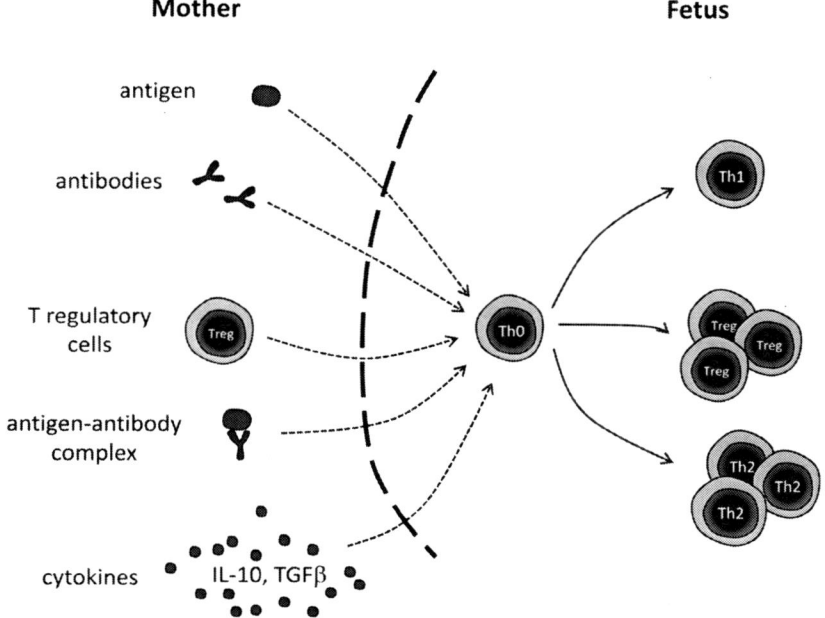

Fig. 11.3 Potential mechanisms by which maternal infections may modulate cellular immune responses in their offspring. The fetal chorion of mammals with hemochorial placentas, including primates and rodents, is in direct contact with maternal blood allowing transplacental transfer of IgG antibodies, antigens, lymphocytes, cytokines, and other immunomodulatory proteins. Adapted from Herz et al. (2000)

(splenocyte proliferative responses) responsiveness (Colley et al. 1999; Montesano et al. 1999b; Montesano et al. 1999a). This in utero sensitization has beneficial effects on offspring as it attenuates hepatic granuloma formation and improves their survival when infected with *S. mansoni* (Montesano et al. 1999a).

Schistosomal antigens also cross the placenta, as measured by antigen tolerance in the offspring of mice infected with *S. mansoni* (Lewert and Mandlowitz 1969), skin test sensitization in humans (Camus et al. 1976; Tachon and Borojevic 1978), and detection of circulating *S. mansoni* soluble antigens in umbilical cord blood (Attallah et al. 2003; Carlier et al. 1980; Hassan et al. 1997; Romia et al. 1992). For example, studies conducted in Egypt detected a 63 kD antigen of *S. mansoni* in 86% of 176 umbilical cord blood samples from infected women (Attallah et al. 2003).

Maternal infection can also modulate immune responses to heterologous antigens. In utero sensitization to *S. haematobium* establishes immunological memory persisting into childhood and biases T lymphocyte cytokine responses to BCG vaccination (King et al. 1998; Malhotra et al. 1997; Malhotra et al. 1999). Prenatal sensitization to schistosomiasis alters immune responses to BCG in children 2–10 years of age. In this group, ex vivo production of IFN-γ by T lymphocytes was diminished relative to control T lymphocytes following stimulation with purified protein derivative (Malhotra et al. 1999). Whether maternal infections modulate immune responses to measles vaccines administered to children is currently unknown, but may become of importance with the development of novel measles vaccines that are administered early in infancy.

11.9   Conclusions

The work discussed herein demonstrates the challenges in interpreting data from studies in the field where multiple factors including nutrition, endemic diseases, ethnicity, and behavior exert varying degrees of influence on the effect of immunization against measles virus. Nonetheless, clear differences between the sexes are evident in how these factors integrate in certain geographical regions and illustrate the need to carefully monitor and evaluate the effect of vaccination programs under different conditions and between the sexes. The fact that these differences are evident in children would suggest that they are likely to be independent of steroid hormone-induced immunomodulation that is normally attributed to such differences in adults. Future studies must consider genetic factors as possible mediators of sex differences in susceptibility to infection in children. In particular, the mosaic of X chromosomal genes in females that are inherited from both mother and father and the consequence of X inactivation should be considered in greater detail as possible mechanisms shaping immunological differences between the sexes (Migeon, 2007).

References

Aaby P (1991) Severity of measles and cross-sex transmission of infection in Copenhagen 1915–1925. Int J Epidemiol 20:504–507

Aaby P (1992) Influence of cross-sex transmission on measles mortality in rural Senegal. Lancet 340:388–391

Aaby P, Lamb WH (1991) The role of sex in the transmission of measles in a Gambian village. J Infect 22:287–292

Aaby P, Leeuwenburg J (1991) Gender and the pattern of transmission of measles infection. A reanalysis of data from the Machakos area, Kenya. Ann Trop Paediatr 11:397–402

Aaby P, Bukh J, Hoff G, Lisse IM, Smits AJ (1986) Cross-sex transmission of infection and increased mortality due to measles. Rev Infect Dis 8:138–143

Aaby P, Burstrom B, Mutie DM (1992a) Measles mortality in same-sex and mixed-sex siblings in Kenya. Lancet 340:923–924

Aaby P, Oesterle H, Dietz K, Becker N (1992b) Case-fatality rates in severe measles outbreak in rural Germany in 1861. Lancet 340:1172

Aaby P, Andersen M, Sodemann M, Jakobsen M, Gomes J, Fernandes M (1993a) Reduced childhood mortality following standard measles vaccination at 4–8 months compared to 9–11 months of age. BMJ 307:1308–1311

Aaby P, Knudsen K, Whittle H, Lisse IM, Thaarup J, Poulsen A, Sodemann M, Jakobsen M, Brink L, Gansted U, Permin A, Jensen TG, Andersen H, Da Silva MC (1993b) Long-term survival after Edmonston-Zagreb measles vaccination in Guinea-Bissau: increased female mortality rate. J Pediatr 122:904–908

Aaby P, Samb B, Simondon F, Seck AMC, Knudsen K, Whittle H (1995) Non-specific beneficial effect of measles immunisation: analysis of mortality studies from developing countries. Brit Med J 311:481–485

Aaby P, Jensen H, Samb B, Cisse B, Sodemann M, Jakobsen M, Poulsen A, Rodrigues A, Lisse IM, Simondon F, Whittle H (2003a) Differences in female-male mortality after high-titre measles vaccine and association with subsequent vaccination with diphtheria-tetanus-pertussis and inactivated poliovirus: reanalysis of West African studies. Lancet 361:2183–2188

Aaby P, Jensen H, Simondon F, Whittle H (2003b) High-titer measles vaccination before 9 months of age and increased female mortality: do we have an explanation? Semin Pediatr Infect Dis 14:220–232

Aaby P, Ibrahim SA, Libman MD, Jensen H (2006a) The sequence of vaccinations and increased female mortality after high-titre measles vaccine: trials from rural Sudan and Kinshasa. Vaccine 24:2764–2771

Aaby P, Jensen H, Walraven G (2006b) Age-specific changes in the female-male mortality ratio related to the pattern of vaccinations: an observational study from rural Gambia. Vaccine 24:4701–4708

Aaby P, Biai S, Veirum JE, Sodemann M, Lisse I, Garly ML, Ravn H, Benn CS, Rodrigues A (2007a) DTP with or after measles vaccination is associated with increased in-hospital mortality in Guinea-Bissau. Vaccine 25:1265–1269

Aaby P, Garly ML, Nielsen J, Ravn H, Martins C, Bale C, Rodrigues A, Benn CS, Lisse IM (2007b) Increased female-male mortality ratio associated with inactivated polio and diphtheria-tetanus-pertussis vaccines: Observations from vaccination trials in Guinea-Bissau. Pediatr Infect Dis J 26:247–252

Albrecht P, Ennis FA, Saltzman EJ, Krugman S (1977) Persistence of maternal antibody in infants beyond 12 months: mechanism of measles vaccine failure. J Pediatr 91:715–718

Atabani S, Landucci G, Steward MW, Whittle H, Tilles JG, Forthal DN (2000) Sex-associated differences in the antibody-dependent cellular cytotoxicity antibody response to measles vaccines. Clin Diagn Lab Immunol 7:111–113

Atabani SF, Byrnes AA, Jaye A, Kidd IM, Magnusen AF, Whittle H, Karp CL (2001) Natural measles causes prolonged suppression of interleukin-12 production. J Infect Dis 184:1–9

Attallah AM, Ghanem GE, Ismail H, El Waseef AM (2003) Placental and oral delivery of *Schistosoma mansoni* antigen from infected mothers to their newborns and children. Am J Trop Med Hyg 68:647–651

Avanzini MA, Pignatti P, Chirico G, Gasparoni A, Jalil F, Hanson LA (1998) Placental transfer favours high avidity IgG antibodies. Acta Paediatr 87:180–185

Babbott FL Jr, Gordon JE (1954) Modern measles. Am J Med Sci 228:334–361

Bawkin H (1929) The sex factor in infant mortality. Hum Biol 1:90–116

Beckford AP, Kaschula RO, Stephen C (1985) Factors associated with fatal cases of measles A retrospective autopsy study. S Afr Med J 68:858–863

Belkaid Y (2007) Regulatory T cells and infection: a dangerous necessity. Nat Rev Immunol 7:875–888

Benn CS, Aaby P, Bale C, Olsen J, Michaelsen KF, George E, Whittle H (1997) Randomised trial of effect of vitamin A supplementation on antibody response to measles vaccine in Guinea-Bissau, west Africa. Lancet 350:101–105

Bennett CL, Christie J, Ramsdell F, Brunkow ME, Ferguson PJ, Whitesell L, Kelly TE, Saulsbury FT, Chance PF, Ochs HD (2001) The immune dysregulation, polyendocrinopathy, enteropathy, X-linked syndrome (IPEX) is caused by mutations of FOXP3. Nat Genet 27:20–21

Bhuiya A, Wojtyniak B, D'Souza S, Nahar L, Shaikh K (1987) Measles case fatality among the under-fives: a multivariate analysis of risk factors in a rural area of Bangladesh. Soc Sci Med 24:439–443

Black FL, Yannet H (1960) Inapparent measles after gamma globulin administration. JAMA 173:1183–1188

Caceres VM, Strebel PM, Sutter RW (2000) Factors determining prevalence of maternal antibody to measles virus throughout infancy: a review. Clin Infect Dis 31:110–119

Camus D, Carlier Y, Bina JC, Borojevic R, Prata A, Capron A (1976) Sensitization to *Schistosoma mansoni* antigen in uninfected children born to infected mothers. J Infect Dis 134:405–408

Carlier Y, Truyens C (1995) Influence of maternal infection on offspring resistance towards parasites. Parasitol Today 11:94–99

Carlier Y, Nzeyimana H, Bout D, Capron A (1980) Evaluation of circulating antigens by a sandwich radioimmunoassay, and of antibodies and immune complexes, in *Schistosoma mansoni*-infected African parturients and their newborn children. Am J Trop Med Hyg 29:74–81

Cherian T, Varkki S, Raghupathy P, Ratnam S, Chandra RK (2003) Effect of Vitamin A supplementation on the immune response to measles vaccination. Vaccine 21:2418–2420

Clarke J (1786) Observations on some causes of the excess mortality of males above that of females. Phil Trans R Soc Lond 16:122

Colley DG, Montesano MA, Freeman GL, Secor WE (1999) Infection-stimulated or perinatally initiated idiotypic interactions can direct differential morbidity and mortality in schistosomiasis. Microbes Infect 1:517–524

Cook IF (2008) Sexual dimorphism of humoral immunity with human vaccines. Vaccine 26(29–30):3551–3555

Coovadia HM, Wesley A, Henderson LG, Brain P, Vos GH, Hallett AF (1978) Alterations in immune responsiveness in acute measles and chronic post-measles chest disease. Int Arch Allergy Appl Immunol 56:14–23

Dabis F, Sow A, Waldman RJ, Bikakouri P, Senga J, Madzou G, Jones TS (1988) The epidemiology of measles in a partially vaccinated population in an African city: implications for immunization programs. Am J Epidemiol 127:171–178

de Moraes-Pinto MI, Farhat CK, Carbonare SB, Curti SP, Otsubo MES, Lazarotti DS, Campagnoli RC, Carneiro-Sampaio MMS (1993) Maternally acquired immunity in newborns from women infected by the human immunodeficiency virus. Acta Paediatr 82:1034–1038

de Moraes-Pinto MI, Almeida ACM, Kenj G, Filgueiras TE, Tobias W, Santos AMN, Carneiro-Sampaio MMS, Farhat CK, Milligan PJM, Johnson CA (1996) Placental transfer and maternally acquired neonatal IgG immunity in Human Immunodeficiency Virus infection. J Infect Dis 173:1077–1084

de Moraes-Pinto MI, Verhoeff F, Chimsuku L, Milligan PJM, Wesumperuma HL, Broadhead RL, Brabin B, Johnson PM, Hart CA (1998) Placental antibody transfer: influence of maternal HIV infection and placental malaria. Arch Dis Child Fetal Neonatal 79:F202–F205

de Swart RL, Yuksel S, Osterhaus AD (2005) Relative contributions of measles virus hemagglutinin- and fusion protein-specific serum antibodies to virus neutralization. J Virol 79:11547–11551

Duke T, Mgone CS (2003) Measles: not just another viral exanthem. Lancet 361:763–773

Eloi-Santos SM, Novato-Silva E, Maselli VM, Gazzinelli G, Colley DG, Correa-Oliveira R (1989) Idiotypic sensitization in utero of children born to mothers with schistosomiasis or Chagas' disease. J Clin Invest 84:1028–1031

Embree JE, Datta P, Stackiw W, Sekla L, Braddick M, Kreiss JK, Pamba H, Wamola I, Ndinya-Achola JO, Law BJ (1992) Increased risk of early measles in infants of human immunodeficiency virus type 1-seropositive mothers. J Infect Dis 165:262

Expanded Programme on Immunization (1990) Global advisory group. Weekly Epidemiol Rec 65:6–11

Expanded Programme on Immunization (1992) Safety of high titre measles vaccines. Wkly Epidemiol Rec 67:357–361

Falagas ME, Mourtzoukou EG, Vardakas KZ (2007) Sex differences in the incidence and severity of respiratory tract infections. Respir Med 101:1845–1863

Farzadegan H, Hoover DR, Astemborski J, Lyles CM, Margolick JB, Markham RB, Quinn TC, Vlahov D (1998) Sex differences in HIV-1 viral load and progression to AIDS. Lancet 352:1510–1514

Fine P (2000) Commentary: an unexpected finding that needs confirmation or rejection. BMJ 321:7–8

Forthal DN, Landucci G, Habis A, Laxer M, Javato-Laxer M, Tilles JG, Janoff EN (1995) Age, sex, and household exposure are associated with the acute measles-specific antibody-dependent cellular cytotoxicity antibody response. J Infect Dis 172(6):1587–1591

France EK, Glanz J, Xu S, Hambidge S, Yamasaki K, Black SB, Marcy M, Mullooly JP, Jackson LA, Nordin J, Belongia EA, Hohman K, Chen RT, Davis R, Team Vaccine Safety Datalink (2008) Risk of immune thrombocytopenic purpura after measles-mumps-rubella immunization in children. Pediatrics 121:e687–e692

Gandhi M, Bacchetti P, Miotti P, Quinn TC, Veronese F, Greenblatt RM (2002) Does patient sex affect human immunodeficiency virus levels? Clin Infect Dis 35:313–322

Gans HA, Arvin AM, Galinus J, Logan L, DeHovitz R, Maldonado Y (1998) Deficiency of the humoral immune response to measles vaccine in infants immunized at age 6 months. JAMA 280:527–532

Garenne M (1994) Sex differences in measles mortality: a world review. Int J Epidemiol 23:632–642

Garenne M, Leroy O, Beau JP, Sene I (1991) Child mortality after high-titre measles vaccines: prospective study in Senegal. Lancet 338:903–907

Good RA, Zak SJ (1956) Disturbances in gamma globulin synthesis as "experiments of nature". Pediatrics 18:109–149

Graunt J (1620–1674) Natural and political observations mentioned in a following index, and made upon the bills of mortality. By John Graunt, citizen of London. With reference to the government, religion, trade, growth, ayre, diseases, and the several changes of the said city. London : printed by Tho: Roycroft, for John Martin, James Allestry, and Tho: Dicas, at the sign of the Bell in St. Paul's Church-yard, MDCLXII. [1662]

Green MS, Shohat T, Lerman Y, Cohen D, Slepon R, Duvdevani P, Varsano N, Dagan R, Mendelson E (1994) Sex differences in the humoral antibody response to live measles vaccine in young adults. Int J Epidemiol 23:1078–1081

Greenberg BL, Sack RB, Salazar-Lindo E, Budge E, Gutierrez M, Campos M, Visberg A, Leon-Barua R, Yi A, Maurutia D (1991) Measles-associated diarrhea in hospitalized children in Lima, Peru: pathogenic agents and impact on growth. J Infect D 163(3):495–502

Griffin DE, Ward BJ (1993) Differential CD4 T cell activation in measles. J Infect Dis 168:275–281

Griffin DE, Cooper SJ, Hirsch RL, Johnson RT, Soriano IL, Roedenbeck S, Vaisberg A (1985) Changes in plasma IgE levels during complicated and uncomplicated measles virus infections. J Allergy Clin Immunol 76:206–213

Griffin DE, Ward BJ, Jauregui E, Johnson RT, Vaisberg A (1990) Natural killer cell activity during measles. Clin Exp Immunol 81:218–224

Halsey NA (1993) Increased mortality after high-titre measles vaccines: too much of a good thing. Pediatr Infect Dis J 12:462–465

Hassan MM, Hassounah OA, Hegab M, Salah K, el Mahrouky L, Galal N (1997) Transmission of circulating schistosomal antigens from infected mothers to their newborns. J Egypt Soc Parasitol 27:773–780

Herz U, Joachim R, Ahrens B, Scheffold A, Radbruch A, Renz H (2000) Prenatal sensitization in a mouse model. Am J Respir Crit Care Med 162:S62–S65

Hirsch RL, Griffin DE, Johnson RT, Cooper SJ, Lindo de Soriano I, Roedenbeck S, Vaisberg A (1984) Cellular immune responses during complicated and uncomplicated measles virus infections of man. Clin Immunol Immunopathol 31:1–12

Holmes CB, Hausler H, Nunn P (1998) A review of sex differences in the epidemiology of tuberculosis. Int J Tuberc Lung Dis 2:96–104

Holt EA, Moulton LH, Siberry GK, Halsey NA (1993) Differential mortality by measles vaccine titer and sex. J Infect Dis 168:1087–1096

Hori S, Nomura T, Sakaguchi S (2003) Control of regulatory T cell development by the transcription factor Foxp3. Sci 299:1057–1061

Karp CL, Wysocka M, Wahl LM, Ahearn JM, Cuomo PJ, Sherry B, Trinchieri G, Griffin DE (1996) Mechanism of suppression of cell-mediated immunity by measles virus. Science 273:228–231

Kemper C, Chan AC, Green JM, Brett KA, Murphy KM, Atkinson JP (2003) Activation of human CD4+ cells with CD3 and CD46 induces a T-regulatory cell 1 phenotype. Nature 421:388–392

King CL, Malhotra I, Mungai P, Wamachi A, Kioko J, Ouma JH, Kazura JW (1998) B cell sensitization to helminthic infection develops in utero in humans. J Immunol 160:3578–3584

Knudsen KM, Aaby P, Whittle H, Rowe M, Samb B, Simondon F, Sterne J, Fine P (1996) Child mortality following standard, medium or high titre measles immunization in West Africa. Int J Epidemiol 25:665–673

Kristensen I, Aaby P, Jensen H (2000) Routine vaccinations and child survival: follow up study in Guinea-Bissau, West Africa. BMJ 321:1–7

Leon ME, Ward B, Kanashiro R, Hernandez H, Berry S, Vaisberg A, Escamilla J, Campos M, Bellomo S, Azabache V, Halsey NA (1993) Immunologic parameters 2 years after high-titer measles immunization in Peruvian children. J Infect Dis 168:109

Lewert RM, Mandlowitz S (1969) Schistosomiasis: prenatal induction of tolerance to antigens. Nature 224:1029–1030

Lisse IM, Aaby P, Knudsen K, Whittle H, Andersen H (1994) Long term impact of high titer Edmonston-Zagreb measles vaccine on T lymphocyte subsets. Pediatr Infect Dis J 13:109–112

Lisse I, Samb B, Whittle H, Jensen H, Soumare M, Simondon F, Aaby P (1998) Acute and long-term changes in T-lymphocyte subsets in response to clinical and subclinical measles. A community study from rural Senegal. Scand J Infect Dis 30:17

Lopez AD, Ruzicka LT (eds) (1983) Sex differentials in mortality. Trends, determinants and consequences. Australian National University Press, Canberra

Malhotra I, Ouma J, Wamachi A, Kioko J, Mungai P, Omollo A, Elson L, Koech D, Kazura JW, King CL (1997) In utero exposure to helminth and mycobacterial antigens generates cytokine responses similar to that observed in adults. J Clin Invest 99(7):1759–1766

Malhotra I, Mungai P, Wamachi A, Kioko J, Ouma JH, Kazura JW, King CL (1999) Helminth- and Bacillus Calmette-Guerin-induced immunity in children sensitized in utero to filariasis and schistosomiasis. J Immunol 162:6843–6848

Martinez AN, Rhee JT, Small PM, Behr MA (2000) Sex differences in the epidemiology of tuberculosis in San Francisco. Int J Tuberc Lung Dis 4:26–31

Migeon B (2007) Females are mosaics: X inactivation and sex differences in disease. Oxford University Press, New York

Montesano MA, Colley DG, Eloi-Santos S, Freeman GL Jr, Secor WE (1999a) Neonatal idiotypic exposure alters subsequent cytokine, pathology, and survival patterns in experimental Schistosoma mansoni infections. J Exp Med 189:637–645

Montesano MA, Colley DG, Freeman GL Jr, Secor WE (1999b) Neonatal exposure to idiotype induces Schistosoma mansoni egg antigen-specific cellular and humoral immune responses. J Immunol 163:898–905

Moraes-Pinto MI, Verhoeff F, Chimsuku L, Milligan PJ, Wesumperuma L, Broadhead RL, Brabin BJ, Johnson PM, Hart CA (1998) Placental antibody transfer: influence of maternal HIV infection and placental malaria. Arch Dis Child Fetal Neonatal Ed 79:F202–F205

Moss WJ, Monze M, Ryon JJ, Quinn TC, Griffin DE, Cutts F (2002a) Prospective study of measles in hospitalized human immunodeficiency virus (HIV)-infected and HIV-uninfected children in Zambia. Clin Infect Dis 35:189–196

Moss WJ, Ryon JJ, Monze M, Griffin DE (2002b) Differential regulation of interleukin (IL)-4, IL-5, and IL-10 during measles in Zambian children. J Infect Dis 186:879–887

Moss WJ, Fisher C, Scott S, Monze M, Ryon JJ, Quinn TC, Griffin DE, Cutts FT (2008) HIV type 1 infection is a risk factor for mortality in hospitalized Zambian children with measles. Clin Infect Dis 46:523–527

Okada H, Sato TA, Katayama A, Higuchi K, Shichijo K, Tsuchiya T, Takayama N, Takeuchi Y, Abe T, Okabe N, Tashiro M (2001) Comparative analysis of host responses related to immunosuppression between measles patients and vaccine recipients with live attenuated measles vaccines. Arch Virol 146:859–874

Owens S, Harper G, Amuasi J, Offei-Larbi G, Ordi J, Brabin BJ (2006) Placental malaria and immunity to infant measles. Arch Dis Child 91:507–508

Permar SR, Moss WJ, Ryon JJ, Monze M, Cutts F, Quinn TC, Griffin DE (2001) Prolonged measles virus shedding in human immunodeficiency virus-infected children, detected by reverse transcriptase-polymerase chain reaction. J Infect Dis 183:5

Perry RT, Halsey NA (2004) The clinical significance of measles: a review. J Infect Dis 189 (Suppl 1):S4–S16

Pison G, Aaby P, Knudsen K (1992) Increased risk of death from measles in children with a sibling of opposite sex in Senegal. BMJ 304:284–287

Riddell MA, Moss WJ, Hauer D, Monze M, Griffin DE (2007) Slow clearance of measles virus RNA after acute infection. J Clin Virol 39:312–317

Romia SA, Handoussa AE, Youseff SA, el Zayat MM (1992) Transplacental transfer of schistosomal antigens and antibodies. J Egypt Soc Parasitol 22:575–582

Ryon JJ, Moss WJ, Monze M, Griffin DE (2002) Functional and phenotypic changes in circulating lymphocytes from hospitalized Zambian children with measles. Clin Diagn Lab Immunol 9:994–1003

Ryon JJ, Moss WJ, Monze M, Quinn TC, Griffin DE (2005) Influence of HIV infection on changes in circulating leukocytes during measles in Zambian children. J Infect Dis 192:1950–1955

Samb B, Whittle H, Aaby P, Seck AMC, Bennett J, Markowitz L, Ngom PT, Zeller J, Michaelsen KM, Simondon F (1995) No evidence of long-term immunosuppression after high-titer Edmonston-Zagreb measles vaccination in Senegal. J Infect Dis 171:506–508

Scott S, Cumberland P, Shulman CE, Cousens S, Cohen B, Brown DWG, Bulmer JN, Dorman EK, Kawuondo K, Marsh K, Cutts F (2005a) Neonatal measles immunity in rural Kenya: the influence of HIV and placental malaria infections on placental transfer of antibodies and levels of antibody in maternal and cord serum samples. J Infect Dis 191:1854–1860

Scott S, Cumberland P, Shulman CE, Cousens S, Cohen BJ, Brown DW, Bulmer JN, Dorman EK, Kawuondo K, Marsh K, Cutts F (2005b) Neonatal measles immunity in rural Kenya: the influence of HIV and placental malaria infections on placental transfer of antibodies and levels of antibody in maternal and cord serum samples. J Infect Dis 191:1854–1860

Scott S, Moss WJ, Cousens S, Beeler JA, Audet SA, Mugala N, Quinn TC, Griffin DE, Cutts FT (2007) The influence of HIV-1 exposure and infection on levels of passively acquired antibodies to measles virus in Zambian infants. Clin Infect Dis 45:1417–1424

Semba RD, Munasir Z, Beeler J, Akib A, Muhilal AS, Sommer A (1995) Reduced seroconversion to measles in infants given vitamin A with measles vaccination. Lancet 345:1330–1332

Seng R, Samb B, Simondon F, Cisse B, Soumare M, Jensen H, Bennett J, Whittle H, Aaby P (1999) Increased long term mortality associated with rash after early measles vaccination in rural Senegal. Pediatr Infect Dis J 18:48–52

Servet-Delprat C, Vidalain PO, Azocar O, Le Deist F, Fischer A, Rabourdin-Combe C (2000) Consequences of Fas-mediated human dendritic cell apoptosis induced by measles virus. J Virol 74:4387–4393

Shann F (2000) Non-specific effects of vaccines in developing countries. BMJ 321:1423–1424

Shohat T, Green MS, Nakar O, Ballin A, Duvdevani P, Cohen A, Shohat M (2000) Gender differences in the reactogenicity of measles-mumps-rubella vaccine. Isr Med Assoc J 2: 192–195

Simister NE (2003) Placental transport of immunoglobulin G Vaccine 21:3365–3369

Tachon P, Borojevic R (1978) Mother-child relation in human schistosomiasis mansoni : skin test and cord blood reactivity to schistosomal antigens. Trans R Soc Trop Med Hyg 72:605–609

Tamashiro VG, Perez HH, Griffin DE (1987) Prospective study of the magnitude and duration of changes in tuberculin reactivity during uncomplicated and complicated measles. Pediatr Infect Dis J 6:451–454

Taylor WR, Mambu RK, Ma-Disu M, Weinman JM (1988) Measles control efforts in urban Africa complicated by high incidence of measles in the first year of life. Am J Epidemiol 127:788–794

Thorne C, Newell ML & European Collaborative Study (2004) Are girls more at risk of intrauterine-acquired HIV infection than boys? AIDS 18:344–347

Veirum JE, Sodemann M, Biai S, Jakobsen M, Garly ML, Hedegaard K, Jensen H, Aaby P (2005) Routine vaccinations associated with divergent effects on female and male mortality at the paediatric ward in Bissau, Guinea-Bissau. Vaccine 23:1197–1204

Vercruysse J, Gabriel S (2005) Immunity to schistosomiasis in animals: an update. Parasite Immunol 27:289–295

Ward B, Boulianne N, Ratnam S, Guiot MC, Couilard M, De Serres G (1995) Cellular immunity in measles vaccine failure: demonstration of measles antigen-specific lymphoproliferative responses despite limited serum antibody production after revaccination. J Infect Dis 172:1591–1595

Washburn TC, Medearis DN Jr, Childs B (1965) Sex differences in susceptibility to infections. Pediatrics 35:57–64

World Health Organization (2007) Progress in global measles control and mortality reduction, 2000–2006. Wkly Epidemiol Rec 82:418–424

Chapter 12
Epilogue: Challenges for the Future

Craig W. Roberts and Sabra L. Klein

Abstract Charles Darwin contemplated the evolution of diversity within species in his book, *On the Origin of Species*. Darwin noted that there are profound differences between the sexes of a single species that evolve through a process he termed sexual selection. We propose that consideration of the adaptive significance as well as the mechanisms underlying why males and females respond differently to infection is fundamental for progress in infectious disease research. Future studies must consider not only the role that hormones play in modulating responses to infection, but also how genes, including sex chromosomal genes, impact responses to infection differentially between the sexes. Finally, by illustrating that sex and reproductive status impact responses to infection it is demonstrated that these factors might also influence responses to therapeutic treatments, including vaccines.

12.1 Natural Selection, Sexual Selection and the Evolution of Sexual Dimorphism

For around 150 years, two theories, both proposed by Charles Darwin, have prevailed to explain the diversity and complexity of life on earth and the differences observed between the sexes of a single species. The first theory, explains how diversity and speciation occurs. Simply put, variations in the diversity of a species that are beneficial for survival are preserved in the proceeding generations. This theory is termed natural selection.

> "I have called this principle, by which each slight variation, if useful, is preserved, by the term Natural Selection"
>
> Charles Darwin, 1859 on the Origin of Species

C.W. Roberts (✉)
Strathclyde Institute of Pharmacy and Biomedical Sciences, University of Strathclyde, 27 Taylor St, G4 0NR Glasgow, Scotland, UK
e-mail: c.w.roberts@strath.ac.uk

The second theory, in some ways, is a refinement of the first as it accounts for differences observed between the sexes of a single species that cannot immediately be explained by natural selection. This theory is termed sexual selection.

"Thus it is rendered possible for the two sexes to be modified through natural selection in relation to different habits of life, as is sometimes the case; or for one sex to be modified in relation to the other sex, as commonly occurs. This leads me to say a few words on what I have called Sexual Selection. This form of selection depends, not on a struggle between individuals but of one sex, generally the males, for the possession of the other sex. The result is not death to the unsuccessful competitor, but few or no offspring. Sexual selection is therefore less rigorous than natural selection. Generally, those most vigorous males, those that which are best fitted for their places in nature, will leave the most progeny."

Charles Darwin, 1859 on the Origin of Species

Any attempt to establish why males and females have differential susceptibility to disease or why their immune systems exhibit differences, should be consistent with these widely accepted theories or should propose a refinement thereof.

Differences in the susceptibility of males and females to infection have been noted in diverse, phylogenetically distinct species of invertebrates and vertebrates (Chap. 1). This book has concentrated on mammals as the functional significance of this is obvious and is of medical and veterinary importance. Consideration, however, has been given to other animals in Chap. 1, in which the evolution of these sex differences is discussed. The rationale for considering all diverse forms of life for deriving theories of the providence of these observations is clear and allows scenarios to be considered that would otherwise be impossible to envisage by examining mammals in isolation. For example, the absence of testosterone in insects demonstrates that the evolutionary origins of the observed differences in humans, although in part mediated by testosterone, are not primarily due to an unfortunate side effect of this hormone. Males respond to infection differently from females in general and this is conserved in diverse life forms by independent signals. This would suggest that the evolutionary reason for male-associated susceptibility to infection is a generality.

12.2   Male-Associated Immuno-incompetence: Limitations of Bateman's Principle and the Immunocompetence Handicap Hypothesis

There are a number of limitations of current theories that attempt to explain "male immuno-incompetence." The evolutionary rationale that has been proposed for this phenomenon combines the idea of Bateman's principle with the immunocompetence handicap hypothesis (Bateman 1948; Rolff 2002; Folstad and Karter 1992). In essence, females are highly discriminate about selecting mates as they are required to invest more energy in progeny, whereas males are less discriminate about mates and expend more energy in gaining as many mates as possible. The energy that males spend on mating is in expense of their immune system and, as a consequence, males are hypothesized to be immunologically handicapped. The combination of these theories is appealing as it provides an overall principle independent of any single mediator that must be conserved in all species. However, it also implies that males have evolved poorer immune systems not due to positive selection of that trait, but due to the adverse effects of positive selection on secondary sexual characteristics that promote the acquisition of mates and reproductive success. However, male immuno-incompetence is not passive in mammals and has been largely attributed to the active immunosuppressive effects of testosterone. Thus, as detailed in Chap. 3, testosterone receptors are present in immune cells and the ligation of these receptors and downstream signaling events actively expend energy in responding to testosterone. Furthermore, males are not more susceptible to infection with all pathogens and female-typic reproductive hormones, including estrogen and progesterone, can alter immune responses and disease pathogenesis. Consequently, theories that attempt to justify "male-immunoincompetence" require refinement to take into account these proximate mechanisms and alternative hypotheses should be considered in the future.

12.3   Beyond the Dogma of Female-Biased Resistance

A notion of "female supremacy" has existed in the literature, as in most cases females are less susceptible to infectious diseases than are males. As outlined in Chap. 7, a number of exceptions exist that illustrate that heightened immunity in females is not always beneficial and can result in development of immune-mediated disease following infection. One obvious mediator of female susceptibility to the development of immune-mediated disease is elevated production of estrogen and progesterone. These hormones are extremely elevated during pregnancy and may account for the increased susceptibility of pregnant females to certain pathogens, as discussed in Chap. 9. Notably, pregnancy increases susceptibility to a number of pathogens irrespective of whether they conform to or contradict the general notion of female-biased resistance to infection. For example, female mice are more susceptible to *T. gondii* infection and this is accentuated during pregnancy (Roberts et al. 1995; Shirahata et al. 1992). In contrast, female mice are more resistant than male mice to *Leishmania* spp., but have increased susceptibility to infection during pregnancy (Alexander 1998; Krishnan et al. 1996).

These results illustrate that exceptions to the notion of female-biased resistance to infection exist and these might be a reflection of hormone-mediated immune alteration necessary for successful pregnancy operating to a reduced extent in a nonpregnant host. This would be consistent with the obvious evolutionary selective pressure of successful pregnancy. As an added complication, however, pathogens also are under evolutionary pressure and have coevolved with their hosts. In some cases, pathogens have evolved to respond to and benefit from host mediators, including hormones. The extent of these interactions has not been adequately explored, but the potential benefits to pathogens are clear in some interactions, such as *Toxocara canis* infection of dogs. The parasite is able to "sense" pregnancy and through hormone-induced immunological changes; *T. canis* can reactivate and infect the offspring (Burke and Roberson 1985).

12.4   Limitations of Current Knowledge About Immune–Endocrine Interactions

The immune system is complex with many redundant and pleiotropic functions that are challenging to understanding in isolation, let alone in combination with the endocrine system. As the chapters contained herein illustrate, the immune system does not function in isolation of the endocrine system and consequently studying each in isolation may provide only a partial understanding of both. Furthermore, the division of each of these systems is purely arbitrary, based on the independent evolution of what were originally two fields of study. Many of the peptide mediators of the immune system, "cytokines," are by classical definition "hormones" and had they been described first by endocrinologists would have been so designated. Conversely, many of the hormones described by endocrinologists are by themselves immunologically active (e.g., prolactin) and could equally as well be considered as cytokines. The borders are further blurred, as steroid hormones, most notably cortisol, have been known for many years to be antiinflammatory and to be induced by immunological events (e.g., IL-1β signaling in the hypothalamus). Less well appreciated by the scientific and medical communities is the notion that sex and pregnancy-associated steroids interact with the immune system, which clearly merits further study. In Chaps. 2 and 3, the distribution of hormone receptors and their action in immune cells are described. The expression of hormone receptors in various cells involved in innate and adaptive immunity and how downstream signaling impacts immunity have not been fully characterized. An approach of utilizing transcriptomics and proteomics will provide novel insights into this emerging area of research. Similarly, characterization of hormone response elements, and their cross reference with immunologically relevant molecules, using the current mammalian genomes will provide a further reference point to study immuno–endocrine interactions. Comparative analysis of mammals using a systems biology approach is also likely to provide information of the evolution of such interactions.

Immunological advances have been made using animals that have been genetically engineered to be deficient in immunologically important genes. Although mice have been engineered to be deficient in certain steroid hormone receptors, their use has been limited, probably due to their extreme phenotypes as many of these hormones are essential for survival or reproduction. The development of tissue specific and conditional gene-deficient mice should be exploited to establish the role of hormone receptors in individual immunological cell lineages as well as at specific developmental stages.

12.5   Sex-Determining Chromosomes in Mammals and Their Influence on Immunity

In mammals, sex is chromosomally determined, with males possessing an X and Y chromosome, but females possessing two X chromosomes. Consequently, genes present on the Y chromosome are unique to males and can influence immune response and disease susceptibility. Furthermore, as males only have a single copy of the X chromosome, they possess only a single allele of any gene thereon, and are consequently vulnerable to X-linked diseases, including immunodeficiencies. This would include XSCID, which occurs as a result of a mutation in the common γ-chain that forms part of the heterodimeric IL-2, IL-4, IL-7, IL-9, IL-15, and IL-21 receptors (reviewed Fish 2008).

Females have two alleles for each X-chromosome-located gene, although only one of these is functional as one of the X chromosomes is randomly inactivated at the late blastula stage. This mosaic pattern can compensate for some X-linked immunodeficiencies, but would logically also provide diversity in immune response within a single heterozygous female. The advantage of X-chromosome mosaicism is clearly evident when looking at the consequence of a rare mutation in *FoxP3* gene present on the X chromosome (reviewed, Wildin and Freitas 2005). This gene encodes a transcription factor that is required for the development of regulatory T cells and in males causes immunodysregulation, polyendocrinopathy, enteropathy, X-linked syndrome (IPEX). This disease manifests in young males as early onset autoimmune diseases, with female mosaicism providing protection (reviewed, Wildin and Freitas 2005).

The two examples provided here are extreme cases that illustrate the vulnerability of males to adverse mutations in genes of the X chromosome. The X chromosome encodes more than 1,100 identified genes, a number of which are known to have obvious immunological functions (Fig. 12.1) (reviewed, Fish 2008). However, the advantage of being a female with a partial "mosaic immune system" remains to be fully determined. There are some cases where it has been postulated that the mosaic system can have adverse effects when one X chromosome is preferentially inactivated. In these individuals, rather than the predicted 50% inactivation of each X chromosome, inactivation can be skewed in certain tissues so severely that few dendritic cells possess one of the X chromosomes. T cells maturation would be tolerated in these individuals to antigens of only one X chromosome and, consequently, T cells could interact with tissue expressing the other X chromosome. This, in turn, could be responsible for certain female-associated autoimmune diseases including systemic lupus erythematosus (reviewed, Stewart 1998). The extent to

a Receptors & associated proteins

AR	Androgen receptor
AGTR2	Angiotensin receptor 2
CSF2RA	Colony-stimulating factor 2 receptor α (granulocyte-macrophage)
GPCR	G-protein coupled receptors 23, 50, 101, 112, 119, 174 and CX-chemokine receptor 3
CYSLTR1	Cysteinyl leukotriene receptor 1
IL-1RAP1	Interleukin-1 (IL-1) receptor accessory protein-like 1
IL-1RAP2	IL-1 receptor accessory protein-like 2
IL-2RG	IL-2 receptor γ-chain
IL-3RA	IL-3 receptor α-chain
IL-9R	IL-9 receptor
IL-13RA1	IL-13 receptor α1-chain
IL-13RA2	IL-13 receptor α2-chain
IRAK	IL-1 receptor-associated kinase
NGFRAP1	Nerve-growth-factor receptor associated protein 1
TLR7	Toll-like receptor 7
TLR8	Toll-like receptor 8

b Immune-response related proteins

XSCID	X-linked severe combined immunodeficiency
ELK1	Involved in B-cell development
EPAG	Early lymphoid activation protein
GATA1	GATA-binding protein 1
GTD	Gonadotropin deficiency
IDDMX	X-linked susceptibility to insulin-dependent diabetes
IGBP1	CD79A, immunoglobulin binding protein 1
IGSF1	Immunoglobulin superfamily member 1
ITGB1BP2	Integrin-β1-binding protein 2
CD99	Also known as MIC2; associated with T-cell function
MTCP1	Mature T-cell proliferation 1
PFC	Properdin P factor, complement
TIMP1	Tissue inhibitor of metalloproteinase 1
CD40L	CD40 ligand
Z39IG	An immunoglobulin superfamily protein

c Transcriptional & translational control effectors

RHOGAP	RAS homologue (RHO) GTPase activating proteins 4, 6
CDC42GEF	Cell-division cycle 42 guanine-nucleotide-exchange factors 6, 9
ETK	Also known as BMX
BTK	Bruton agammaglobulinaemia tyrosine kinase
CDX4	Caudal homeobox transcription factor 4
TRAP170	A co-factor for SP1 transcription factor activation
DUSP	Dual specificity phosphatases 9, 21
EEF	Eukaryotic translation elongation factors 1α3, β4
EIF	Eukaryotic translation initiation factor 1A*, 2a
FOXP3	Forkhead box P3 (associated with the development and function of regulatory T cells)
GAB3	Growth-factor-receptor-bound protein 2-associated binding protein 3
HDAC	Histone deacetylases 6, 8
IKKγ	IκB kinase; also known as NEMO
MAPKKK15	Mitogen-activated protein kinase kinase kinase 15
NFκBRF	Nuclear factor-κB (NF-κB) repressing factor
NRK	NF-κB-inducing kinase-related kinase
NXF	Nuclear RNA export factors 2, 3, 4, 5
PAK3	p21 (also known as CDKN1A)-activated kinase 3
PPP	Protein phosphatases 1, 2*, 6
PRKCI	Protein kinase Ci
S6K	Ribosomal protein S6 kinase
SWI/SNF	SWI/SNF-related, matrix associated, actin-dependent regulator of chromatin
STK9	Serine/threonine kinase 9
TAF1	TATA-box-binding protein-associated factor 1, TFIID subunit
UBE1	Ubiquitin-activating enzyme E1
UBE2A	Ubiquitin-conjugating enzyme E2A
USP	Ubiquitin-specific proteases 9*, 11, 26, 27, 51l
WASP	Wiskott–Aldrich syndrome protein

Fig. 12.1 Genes on the X chromosome with the potential to influence immunocompetence. Several proteins encoded by genes that are found on the X chromosome might underlie sex-based differences in immune responses. The proteins listed were selected from more than 1,100 identified genes on the X chromosome, and have been grouped according to their associated function as receptors and associated proteins (**a**), proteins related to the immune response (**b**), or proteins involved in transcriptional and translational control (**c**). Proteins with their definition and/or known function are listed. The protein marked with the asterisk indicated those encoded by genes also found on Y-chromosome. Adapted from Fish, 2008. Reprinted by permission from Macmillan Publishers Ltd: Nature Reviews Immunology (Fish 2008), copyright (2008)

which sex chromosomal genes impact infectious disease pathogenesis requires consideration.

12.6 Sex Differences in Vaccinology and Therapeutics

Throughout this book, we have illustrated that the outcome of infection differs between males and females and among females in different reproductive conditions. Whether the sexes respond differentially to vaccines and other therapeutic treatments is considerably less well characterized, as discussed in Chap. 11.

Sex differences are reported in response to both childhood and adult vaccination. For example, sex differences are reported in the humoral immune responses of adults to vaccines against influenza, hepatitis A, hepatitis B, pneumococcal polysaccharide, and diphtheria (Yu et al. 2004; Cook et al. 2006; Volzke et al. 2006; Cook 2008). One notable example is in response to the intramuscular trivalent-inactivated influenza vaccine, women generate a more robust antibody response and develop more severe side effects (e.g., headache, other pain, and fatigue) following vaccination than do men (Engler et al. 2008). Importantly, the antibody response of women to a half dose of the influenza vaccine is equivalent to the antibody response of men to the full dose, suggesting that sex should be considered in guidelines for influenza vaccine dosages (Engler et al. 2008). Vaccination of adults against herpes simplex virus type 2 (HSV-2) also results in disproportionately greater protection in women than men. In fact, the HSV-2 vaccine seems to work exclusively in women and provides little to no protection in men (Stanberry et al. 2002). Men and women also differ in adverse reactions to vaccines. Females develop more adverse side effects following vaccination against anthrax, rubella, diphtheria–tetanus–pertussis, and measles, whereas males develop more severe encephalitis in response to the yellow fever vaccine (Knudsen et al. 1996; Monath et al. 2002; Pittman et al. 2002; Cook et al. 2006; Aaby et al. 2007; McNeil et al. 2007; Cook 2008). Whether the dose of vaccine should be tailored to men and women should be considered. Future studies also should continue to consider whether antibody responses, the activity of memory B cells, or cytokine gene polymorphisms underlie sex differences in the efficacy of both childhood and adult vaccination (Baynam et al. 2008; Cook 2008; Sanchez-Ramon et al. 2008).

In addition to vaccines, the pharmacokinetics and pharmacodynamics of therapeutic medications also differ between the sexes, which may impact effectiveness of treatments of infectious diseases. Men and women differ in gastric emptying, body weight, body fat, organ blood flow, hepatic enzymes that metabolize drugs, and renal clearance – all of which influence the efficacy of drug treatments (Wizemann and Pardue 2001; Fish 2008). Of all the drugs that the Federal drug administration in the United States withdrew in 2005, the reason for 80% of these was due to adverse effects in women (Simon 2005). The under-representation of women in clinical studies has impacted our understanding and treatment of infectious diseases (Fish 2008). Even though current policies in the United States require women to be included in studies on the safety and effectiveness of drugs, there is still inadequate compliance (Fish 2008). When women are included, clinical studies often do not analyze data in a manner that allows for distinctions between men and women in responses to therapeutic treatments. If the sexes are compared, this is often conducted post hoc using data from studies not explicitly designed to compare responses between men and women using subgroup analyzes with adequate statistical power (Uhl and Marts 2008).

As the concept of "personalized medicine" continues to gain momentum in the scientific and medical communities, understanding how and why the immunology of men and women differs both in response to infectious agents as well as in response to treatments of infection will result in a paradigm shift in how basic scientists, epidemiologists, clinical researchers, and healthcare providers evaluate disease and therapeutic treatment. Thus, the data presented in this book could have a fundamental impact on the evaluation and treatment of infectious diseases. Rather than focus on the need for inclusion of women in studies of infectious diseases, this book has emphasized the importance of examining responses in men and women to improve our understanding about the manifestation and treatment of disease in both sexes. An extension of this is that the treatment of at least certain diseases would be optimized if specifically tailored for hormonal status taking into account both age and sex.

References

Aaby P, Garly ML, Nielsen J, Ravn H, Martins C, Bale C, Rodrigues A, Benn CS, Lisse IM (2007) Increased female–male mortality ratio associated with inactivated polio and diphtheria–tetanus–pertussis vaccines: observations from vaccination trials in Guinea-Bissau. Pediatr Infect Dis J 26:247–252

Alexander J (1998) Sex differences and cross-immunity in DBA/2 mice infected with *L. mexicana* and *L. major*. Parasitology 96:297–302

Bateman AJ (1948) Intra-sexual selection in *Drosophila*. Heredity 2:349–368

Baynam G, Zhang G, Khoo SK, Sly P, Holt P, Goldblatt J, Le Souef PN (2008) Gender-specific effects of cytokine gene polymorphisms on childhood vaccine responses. Vaccine 26:3574–3579

Burke TM, Roberson EL (1985) Prenatal and lactational transmission of *Toxocara canis* and *Ancylostoma caninum*: experimental infection of the bitch before pregnancy. Int J Parasitol 15:71–75

Cook IF (2008) Sexual dimorphism of humoral immunity with human vaccines. Vaccine 26:3551–3555

Cook IF, Barr I, Hartel G, Pond D, Hampson AW (2006) Reactogenicity and immunogenicity of an inactivated influenza vaccine administered by intramuscular or subcutaneous injection in elderly adults. Vaccine 24:2395–2402

Engler RJ, Nelson MR, Klote MM, VanRaden MJ, Huang CY, Cox NJ, Klimov A, Keitel WA, Nichol KL, Carr WW, Treanor JJ (2008) Half- vs full-dose trivalent inactivated influenza vaccine (2004–2005): age, dose, and sex effects on immune responses. Arch Intern Med 168:2405–2414

Fish EN (2008) The X-files in immunity: sex-based differences predispose immune responses. Nat Rev Immunol 8:737–744

Folstad I, Karter A (1992) Parasites, bright males, and the immunocompetence handicap. Am Nat 139:603–622

Knudsen KM, Aaby P, Whittle H, Rowe M, Samb B, Simondon F, Sterne J, Fine P (1996) Child mortality following standard, medium or high titre measles immunization in West Africa. Int J Epidemiol 25:665–673

Krishnan L, Guilbert LJ, Russell AS, Wegmann TG, Mosmann TR, Belosevic M (1996) Pregnancy impairs resistance of C57BL/6 mice to *Leishmania major* infection and causes decreased antigen-specific IFN-gamma response and increased production of T helper 2 cytokines. J Immunol 156:644–652

McNeil MM, Chiang IS, Wheeling JT, Zhang Y (2007) Short-term reactogenicity and gender effect of anthrax vaccine: analysis of a 1967–1972 study and review of the 1955–2005 medical literature. Pharmacoepidemiol Drug Saf 16:259–274

Monath TP, Nichols R, Archambault WT, Moore L, Marchesani R, Tian J, Shope RE, Thomas N, Schrader R, Furby D, Bedford P (2002) Comparative safety and immunogenicity of two yellow fever 17D vaccines (ARILVAX and YF-VAX) in a phase III multicenter, double-blind clinical trial. Am J Trop Med Hyg 66:533–541

Pittman PR, Kim-Ahn G, Pifat DY, Coonan K, Gibbs P, Little S, Pace-Templeton JG, Myers R, Parker GW, Friedlander AM (2002) Anthrax vaccine: immunogenicity and safety of a dose-reduction, route-change comparison study in humans. Vaccine 20:1412–1420

Rolff J (2002) Bateman's principle and immunity. Proc Biol Sci 269:867–72

Roberts CW, Cruickshank SM, Alexander J (1995) Sex-determined resistance to *Toxoplasma gondii* is associated with temporal differences in cytokine production. Infect Immun 63:2549–2555

Sanchez-Ramon S, Radigan L, Yu JE, Bard S, Cunningham-Rundles C (2008) Memory B cells in common variable immunodeficiency: clinical associations and sex differences. Clin Immunol 128:314–321

Shirahata T, Muroya N, Ohta C, Goto H, Nakane A (1992) Correlation between increased susceptibility to primary *Toxoplasma gondii* infection and depressed production of gamma interferon in pregnant mice. Microbiol Immunol 36:81–91

Simon V (2005) Wanted: women in clinical trials. Science 308:1517

Stanberry LR, Spruance SL, Cunningham AL, Bernstein DI, Mindel A, Sacks S, Tyring S, Aoki FY, Slaoui M, Denis M, Vandepapeliere P, Dubin G (2002) Glycoprotein-D-adjuvant vaccine to prevent genital herpes. N Engl J Med 347:1652–1661

Stewart JJ (1998) The female X-inactivation mosaic in systemic lupus erythematosus. Immunol Today 19:352–357

Uhl K, Marts S (2008) Assessing sex differences: methodological considerations. Exp Rev Clin Pharmacol 1:585–587

Volzke H, Kloker KM, Kramer A, Guertler L, Doren M, Baumeister SE, Hoffmann W, John U (2006) Susceptibility to diphtheria in adults: prevalence and relationship to gender and social variables. Clin Microbiol Infect 12:961–967

Wildin RS, Freitas A (2005) IPEX and FOXP3: clinical and research perspectives. J Autoimmun 25(Suppl):56–62

Wizemann TM, Pardue M (eds) (2001) Exploring the biological contributions to human health: does sex matter? National Academy Press, Washington DC

Yu AS, Cheung RC, Keeffe EB (2004) Hepatitis B vaccines. Clin Liver Dis 8:283–300

Index

Lightning Source UK Ltd.
Milton Keynes UK
19 February 2010

150364UK00001B/52/P